Reconsidering Sputnik

Studies in the History of Science, Technology and Medicine
Edited by John Krige, CRHST, Paris, France

Studies in the History of Science, Technology and Medicine aims to stimulate research in the field, concentrating on the twentieth century. It seeks to contribute to our understanding of science, technology and medicine as they are embedded in society, exploring the links between the subjects on the one hand and the cultural, economic, political and institutional contexts of their genesis and development on the other. Within this framework, and while not favouring any particular methodological approach, the series welcomes studies that examine relations between science, technology, medicine and society in new ways, e.g., the social construction of technologies and large technical systems.

Volume 1
Technological Change: Methods and Themes in the History
of Technology
Edited by Robert Fox

Volume 2
Technology Transfer out of Germany after 1945
Edited by Matthias Judt and Burghard Ciesla

Volume 3
Entomology, Ecology and Agriculture: The Making of Scientific Careers
in North America, 1885–1985
Paolo Palladino

Volume 4
The Historiography of Contemporary Science and Technology
Edited by Thomas Söderqvist

Volume 5
Science and Spectacle: The Work of Jodrell Bank in Post-war
British Culture
Jon Agar

Volume 6
Molecularizing Biology and Medicine: New Practices
and Alliances, 1910s–1970s
Edited by Soraya de Chadarevian and Harmke Kamminga

See the back of this book for other titles in Studies in the History of Science, Technology and Medicine.

Reconsidering Sputnik

Forty Years Since the Soviet Satellite

Edited by

Roger D. Launius

National Aeronautics and Space Administration
Washington, DC, USA

John M. Logsdon

George Washington University
Washington, DC, USA

and

Robert W. Smith

University of Alberta
Edmonton, Alberta, Canada

Routledge
Taylor & Francis Group
New York   London

Transferred to Digital Printing 2002

by Routledge

2 Park Square, Milton Park, Abingdon, Oxfordshire OX14 4RN
711 Third Avenue, New York, NY 10017
First issued in paperback 2014

*Routledge is an imprint of the Taylor and Francis Group,
an informa business*

British Library Cataloguing in Publication Data

Reconsidering Sputnik : forty years since the Soviet
    satellite. — (Studies in the history of science, technology
    and medicine ; v. 11 — ISSN 1024-8048)
    1. Outer space — Exploration — History 2. Outer space —
    Exploration — Political aspects — Soviet Union 3. Outer
    space — Exploration — Political aspects — United States
    I. Launius, Roger D., 1954– II. Losdgon, John III. Smith,
    Robert W. (Robert William), 1952–
    629.4′1′0947

ISBN 978-90-5702-623-2 (hbk)
ISBN 978-1-138-01224-0 (pbk)

Contents

List of Figures   vii

Preface and Acknowledgments   ix
Roger D. Launius

Introduction: Was Sputnik Really a Saltation?   xv
Walter A. McDougall

Part 1—Space Flight in the Soviet Union   3
Roger D. Launius

1.   Rising from a Cradle: Soviet Public Perceptions of
     Space Flight before Sputnik   11
     Peter A. Gorin

2.   Korolev, Sputnik, and the International Geophysical
     Year   43
     Asif A. Siddiqi

3.   Korolev's Triple Play: *Sputniks 1, 2,* and *3*   73
     James J. Harford

4.   Sputnik and the Creation of the Soviet Space Industry   95
     William P. Barry

Part 2—A Setting for the International Geophysical Year   117
Robert W. Smith

5.   The Sputniks and the IGY   125
     Rip Bulkeley

6.   Cover Stories and Hidden Agendas: Early American
     Space and National Security Policy   161
     Dwayne A. Day

7.   Before Sputnik: National Security and the Formation
        of U.S. Outer Space Policy   197
        Kenneth A. Osgood

8.   *Orbiter*, Overflight, and the First Satellite: New Light
        on the *Vanguard* Decision   231
        Michael J. Neufeld

Part 3—Ramifications and Reactions   259
John M. Logsdon

9.   The First Earth Satellite: A Retrospective View from
        the Future   267
        Sergey Khrushchev

10.   Building a Third Space Power: Western European
        Reactions to Sputnik at the Dawn of the Space Age   289
        John Krige

11.   Organizing the U.S. Government for Outer Space,
        1957–1958   309
        Eilene Galloway

12.   A Certain Future: Sputnik, American Higher
        Education, and the Survival of a Nation   327
        John A. Douglass

13.   Sputnik: A Political Symbol & Tool in 1960 Campaign
        Politics   363
        Gretchen, J. Van Dyke

Epilogue: Sputnik and Technological Surprise   401
Glenn P. Hastedt

List of Contributors   425

Index   427

Figures

4.1   The Council of Chief Designers (Prior to 1953)   99

12.1   Total Federal Funding for R&D and for Basic Research: 347
1957–1997

12.2   Source of U.S. Expenditures for Basic Research: 348
1957–1997

12.3   U.S. Basic Research Expenditures: 1957–1997   349

12.4   Percentage of Federal Funds in Higher Education 351
Operating Budgets: 1951–1973

12.5   Percentage of UC Campus Income from State and 352
Federal Sources: 1947–1995

12.6   University Research by Type: 1957–1997 (est.)   357

Preface and Acknowledgments

Roger D. Launius

The contours of space exploration in the latter half of the twentieth century owe much to the seminal but hardly singular event of Sputnik. To a very real extent, history changed on October 4, 1957, when the former Soviet Union successfully launched *Sputnik 1*. The world's first artificial satellite was about the size of a basketball, weighed only 183 pounds, and took about 98 minutes to orbit the Earth on its elliptical path. That launch ushered in new political, military, technological, and scientific developments. While the Sputnik launch was a single event, it marked the start of the space age and the U.S.–USSR space race.

That story, however, really begins in 1952, when the International Council of Scientific Unions decided to establish July 1, 1957, to December 31, 1958, as the International Geophysical Year (IGY) because the scientists knew that the cycles of solar activity would be at a high point then. In October 1954, the council adopted a resolution calling for artificial satellites to be launched during the IGY to map the Earth's surface.

Responding to this, in July 1955, the White House announced plans to launch an Earth-orbiting satellite for the IGY and solicited proposals from various Government research agencies to undertake development. In September 1955, the Naval Research Laboratory's Vanguard proposal was chosen to represent the U.S. during the IGY.

The Sputnik launch changed everything. As a technical achievement, Sputnik caught the world's attention and the American public off-guard. Its size was more impressive than Vanguard's intended 3.5-pound payload. In addition, the public feared that the Soviets' ability to launch satellites also translated into the capability to launch ballistic missiles that could carry nuclear weapons from Europe to the U.S. Then the Soviets struck again; on November 3, *Sputnik 2* was launched, carrying a much heavier payload, including a dog named Layka.

Immediately after the *Sputnik 1* launch in October, the U.S. Defense Department responded to the political furor by approving funding for another U.S. satellite project. As a simultaneous alternative to Vanguard, Wernher von Braun and his Army Redstone Arsenal team began work on the Explorer project.

On January 31, 1958, the tide changed, when the United States successfully launched Explorer I. This satellite carried a small scientific payload that eventually discovered the magnetic radiation belts around the Earth, named after principal investigator James Van Allen. The Explorer program continued as a successful ongoing series of lightweight, scientifically useful spacecraft.

The Sputnik launch also led directly to the creation of National Aeronautics and Space Administration (NASA). In July 1958, Congress passed the National Aeronautics and Space Act (commonly called the "Space Act"), which created NASA as of October 1, 1958, from the National Advisory Committee for Aeronautics (NACA) and other government agencies.

The success of Sputnik, exacerbated as it was by later Soviet accomplishments in space flight, would naturally create stress in the United States. For much of American history, and certainly throughout the twentieth century, if there is one hallmark of the American people, it is their enthusiasm for technology and what it can help them to accomplish. Historian Perry Miller wrote that the Puritans of New England "flung themselves in the technological torrent, how they shouted with glee in the midst of the cataract, and cried to each other as they went headlong down the chute that here was their destiny" as they used technology to transform a wilderness into their "City upon a hill." Since that time the United States has been known as a nation of technological system builders who could use this ability to create great machines and the components of their operation, of wonder.

Perceptive foreigners might be enamored with American political and social developments, with democracy and pluralism, but they are more taken with U.S. technology. The United States is not just the nation of George Washington, Thomas Jefferson, Abraham Lincoln, Frederick Douglas, and Elizabeth Cady Stanton, but also of Thomas Edison, Henry Ford, the Tennessee Valley Authority, and the Manhattan Project. These people and achievements reinforced the belief among Americans that their nation was the technological giant of the world. The Soviet success with *Sputnik 1* raised in a very fundamental way the question of American technological virtuosity, and questioned American capability in so many other areas already underway that setbacks in this one was all the more damaging to the American persona.

At the same time, Sputnik ensured a hubris among the citizens of the Soviet Union not seen to the same degree since the end of World War II in 1945 or to be experienced again in that empire's history. It represented a high-water mark of success and Nikita Khrushchev's leadership exploited it to the fullest for the next decade. Thereafter, with high priorities given the effort by the Soviet leadership, the communist state's rocketeers led the way in one stunning success after another:

— The first living thing in orbit, the dog Layka launched on *Sputnik 2* in November 1957.

— The first human-made object to escape the Earth's gravity and to be placed in orbit around the Sun, *Luna 1* in January 1959.

— The first clear images of the Moon's surface in September 1959 from *Luna 2*.

— The first pictures of the far side of the Moon, October 1959, taken by the Soviet Union's *Luna 3*.

— The first return of living creatures from orbital flight, two dogs sent into space by the Soviet Union in August 1960 aboard *Sputnik 5*.

— The first human in space, Soviet Cosmonaut Yuri Gagarin, who flew a one-orbit mission aboard the spacecraft *Vostok 1*, April 1961.

— The first day-long human space flight mission, August 1961, made by *Vostok 2* with cosmonaut Gherman Titov aboard.

— The first long duration space flight, Cosmonaut Andrian Nicolayev spent four days in space aboard *Vostok 3*, August 1962.

— The first woman in space, Soviet cosmonaut Valentina Tereshkova, flew 48 orbits aboard *Vostok 6* in June 1963.

— The first multi-person mission into space, *Voskhod 1* carrying cosmonauts Komarov, Yegorov, and Feoktistov in October 1964.

— The first spacewalk or extravehicular activity, March 1965, by Alexei Leonov during the *Voskhod 2* mission.
— The first soft landing on the Moon, the Soviet Union's *Luna 9* returns the first photos to Earth from the lunar surface in February 1966.
— The first landing on another planet, *Venera 3* crash landed on Venus in March 1966.

It looked as if the Soviet Union did everything right in space flight, and the United States appeared at best a weakling without the kind of capabilities that the command economy of the "workers' state" in the Soviet Union had been able to muster. The result was that the United States mobilized to "catch up" to the apparent might of its Cold War rival. As surely as the several crises in Berlin—the blockade and airlift, the wall—and the other flashpoints of competition, Sputnik served to fuel the antagonism and steel the resolve of both sides.

Sputnik was a watershed, no doubt. Almost immediately after the event, two phrases entered the international lexicon to define time, "pre-Sputnik" and "post-Sputnik." The other phrase that soon replaced earlier definitions of time was "Space Age," for with the launch of *Sputnik 1* the space age had been born and the world would be different ever after. Such a demarcation may be over-wrought, but the difference between the world a decade before and a decade after the satellite is remarkable.

At the fortieth anniversary of the launch of that critically important Soviet satellite, the NASA History Office, the National Air and Space Museum of the Smithsonian Institution, the Space Policy Institute of The George Washington University, the Kennan Institute for Advanced Russian Studies of the Woodrow Wilson Center for International Scholars, and the D.C. Space Grant Consortium teamed to co-sponsor a well-received conference that attracted much media attention, "Reconsidering Sputnik: 40 Years Since the Soviet Satellite," on September 30 and October 1, 1997. The symposium involved more than 150 participants drawn from throughout the world, and some thirty media. Because the Cold War has now ended, and because of the resultant opening of archives on both sides, the meeting was especially significant because of the new perspectives that could now be brought to light.

Because of the success of that symposium, the organizers have assembled selected essays first presented at the meeting, added additional research essays, and assembled the set into a cohesive and comprehensive collection that explores the planning for, the immediate aftermath of, and the long-term consequences of the launch of Sputnik. The collection is introduced by Walter A. McDougall, professor of history at the University of Pennsylvania and author of the Pulitzer Prize winning book, ...*The Heavens and the Earth: A Political History of the Space Age* (1986), who takes a long view of the themes explored in the various essays. There follows essays in three major parts, "Space Flight in the Soviet Union, A Setting for the International Geophysical Year," and "Ramifications and Reactions."

In the first part, historians of the Soviet Union explore the reasons for the Soviet Union's stunning success with Sputnik. They describe the long-standing scientific basis for space flight from the work of Konstantin Tsiolkoskiy of Tsarist Russia through the pioneering technical work of Sergey Korolev and the high priority given the Soviet space program by national leaders as the 1950s emerged with its promise of ballistic missiles as a weapons delivery system in the offing. These efforts ensured the development of a powerful and significant technical community, what might be termed the Soviet equivalent of the aerospace industry, that held considerable sway in the policymaking arena of national defense.

In the second part, two historians and two political scientists explore the origins and development of the International Geophysical Year and how it related to the larger questions of national security, international prestige, and scientific advancement.

Finally, in the last part, five scholars investigate the long-term ramifications of the satellite. Sergey Khrushchev, son of the Soviet premier and an aerospace engineer, discusses how the Soviet space program evolved as a result of its early success. Other scholars look at the results of the crisis from the vantage point of western Europe, the American effort to keep the heavens unarmed, and the importance Sputnik had for the educational community in the United States. Fittingly, the collection is brought to a conclusion with an essay on the role of technological surprise in history and an assessment of why Sputnik held such surprise for America and what it has meant thereafter.

Whenever historians take on a project of historical synthesis such as this, they stand squarely on the shoulders of earlier investigators and incur a good many intellectual debts. I would like to acknowledge the assistance of several individuals who aided in the preparation of this overview of space exploration history. Lee D. Saegesser, NASA History Office Archivist, was instrumental in obtaining documents used in the preparation of this book and in providing other archival services; Stephen D. Garber, assistant historian in the NASA History Office, provided important help in the preparation and critiquing of the text; Nadine Andreassen helped with proofreading and other editorial details, as Louise Alstork; the staffs of the NASA Headquarters Library and the Scientific and Technical Information Program provided assistance in locating materials; and archivists at various presidential libraries, the National Archives and Records Administration, and in other research centers aided with research efforts. Several individuals read portions of the manuscript or talked with the presenters and editors, in the process helping us more than they could ever know. These include Brian Balogh, Donald R. Baucom, Linda Billings, Roger E. Bilstein, Andrew J. Butrica, Michael L. Ciancone, Tom D. Crouch, Robert Dallek, Virginia P. Dawson, Henry C. Dethloff, David H. DeVorkin, Robert A. Divine, Deborah G. Douglas, Andrew Dunar, Donald C. Elder, Robert H. Ferrell, Michael H. Gorn, Charles J. Gross, R. Cargill Hall, Richard P. Hallion, James R. Hansen, Gregg Herken, Norriss S. Hetherington, Robin Higham, Francis T. Hoban, Joan Hoff, J.D. Hunley, Karl Hufbauer, W.D. Kay, Sylvia K. Kraemer, W. Henry Lambright, Howard E. McCurdy, Pamela E. Mack, John E. Naugle, Allan E. Needell, Lyn Ragsdale, Joseph N. Tatarewicz, Stephen P. Waring, and Mike Wright. All of these people would disagree with some of the essays included, but such is both the boon and the bane of historical inquiry.

Introduction

Was Sputnik Really a Saltation?

Walter A. McDougall

Nineteen years ago I came to the history of space technology as a novice, and if I accomplished something worthwhile it was due to the support I received from Monte Wright, Alex Roland, and Lee Saegesser of the NASA History Office, from the Woodrow Wilson Center and Kennan Institute, from the Smithsonian's Air and Space Museum, and from John Logsdon of the Space Policy Institute—in short, all the sponsors of this week's celebration. As some of you know, that phase of my career was fraught with personal and professional crises, but I always remember those years with nostalgia, because of the unfailing generosity and collegiality which the space history and policy community showed me, even though I was something of a poacher on your reservation. And since I see that many of the topics I wrote about are on this week's agenda, let me hasten to quote what I wrote in the preface to ...*The Heavens and the Earth*: "Future historians will have much more to say, and will surely correct much of what I write here." I'm sure some of you intend to do just that—pray, do so gently!

Something else that will always bind me to you is our common enthusiasm for the exploration of outer space and the technologies that make it happen. But the passion we share also means that most of us probably look back on the four decades of the Space Age with a sense of unfulfilled promise and profound disappointment. I do not mean that tremendous strides have not been made. Arthur C. Clarke's vision of a world united through satellite communications has been reified, and earth sensors have revolutionalized meteorology, oceanography, pollution control, extractive industries, even archaeology. Robotic explorers have increased our understanding of the solar system, and astronomical satellites and telescopes our knowledge of the universe, by several orders of magnitude, as a mere glance at my 1964 college Astronomy textbook brought home to me. The strides made in the military applications of orbiting systems were made so abundantly clear in the Gulf War that strategic thinkers from the Beltway to Baikonur, Bagdad, and Beijing now routinely debate how to adjust to the so-called Revolution in Military Affairs.

But how many of our most exciting hopes have been dashed? Consider, for instance, that in the months and years after the great Sergey Korolev oversaw launch of the first artificial earth satellite, experts confidently testified before Congress or wrote in the press that by the year 2000 the United States and Soviet Union would found lunar colonies and elaborate space stations, and compete for the high ground with fleets of laser-armed spaceships. The Stanley Kubrick film based on Clarke's novel *2001* depicted Hilton hotels on the Moon and a manned mission to Jupiter. And how many remember that the birthday of the on-board supercomputer "Hal" in *2001* was January 12, 1992?

NASA promoters were more cautious, but by 1970 they forecast reusable spacecraft ascending and descending like angels on Jacob's ladder, permanent space stations, and human missions to Mars—all to be achieved, if the money was there, within a decade or two. By the mid-1970s space visionaries imagined using the Space Shuttle to launch into orbit huge solar panels to beam unlimited, non-polluting energy to earth, hydroponic farming in space to feed the earth's exploding population, and colonies at the libration points in the terralunar system. In the 1980s the space station idea revived, again to be completed "within a decade," and the Strategic Defense Initiative promised directed-energy weapons in orbit to render ballistic missiles obsolete, and the space telescope was to unlock the last secrets of the universe. By 1990, a manned mission to Mars by the year 2010 and an aerospace plane (the Orient Express) were again on the President's wish list.

None of it came to pass. Instead, the dream of limitless progress through government-sponsored research and development began to fade even before astronauts first stepped on the Moon. NASA's budget began to decline in 1966, and by the time Apollo wound down many of its best engineers had retired or left for the private sector, its rocket and spacecraft design teams broke up, and the agency lost the creative synergy and "institutional charisma" of its early heroic years, just as Max Weber would have predicted. The

Nixon Administration then chose to discard the incomparable Saturn/Apollo system and start from scratch on a reusable launch vehicle that was promptly oversold and underfunded. Perhaps NASA administrators in the 1970s were remiss when they promised to do the job on a shoestring, but however one apportions the blame, the result was that when the Shuttle fleet finally began flying in 1981 it was late, way over budget, troubled by bugs, capable of just four to six missions per year instead of the 24 promised, and—far from cutting the cost-per-pound of launching payloads "by a factor of ten"—increased the cost several times over that of the old Saturn V. What is more, the Shuttle ate up so much of the space program budget that robotic exploration, science, and space applications went begging. The *Challenger* accident was a revelation not because a spacecraft was lost (which was bound to happen sooner or later), but because the nation witnessed an agency, so recently praised as a model of squeaky-clean efficiency, in such turmoil that an outsider, Dr. Richard Feynman, had to be called in to discover where NASA went wrong.

The story of the space station has been equally frustrating, and the Hubble Space Telescope plagued from the start. But the foibles of Space Age technocracy have been most strikingly exposed in the fate of the regime that made technocracy its founding principle. Not just its space program, but the Soviet Union itself crashed and burned. And when the Mir finally re-enters the atmosphere, aborigines some- where on earth, like those who chanted for the Mercury capsule in Tom Wolfe's *The Right Stuff,* will stare in wonder at the fire from space and declare it an augury that somewhere an empire has fallen, somewhere a god has just died.

Perhaps the litany of unfulfilled promises is what prompted John Logsdon to ask whether I still believe Sputnik was a saltation, a truly discontinuous leap forward in the otherwise incremental evolution of human technology and its relationship to the State. When I sat down to write the book I was surrounded by documents that showed how reti- cent Truman and Eisenhower had been to embrace government-funded and -directed technological revolution—what I called technocracy—and how eager Kennedy and Johnson were to promote it. To be sure, I knew that the story of government support for R&D was an old one, mostly revolving around war, and that the Manhattan Project was the most obvious antecedent to the missile and space programs. But the explosion of quantitative support for science and technology in the late 1950s and 1960s, and especially the qualitative shift that labeled government R&D no longer a necessary evil but a positive boon of unlimited promise, seemed to me so rapid and thorough as to amount to a saltation.

Only twelve years have passed since the book first appeared, but during that time government-sponsored technological revolution fell increasingly into disfavor for a number of reasons: the cultural critiques of the 1960s that damned the military-industrial complex, the Cold War arms race, the technocratic mentality behind the Vietnam War, the environmental and feminists movements that identified big technological systems with pollution and a kind of macho assault against Mother Nature, the letdown after Apollo, the energy crisis and economic malaise of the 1970s, the growing demand that scarce resources be deployed against problems on earth, the conservative reaction against big government in the 1980s, and finally the end of the Cold War and the 1990s campaign to balance the budget. One hammer blow after another fell on those who yearned to revive the spirit of the Kennedy–Johnson years, or even the "slow but steady" approach to space favored by Eisenhower. Today, privatization and downsizing are the watchwords in the public and private sectors alike, at home and abroad, and they make the rhetoric of a Kennedy or Khrushchev and the methods of a James Webb seem shockingly dated.

In retrospect, therefore, the post-Sputnik burst of enthusiasm for state-directed technological revolution seems to have been an ephemeral episode in the larger history of the Cold War, rather than the Cold War having been an episode in the larger story of the march of technocracy. Curiously, my own arguments in the book hinted at just that conclusion. As a diplomatic historian I held that international competition was the first cause driving governments to mobilize ever greater resources in an effort to invent and control the future. I even anticipated somewhat the Thatcher/Reagan/Gorbachev retreat from big government when I echoed Eisenhower's suspicions about the politicization and bureaucratization of science, technology, and education. But I tripped up as an historian when I failed to realize that if Sputnik was really a "technological Pearl Harbor," then the real watershed in the role of the U.S. government must still have been Pearl Harbor! And I tripped up as a prophet because, I now realize, I still wanted to believe that the Space Shuttle would usher in an exciting new era, that public excitement about space would rekindle after the post-Vietnam funk, and that in any case international competition—whether military or commerical—would drive the American space program forward in spite of Mugwump Senators like Mondale and Proxmire. And so, however heady my historical account, I let my heart have the last word.

So the answer to your question, John, is, no, I no longer think that *saltation* was the right label for the chain of events kicked off by Sputnik. Indeed, far from being too pessimistic about technocracy, it seems that in the short run I was not pessimistic enough. But that is hardly the keynote I wish to strike for this conference. So let me suggest instead that in the larger story and longer time frame the metaphor may yet prove apt.

If you've read my book you may recall that I was inspired to use the word *saltation*—indeed, I learned the word—when by chance I passed by an exhibit at the Smithsonian's Natural History Museum. It depicted the beast Eusthenopteron, believed to be the first amphibian, crawling out of a stream bed rendered muddy and viscous by climatic change, in hopes that the embryonic lungs his ancestors had mutated would allow him to survive on the land. In the year 1961 humanity repeated the amphibian's feat when Yuri Gagarin left the earth and lived, for 108 minutes, in outer space.

Now, we certainly hope that human beings will never need to go into space because the earth has been rendered uninhabitable. But whatever the mix of our motives, and whatever long periods of inertia may slow our advance, human beings will no more abandon their final frontier anymore than amphibians turned back into fishes when the next geological era turned wet. We should, in fact, be careful not to extrapolate the future from today's present gloom anymore than I should have extrapolated on the basis of the mood of the 1960s. For you who keep the flame burning, and work for years and even careers on reduced life support, trying to push the envelope of science and engineering, will someday be rewarded. For some year, somewhere, some human brain will take all we have learned from our triumphs and failures and transcend them in a burst of true creativity. It will not be the sort of pedestrian creativity that characterizes the normal science most of us do, whereby we take a good idea from one branch of learning and apply it by analogy to another. Nor will it be the sort of dialectical creativity that combines existing ideas in some novel way. Rather, it will be that stroke of true genius that uncovers some unknown and perhaps unsuspected principle of physics, something that makes possible an entirely new launch technology that is truly cheap, reliable, and safe: that liberates humanity not only from the clumsy chemical rocket, but from Konstantin Tsiolkovsky's "prison of terrestrial gravity." It may not happen in the labs of government agencies or big aerospace firms. It may not even happen in some science nerd's garage in L.A., as I used to joke. And the spark of genius may not occur in the United States at all, but in Kazakhstan, Japan, or Sri Lanka. But when that day comes, and some of you may live to see it, then space flight will indeed deserve the analogy Bruce Mazlish drew between the rocket and the railroad of the nineteenth century, human beings will quickly become a race of romping cosmic amphibians, and space flight will indeed merit the label *saltation*.

Part 1
Space Flight in the Soviet Union

Space Flight in the Soviet Union

Roger D. Launius

With the launch of Sputnik on October 4, 1957, a scrambling for explanations ensued about how the Soviets had suddenly bested the United States, arguably the most technically advanced civilization the world had ever known. In fact, as the essays in this part of this book demonstrate, the dream of space flight had enjoyed a long tradition in the Soviet Union. Beginning near the turn of the twentieth century two important developments converged in Russia that made possible Sputnik's extraordinary success: science fiction literature sparked the enthusiasm of a generation of engineers and technicians who longed for the possibility of actually going into space and exploring it firsthand, even as rocket technology began to mature and thereby brought a convergence of dream with likelihood.[1]

Peter Gorin traces the relationship of these two elements during the pre-Sputnik era in the Soviet Union. He begins with the role of a small group of pioneers who developed the theoretical underpinnings and the technical capabilities of rocketry in the first half of this century. Four towering figures have been held up as the godfathers of modern space exploration, largely because of their work on rockets. Gorin finds that the the most significant for the Soviet Union, if not the earliest, was Konstantin E. Tsiolkovskiy. The others included the German Hermann Oberth, the American Robert H. Goddard, and the Frenchman Robert Esnault-Pelterie. Collectively, these men developed theories of rocketry for space exploration, experimented with their own rockets, and inspired others to follow in their footsteps.

Tsiolkovskiy, born on September 17, 1857, in the village of Izhevskoye, south of Moscow, had become enthralled with the possibilities of interplanetary travel as a boy by reading science fiction literature, and at age fourteen started independent study using books from his father's library on natural science and mathematics. He first started publishing on rocketry in 1903, and over the course of the next thirty years he described in depth the use of rockets for launching orbital spacecraft. Virtually unknown until after the Bolshevik revolution of 1917, in the 1920s the Soviet Union glorified his accomplishments in the theory of space flight as an offset to accomplishments in other nations around the world. After his death in 1935 Tsiolkovskiy's theoretical work came to influence later rock-eteers mostly in his native land. While much less well known in the United States, Tsiolkovskiy's work enjoyed broad study in the 1950s and 1960s as Americans sought to understand how the Soviet Union had accomplished such unexpected success in its early efforts in space flight. American space scientists realized upon reading Tsiolkovskiy that his theoretical foundations had provided an important basis for the development of practical rocketry.[2]

Gorin posits the development of a "space boom" in the Soviet Union in the 1920s. The result of seeds about the possibilities of space flight sown earlier by such theorists as Tsiolkovskiy, coupled with the end of civil war and an opening of borders to outside influences, this boom influenced several engineers to explore the possibilities for themselves. Frederik Tsander, Nikolay A. Rynin, Sergey Korolev, and Valentin Glushko all caught the space flight "bug" during this period of open-ness. The first space club in the Soviet Union, the Society for the Study of Interplanetary Communications under the nominal aegis of the Moscow Air Fleet Academy, was organized during this period, and as with rocket societies in other nations it exerted an important influence on the space flight imperative in the nation.[3]

Gorin also discusses the overwhelmingly significant development of the Reactive Scientific Research Institute (RNII), which directed much of the rocketry research in the Soviet Union beginning in the 1930s. Funded by the Red Army, this organization became the spawning ground for Sputnik-era aerospace leaders such as Korolev and Glushko. Hit hard in Stalin's purges of the late 1930s, the RNII barely survived. Korolev, for instance, spent nearly a decade in a Gulag. At no time, Gorin concluded, did he give up on his goal of space flight and, with the coming of World War II and the advent of the missile era wrought by the V-2, Stalin put him and his colleagues to work on missile development. The result was that after the war the Soviet Union was well-placed to enter practical space travel.

Indeed, Gorin asserts that the twin development of science fiction, which fueled the Soviet imagination, and rocketry, which made possible actual flight, combined to prepare a people for Sputnik. Indeed, the decade following the war brought a sea change in perceptions, as most Soviets went from skepticism about the probabilities of space flight to an acceptance of it as a near-term reality. An important shift in perceptions took place during that era, and it was largely the result of well-known advances in rocket technology coupled with a public relations campaign based on the real possibility of space flight.[4]

Gorin appropriately concluded that the combination of technological and scientific advance, political competition with the United States, and changes in popular opinion about space flight came together in a very specific way in the 1950s to affect public policy in favor of an aggressive space program. This found tangible expression in the 1950s when the International Council of Scientific Unions (ICSU) made a satellite a desired objective of the International Geophysical Year (IGY). The Soviet Union immediately announced plans to orbit an IGY satellite, virtually assuring that the United States would respond, and this, coupled with the military satellite program, set both the agenda and the stage for most space efforts through 1958.[5]

In the second essay in this section, Asif Siddiqi explores in depth the decisionmaking process that led to the Soviet support of a satellite effort for the IGY and of the place Sergey Korolev held in bringing it to fruition. Siddiqi describes for the first time the inner-workings of the policy-making apparatus in the Soviet Union that led to the decision to launch a satellite as part of the IGY. He finds that a similar approach to that seen in the United States took place as the emphasis bubbled up from individual interest groups, each with something to gain by supporting it. Essentially five major interest groups, broadly considered, pushed toward an aggressive effort in space. The first of these groups was a loose confederation of physicists, geodesists, and astronomers interested in advancing basic research and obtaining the funding support required to do so. The second group consisted of leaders of the armed services who wanted to use scientific understanding of space for military ends. The third group included engineers working in the state design bureaus who foresaw the opportunity to develop their fortunes. Fourth, there were the space flight enthusiasts who were eager to pursue exploration beyond the atmosphere because of the adventure it would provide and the rather protean idea of human destiny that it invoked. Finally, there were policymakers and government executives who pursued space activities because of a myriad of political ends that might be fostered. Those ends ranged from large-scale strategic victories in the cold war with the Unites States to the potential for garnering greater political power.

These major groups of advocates were by no means either monolithic or mutually exclusive, and members of one were often also members of another of the remaining four. Sometimes dedicated scientists, such as Academician Mstislav V. Keldysh, were also government leaders. Some advocates from the armed forces, such as army Col.-Engineer Aleksandr G. Mrykin, were also space flight enthusiasts. Technical people, regardless of their membership in other interest groups, were often adamant cold warriors. All of these groups came together in a loose coalition aimed at decisive space operations. The complex process of decision-making involved in this arena informs the Siddiqi study.

While present in the two essays already presented, Sergey Korolev takes center stage in the study by James Harford as the central actor in the accomplishment of the early Sputnik successes and how they affected the Soviet space program to move in a certain direction thereafter. Harford describes the details of the development of the first three Sputnik spacecraft. The Soviet Union's R-7 ballistic missile was the carrier for all three of these spacecraft, and its superior capability to hurl several tons into orbit ensured that Korolev could achieve his goals. He and other Soviet scientists planned to launch into orbit for the IGY a huge, sophisticated satellite with instruments that would greatly enhance understanding of the geophysical nature of the Earth. But this satellite fell behind schedule—either because of foot-dragging on the part of military officers or others who thought the IGY effort less important than other national priorities or because of normal technical difficulties. Accordingly, as 1957 entered the mid-year point, Korolev shelved this large spacecraft in favor of a more simple design that could be built and launched within a short period of time, regardless of the marginal scientific value of the data returned from it.

Harford quoted Georgiy Grechko, one of the Korolev design bureau engineers, about this matter. His comments would have been impossible to obtain just a few years ago, but provide a telling point about the Soviet effort. "These devices were not reliable enough so the scientists who created them asked us to delay the launch month by month," he said. "We thought that if we postponed and postponed we would be second to the U.S. in the space race so we made the simplest satellite, called just that—*Prostreishiy Sputnik*, or 'PS'. We made it in one month, with only one reason, to be first in space."[6] Not until May 15, 1958, did the Soviet Union launch the most sophisticated of these satellites, now named *Sputnik 3*, a weighty spacecraft of 1,327 kilograms.

In the interim Korolev had, perhaps through sheer force of will, launched *Sputnik 1*, a spherical satellite containing only a radio transmitter, batteries, and temperature measuring instruments, on November 3, 1957, and *Sputnik 2* containing the dog Layka on November 3, 1957. These launches took place atop a rocket that had failed in five of its first six launch attempts. Korolev had made a nervy decision and it had paid off spectacularly for the Soviet Union. Soviet Premier Nikita Khrushchev, probably surprised at the reaction to *Sputnik 1*, nonetheless seized the initiative and demanded more of Korolev. Harford recounts the words of Khrushchev to Korolev: "We never thought that you would launch a Sputnik before the Americans. But you did it. Now please launch something new in space for the next anniversary of our revolution." He did so, *Sputnik 2* went up in commemoration of the 40th anniversary of the Bolshevik revolution.[7]

The essays of both Siddiqi and Harford provide detailed internal information that was not available before the end of the cold war. They show the complex decision-making process inside the Soviet bureaucracy and the nature of Krushchev's ad hoc leadership as he seized an opportunity. They also demonstrate that much of what we have believed about Soviet space policy has been in error. It was not, at least in the Sputnik era, rigorously thought out and formally thought through. Instead, it appears much more reactive than provocative and represents what some have called the art of muddling through. In this particular instance, the Soviet's appeared masterful to the outside world, thereby steeling an image of ominous technical capability that was an illusion.

To sustain that illusion the Soviet Union invested heavily in space flight technology, as the essay by William Barry demonstrates. He seeks to plow the depths of what he refers to as "an impenetrable bureaucratic maze of design bureaus, state research and production centers, science-production associations, and other 'space corporations'."[8] In every case, this capability was built squarely on the national defense establishment, something that military leaders did not necessarily view as desirable since it took needed resources away from weapons development. Korolev, Barry finds, was delighted with the emphasis on space exploration, but this set him at odds with the military and he had to fight a constant rear-guard against their raids on his resources. Korolev enjoyed the support of Khrushchev, but with his ouster in the mid-1960s Korolev's star fell never to return. Korolev's death in 1966 was probably a blessing in that he did not have to suffer the bureaucratic attacks that most assuredly became more vicious as the decade progressed.[9]

Barry's conclusion that the Soviet space industry was an unusual hybrid of civil and military industrial capability with the two locked like Yin and Yang in an eternal contest provides a powerful counterpoint to the American experience. What support it received came largely because it provided the opportunity for propaganda, overseen and coordinated at the very highest levels of government. He concludes that this "military-political duality prevented the creation of a separate civil space organization on the model of NASA. Although the USSR tried to create the appearance of having a civil space agency (pretending first, that the Academy of Sciences ran the space program, and later setting up Glavkosmos), the Soviet space industry remained subordinated to defense industrial ministries."[10]

Collectively, these four essays represent a modern synthesis of the actions leading up to the Sputnik launch in the fall of 1957 and its immediate ramifications. They complement each other in showing the duality of military and civil priorities for space, the relationship of popular culture and public policy, and the intense nastiness of the cold war struggle then in full swing. Forty years after the fact, the

Soviet actions in launching that first satellite are still present as it shapes the space age.

1. Roger D. Launius, "Prelude to the Space Age," in John M. Logsdon, General Editor, with Linda J. Lear, Jannelle Warren-Findley, Ray A. Williamson, and Dwayne A. Day, *Exploring the Unknown: Selected Documents in the History of the U.S. Civil Space Program, Volume I, Organizing for Exploration* (Washington, DC: NASA Special Publication 4407, 1995), pp. 1–4.

2. Konstantin E. Tsiolkovsky, *Aerodynamics* (Washington, DC: NASA, TT F-236, 1965); Konstantin E. Tsiolkovsky, *Reactive Flying Machines* (Washington, DC: NASA, TT F-237, 1965); Konstantin E. Tsiolkovsky, *Works on Rocket Technology* (Washington, DC: NASA, TT F-243, 1965); Arkady Kosmodemyansky, *Konstantin Tsiolkovskiy* (Moscow, USSR: Nauka, 1985).

3. The standard work on the rocket societies is Frank H. Winter, *Prelude to the Space Age: The Rocket Societies, 1924–1940* (Washington, DC: Smithsonian Institution Press, 1983).

4. These same took place in the United States, as documented in Howard E. McCurdy, *Space and the American Imagination* (Washington, DC: Smithsonian Institution Press, 1997).

5. A good account of the IGY satellite projects can be found in Rip Bulkeley, *The Sputniks Crisis and Early United States Space Policy* (Bloomington: Indiana University Press, 1991), pp. 89–122.

6. Georgiy Grechko interview with author, Moscow, May 16, 1993, quoted in James H. Harford, "Korolev's Triple Play: *Sputniks 1, 2, and 3.*"

7. Quoted in Ibid.

8. William P. Barry, "Sputnik and the Creation of the Soviet Space Industry."

9. The standard work on Korolev is James J. Harford, *Korolev: How One Man Masterminded the Soviet Drive to Beat America to the Moon* (New York: John Wiley & Sons, 1997).

10. Barry, "Sputnik and the Creation of the Soviet Space Industry."

CHAPTER 1

Rising from the Cradle: Soviet Perceptions of Space Flight Before Sputnik

Peter A. Gorin

PRECURSORS OF SOVIET SPACE EXPLORATION

Long before the launch of *Sputnik 1* and the dawn of the space age in 1957, the Russian people had been prepared for its eventual coming. Public perception of space travel began forming in Russia near the beginning of the twentieth century under the same conditions as in other developed nations. Major advances in technology, new discoveries in physics, chemistry, and astronomy in the second half of the nineteenth century caused a dramatic reassessment of philosophical views on the place of human civilization in the universe. In the Russian Empire, where the Orthodox Christian philosophy was arbitrarily imposed on the public by the imperial government, that reassessment was not an easy task. Many Russian philosophers struggled to find a compromise between their strong religious beliefs and new scientific facts. Yet, even the most hard-line religious proponents eventually had to accept the notion that the human race was not the center of God's creation but just one of perhaps many inhabited worlds in the Universe. This laid a foundation for philosophical justification of the idea of space flight.

There was, however, a deep gap between these new scientific discoveries and the public knowledge of them. Like other countries, the Russian scientific community was not very cooperative in distributing its knowledge among the "non-qualified public masses." With few exceptions, professional scientists did not go beyond specialized journals and did not bother to use common language in their writings.

The man who really made a difference on this score in Russia was a young physicist from the city of St. Petersburg, Yakov I. Perelman (1882–1942). In 1913, he published a book, *Zanimatelnaya fizika* (*Entertaining Physics*), that explained many physical phenomena in the form of entertaining stories and in simple language. This immensely popular book outlived its author for decades and enjoyed eighteen printings in millions of copies. Perelman actually founded public science literature in Russia. His next book *Mezhplanetnye puteshestviya* (*Interplanetary Travel*) became the first Russian popular scientific account of space flight. Printed ten times between 1915 and 1935, it played a major role in publicizing the idea of space exploration.[1]

Another component of the public perception of space flight was science fiction. Stories about travel to another planets had existed for centuries, but only in the second half of the nineteenth century did they become scientifically oriented. With the appearance of talented and scientifically-minded writers like Jules Verne (1828–1905) in France, science fiction became a widely popular literary genre. Verne's *From the Earth to the Moon* (1865) was probably the first fiction novel on space flight with a solid scientific foundation. That novel (along with other Verne writings) was translated into Russian and became widely popular. Generations of young people in Russia and the Soviet Union owed their interest in space travel to Verne's novels. One of those whose inspiration from *From the Earth to the Moon* became a lifetime dedication was a young boy from the city of Odessa, Valentin P. Glushko (1908–1989). He read the novel in 1921 and in 1929 joined a military rocket research group, the GDL. Twenty-eight years later, the rocket engines developed under Glushko's leadership propelled Sputnik into space.[2]

Interpretation of the newest trends in astronomy through science fiction also boosted the growing popularity of the space theme. For example, the misconception of "Martian channels" led to a wide-spread assumption in Russia of a civilization on Mars. That made Mars, along with the Moon, the most "frequently visited" celestial body in science fiction writings. Along with the translations of foreign authors, such as H. G. Wells (England) and Camille Flammirion (France), the Russian public was entertained by a number of native writers who tried to explore the universe in their fantasy. Russia was in a deep political and economic crisis in those years, but even politically active intellectuals paid attention to such a popular subject. One of the novels on visitors from Mars, *Krasnaya zvezda* (*Red Star*), was published in 1908 by Aleksandr A. Bogdanov (Malinovskiy, 1873–1928)—a medical doctor, philosopher, and a member of a dissident group in the Bolshevik party. Unlike the blood-thirsty aliens from H. G. Wells' classic *War of the Worlds*, Bogdanov's "Martians" were seeking cooperation with humans.[3]

The demise of a rocket as a weapon in the second half of the nineteenth century did not discourage inventors. Many of them continued to seek and perfect immediate applications of jet propulsion. Some projects went far ahead of that time and level of technology. For example, rocket-like airplanes powered by steam thrusters were independently proposed by Russian engineers Nikolay A. Teleshov (1828–1895)[4] and Fedor R. Geshvend (1838–1890)[5] in 1867 and 1887.

These and other flying machine studies did not have a direct connection with the idea of space travel. At the same time, the philosophers who dreamed about voyages to the cosmos, did not pay much attention to the possible means of transportation. Konstantin E. Tsiolkovskiy (1957–1935) was the first Russian researcher who combined rocket technology with the idea of space flight. A school teacher of mathematics from a small provincial town of Kaluga, Tsiolkovskiy devoted all his spare time to theoretical research in aviation and space. He not only outlined the technical possibility of space flight, but also stressed the necessity of conducting space exploration. He wrote: "Humankind will never stay on Earth forever but in pursuit of flight and living in space it will, cautiously at first, go beyond the atmosphere and then conquer for itself the whole Solar System."[6] By his own account, he also was inspired by the book *From the Earth to the Moon*. At the same time, he conducted a thorough analysis of Verne's novel. Tsiolkovskiy calculated that Verne's huge cannon for sending a piloted projectile into space could not have achieved escape velocity and would have killed its passengers. His search for an alternative and scientifically feasible way of space transportation led to the idea of using rockets. In Tsiolkovskiy's studies, science fiction became a starting point for space science.

Tsiolkovskiy was not the only Russian researcher to consider a rocket for space flight. Such ideas appeared in the nineteenth century in several publications. For example, the engineer Aleksandr P. Fedorov (1872–?) published a monograph *Novye printsipy vozdukhoplavaniya* (*New Principles of Air Flight*, 1896).[7] Although not particularly space-oriented, it clearly indicated that a rocket did not require any support from the atmosphere. Tsiolkovskiy was, however, the first to prove mathematically that only a rocket could be used for space travel. He formulated the "Tsiolkovskiy Equation" which demonstrated that the final acceleration of the rocket would depend on factors such as initial masses of the rocket and propellant, and the speed of exhaust gases from rocket engines. Tsiolkovskiy envisaged a rocket not only as a propulsion device, but as a complex of engines, life-support systems, scientific instruments, etc. Being one of the first to consider liquid explosives instead of a gunpowder for propulsion, he proposed liquid oxygen and liquid hydrogen as propellant components. From his point of view, along with feeding the engines, these propellants would provide air and water for the crew.

Starting in 1883, Tsiolkovskiy wrote a number of books where he predicted artificial satellites, manned orbital stations, and liquid pro-pellant rockets for lifting humans from the Earth into orbit. At first he presented his ideas in the popular form of science fiction. For him, science fiction was not just thrilling story-telling—it was a method of introducing the most revolutionary scientific ideas without fear of ridicule. Tsiolkovskiy's major scientific monograph *Issledovaniya mirovykh prostranstv reaktivnymi priborami* (*Exploration of Space by Reactive Devices*) was published in portions from 1903 to 1914.[8] Unfortunately his books were not widely known to the public due to a very small publication run (he paid for the publications from the scarce resources of his family). Nevertheless, Tsiolkovskiy's ideas slowly reached an audience. One reader was a young engineer of German/Russian origin, Fredrik A. Tsander (1887–1933). He first read Tsiolkovskiy's work in 1905 and as a result began a lifetime quest for human space flight.[9]

Despite the activities of Perelman, Tsiolkovskiy, and other space enthusiasts, the idea of space travel in Russia did not leave the realm of science fiction in the first two decades of the twentieth century. Its sheer novelty made it difficult to comprehend for a Russian public preoccupied with raging economic and social crises. These problems, aggravated by World War I, resulted in the Bolshevik revolution of October 1917 and a bloody Civil War (1918–1921). Millions of people were killed and the country's economy was ruined. It is difficult to imagine that anybody would have been interested in space flight in those years of intellectual degradation, lawlessness, and overwhelming poverty.

Remarkably, the space pioneers continued their work even under those conditions. For example, Fredrik Tsander started experiments in an "air-light laboratory" in his apartment in Moscow. He researched the possibility of growing various plants and vegetables in liquid solutions instead of soil—hydroponics, now routinely employed by cosmonauts aboard *Mir*—to save weight in food production for space travel. Those first experiments were conducted in a city which lacked heat, food, and all other basic supplies by a lonely enthusiast who did not have enough money even to feed himself properly.

Perhaps the most dramatic story of all was one of another space pioneer, whose dedication to the idea of space travel helped him survive the misery of civil war. Aleksandr I. Shargey (1897–1942) became interested in space travel in high school in the Ukrainian city of Poltava in 1914. He started writing his research diary where he evaluated ideas and calculations related to space flight. Hoping to continue his research as an engineer, he entered the Poly-Technical Institute in Petrograd (St. Petersburg) in 1916. World War I, however, prevented him from pursuing his interests. He was forced into a different kind of study: a crash-course for military officers. Sitting in front-line trenches instead of engineering laboratories, Shargey nevertheless continued writing his research diary.

After a separate peace treaty between Russia and Germany and the disintegration of the imperial army in 1918, *Praporshchik* (Junior Lieutenant) Aleksandr Shargey tried to return home to a peaceful life. His mind was preoccupied with the ideas of acceleration of spacecraft by the gravitational fields of the planets, of reentry from space using the atmosphere for air-braking and gliding to a landing site, and of using a separate lander for an expedition to the Moon's surface while the main spacecraft would stay in orbit. But the reality of the civil war soon returned him from the skies to the ground. The Bolshevik Red Army and anticommunist White Guards Movement fought each other across the vast remains of the Russian Empire. To gain much-needed cannon-fodder, both sides employed the same dirty tactics: recruit anybody capable of carrying a weapon. Those who disagreed had only one alternative—execution by firing squad. Twice Shargey was forced as a private under the banners of the so-called Voluntary White Army of Lieutenant-General Anton I. Denikin and twice he deserted, unwilling to fight against his own people.

After the last escape, he had to keep a low profile for three years to avoid any attention from the retreating White Guards and advancing Soviet troops. In hiding during 1919, he summarized his ideas in a single handwritten draft entitled *Tem, kto budet chitat, chtoby stroit* ("To Those Who Will Read To Build"). Not knowing about Tsiolkovskiy's works, Shargey independently came to the same conclusions on the mechanics of space flight and the use of liquid-propellant rockets. Unfortunately, it was impossible to publish his work at that time, but that draft became a backbone of his future studies.[10]

The end of civil war and the establishment of Soviet power in the Ukraine did not bring relief to Aleksandr Shargey. Once outlawed by the White Guards, he also became an outlaw for the new regime since he had been an officer of the imperial army and twice a "volunteer" of the White Movement. This was more than enough for the political police, the ChK (Extraordinary Commission), to imprison or execute him without a trial. The only way out was to disappear, to change his identity. Thus, early in 1921, Aleksandr Shargey assumed the name of Yuriy V. Kondratyuk—a name that would make him suffer for the rest of his life but immortalize his contribution to space flight theory.[11]

THE "SPACE BOOM"

The 1920s saw a period of slow recovery in Russia from the devastation of the civil war. To stabilize the social situation and restore the economy, the Bolshevik government introduced in 1921 the "New Economic Policy" (NEP) based on a free market. Along with more favorable conditions for economic growth, NEP brought much greater flow of information and ideas across the international borders than had existed immediately after the 1917 revolution.

Foreign specialists were invited to Russia, and foreign books, including those on space flight, were translated and published. The ideas of space travel started to spread among the Russian population once again.

The growing popularity of space travel in Europe, the establishment of interplanetary exploration societies abroad, theoretical works and practical rocket experiments in the United States, Germany, and France were all known inside Soviet Russia and found fertile soil in the minds of the Russian public. For example, the rumors spread by the Soviet press in 1924 that American scientist Robert H. Goddard allegedly was going to send a rocket to the Moon sparked intensive public discussions. Space enthusiasts naturally used that public interest to educate people about space travel.

At first only a few events marked the upcoming "Space Boom." In 1921, Fredrik Tsander reported on his idea of a hybrid between an airplane and a spacecraft at the Moscow Regional Conference of Inventors. According to Tsander, it would be more efficient to use atmosphere for the initial acceleration of the space rocket, using wings and aviation engines. A spacecraft would take off from a conventional airfield like a plane. Then, outside the atmosphere, the rocket engine would take over while wings and other expendable parts would be discarded. Today similar ideas are incorporated in the "Single Stage-to-Orbit" reusable spacecraft under development for the twenty-first century. Tsander's report did not produce noticeable consequences but was probably the first public presentation on space technology in post-revolutionary Russia.

By 1924, the popularity of space travel became widespread. Public lectures and discussions on the subject were common and supporters made attempts to unify their growing numbers. That year, a group of space enthusiasts in the Moscow Air Fleet Academy organized the Section of Interplanetary Communications—the first space club in the USSR, which was later transformed into the Society for the Study of Interplanetary Communications. This society boasted a roster of about 200 members and started a campaign of public lectures. It was in constant contact with the Soviet space flight patriarch, Konstantin Tsiolkovskiy. Nikolay A. Rynin (1877–1942), an aviation specialist and longtime space enthusiast, organized a similar "Interplanetary

Section" in the Leningrad Transportation Institute in 1928. Like their Moscow colleagues, the members of the Leningrad section engaged in public presentations and promotions. Interplanetary clubs soon appeared in many other large cities.[12]

In the years after the October revolution, Tsiolkovskiy temporarily turned his attention from space flight to more feasible projects: jet planes and metal airships. But new developments forced him to return to the subject of space travel. In October 1923, an article in the *Izvestiya* newspaper praised the newly published book of German space pioneer Hermann Oberth (1894–1989), *The Rocket Into Space*, as the first and most complete theoretical work on space flight. Although Oberth's book was clearly outstanding, Tsiolkovskiy had already published similar ideas in 1903. Yet, the Soviet newspaper did not even mention his name. Deeply offended, Tsiolkovskiy published at his own expense a brochure where he outlined his space research. He intentionally gave it the same title, *Raketa v kosmicheskoye prostranstvo* (*The Rocket Into Space*, 1924).[13] This soon became known to Oberth and he established a regular correspondence with Tsiolkovskiy. He wrote to him in 1929: "For sure, I would be the last person to challenge your pre-eminence and your contribution in rocket technology. I regret only that I did not know about you before 1925."[14] Oberth not only acknowledged Tsiolkovskiy as his predecessor but started popularization of his works in the West. Tsiolkovskiy's name was first popularized outside the Soviet Union from Oberth's writings.[15] This notoriety boosted Tsiolkovsky's ideas and his status, soon receiving hundreds of letters and dozens of visitors inspired by his writings.

Popular science and specialized literature on rocket technology and its application in space was numerous in those years. Tsiolkovskiy continued his productivity, and from 1917 to 1935 he wrote and published 70 percent of all his works, including a revised and updated edition of his main monograph *Issledovaniya mirovykh prostranstv reaktivnymi priborami* (*Exploration of Space by Reactive Devices*, 1926).[16] In the last years of his life he researched the idea of a multistaged rocket called the "rocket train" or the "rocket squadron." Tsiolkovskiy's concept was a starting point for research into the Soviet "rocket packet" in the late 1940s that led to the development of the R-7 ballistic missile.[17] On October 4, 1957, the R-7 became the first space launcher, propelling Sputnik into orbit.

Other space pioneers were active as well. In 1924, Fredrik Tsander published his first article, "Flights to Another Planet," in which he outlined the controversial idea of using expendable metal structures of a spacecraft as a propellant. Ideal in terms of weight efficiency, this proposal is extremely difficult to implement even by modern standards. Not knowing about works of Kondratyuk/Shargey, Tsander in later books also proposed the acceleration of spacecraft by planetary gravitation and air-braking in the atmosphere during spacecraft reentry.

Meanwhile, Yuriy Kondratyuk quietly worked as a designer of grain depository facilities in Siberia, spending his spare time writing a monograph on space flight. In 1929, after a favorable assessment of the draft by Vladimir P. Vetchinkin (1880–1950), an authority in aerodynamics and a space flight proponent, Kondratyuk finally published his book *Zavoevanie mirovykh prostranstv* (*Conquest of Interplanetary Space*). Although in many ways his work repeated what had already been published by Tsiolkovskiy, Kondratyuk was immediately recognized as a leading specialist in the field of space flight theory. Some of his original ideas have been realized: *Apollo* used a special lander separate from the orbiting spacecraft for landing expeditions to the lunar surface; acceleration by planetary gravitation fields was utilized on the American *Pioneer-10* and *-11* and *Voyager-1* and *-2* missions to distant planets as well as the Soviet *Vega-1 and -2* missions to Halley's Comet; and the Space Shuttle today uses aerodynamic braking and gliding in the atmosphere.[18]

Yakov Perelman also prepared new editions of his classic *Mezhplanetnye puteshestviya* (*Interplanetary Travel*), updating it with new materials from the work of Soviet and foreign space theorists. The 1929 edition of this bestseller was published in 150,000 copies—more than all English-language space literature combined prior to World War II.[19]

The desire to summarize for educational purposes all the experience gained by that time in rocket technology and the theory of space flight in the Soviet Union and abroad led Dr. Nikolay Rynin to the idea of a space encyclopedia. That fundamental nine-volume book, called *Mezhplanetnykh soobshcheniy* (*Interplanetary Communications*), was published between 1928 and 1932 and became a unique reference for decades to come.[20] It not only outlined all major aspects of the subject, but also provided a historical overview and an extensive bibliography. The next attempt to create a space encyclopedia was undertaken in the USSR in the mid-1960s under Valentin Glushko. In this process, Georgy S. Vetrov, the Russian space engineer and space historian, found out that there were about 535 various space flight publications in the Soviet Union prior to World War II.[21]

One of the milestones of the "Space Boom" in Russia was the "Exhibition of Interplanetary Machines and Mechanisms" organized in April-June 1927 in Moscow by the Association of Inventors (AIIZ). It put on display numerous models and drawings of various rockets and spacecraft from science fiction literature, as well as from the works of Konstantin Tsiolkovskiy, Fredrik Tsander, Robert Goddard, Hermann Oberth, Robert Esnault-Pelterie, Max Valier, and other space researchers. The exhibition attracted great attention from the general public and the media. In two months about 12,000 people visited, and a book was made available for visitors to sign-up as volunteers for a "future mission to the Moon." That exhibition apparently was the first of its kind in the world.[22]

Radical social changes in Russia and growing interest in space flight were also reflected in science fiction writings. The first Soviet space fiction novel was published in 1923 by a well-known prose writer Aleksey N. Tolstoy (1883–1945). Called *Aelita*, the novel described the adventures of two Russian astronauts on the planet Mars under the yoke of a cruel Martian empire. The first of the astronauts, named Los—a technical genius but a politically indifferent engineer—fell in love with the Martian princess Aelita. Meanwhile, his companion— a former Red Army soldier named Gusev—staged a plot to over- throw imperial rule on Mars. The western-like "cowboy style" and "political correctness" of the novel pleased the Soviet authorities and the story of "Martian revolution" was put on the "silver screen." Released in the mid-1920s, the movie *Aelita* became the first Russian space feature film. Naive and somewhat primitive by modern standards, that novel nevertheless was very popular and was reprinted many times. Its title became so deeply engraved in the Soviet psyche as something associated with Mars that a top secret Soviet plan of 1969 to outperform the United States in space after the Apollo program by landing Soviet cosmonauts on Mars by the 1980s was code-named *Aelita*.

Another Soviet science fiction writer, Aleksandr R. Belyayev (1884–1942), took a much less politicized approach. He based his writings on realistic projects advertised by space proponents. The novel *Pryzhok v nichto* (*Jump Into Nowhere*, 1934) described a scientific expedition to Mars in a liquid-propellant rocket. In a second novel, *Zvezda K.E.Ts.* (*Star K.E.Ts.*, 1936), Belyaev paid tribute to the late Konstantin Tsiolkovskiy. He painted a broad picture of space exploration based on Tsiolkovskiy's ideas ("K.E.Ts." stood for Tsiolkovskiy's initials): regular launches of reusable rockets, construction of a large orbital space station, and a human landing expedition to the Moon. That novel is especially interesting because Belyayev attempted to describe in detail day-to-day life in space, including the harmful consequences of exposure to weightlessness and space radiation.

It is not known if the classic German film *Frau im Mond* (*Woman on the Moon*, 1929) was released in the Soviet Union. Certainly, that would not have been unusual since many foreign films, especially American westerns, were widely available in the USSR in those years. In 1930, the Moscow *Mosfilm* studio apparently decided to produce a work just as spectacular as the German contribution to the genre. For professional help and advice, the film director V. N. Zhuravlev turned to Konstantin Tsiolkovskiy. The silent movie *Kosmicheskoye putechestviye* (*Space Journey*, 1935), depicted a human flight to the Moon. At least from a technical standpoint, that movie was more accurate than its German counterpart since the astronauts wore spacesuits on the Moon. It also was the first Soviet film which showed weightlessness in action. The actors were suspended by thin cables, invisible to the camera.

The Soviet government, not really interested in "crazy space fantasies," nevertheless realized some usefulness from them for the image of the new political regime. In 1918, Nikolay Rynin published an article on an unusual project found in the archives of the former imperial security service. It described a rocketpropelled flying machine with a steerable gimballed engine and an automatic feeder for gunpowder cartridges into a combustion chamber. Its author, Nikolay I. Kibalchich (1853–1881), outlined the project in a prison cell while awaiting execution. Convinced that imperial rule in Russia could be destroyed just by an assassination of the Emperor, Kibalchich had become a member of a conspiracy and a designer of the bomb that killed Tsar Aleksandr II. His project was not space oriented and was no more realistic nor detailed than other contemporary engineering proposals of rocket-propelled devices (such as those of Nikolay Teleshov and Fedor Geshvend). Kibalchich's connection to the revolutionary movement, however, attracted the interest of Soviet ideologists. As the "Space Boom" progressed in Russia, Kibalchich was advertised by official propaganda as a great rocket designer and pioneer of rocket technology. Such politically oriented exploitation of the designer's name continued through the years of Soviet power and was reflected in many space history books in the Soviet Union and abroad.

Prior to the Bolshevik Revolution, Konstantin Tsiolkovskiy spent decades in fruitless efforts to interest imperial authorities and scientists in his ideas. Recognition came to him only at the end of his life when the Soviet government used Tsiolkovskiy as a living example of the popular propaganda stereotype: "an ordinary person could become a great scientist, being freed from the oppression of the bourgeoisie." Because of this, beginning in 1921 Tsiolkovskiy received a personal government pension, which allowed him better living conditions and the opportunity to continue his research. The use of Tsiolkovskiy's name by official propaganda during the "Space Boom" was also intended to establish the preeminence of the USSR in rocket technology and to create an image of scientific prosperity after thousands of leading scientists and other intellectuals had been forced out of Soviet Russia in 1921–1922. The hypocrisy of these Soviet authorities was demonstrated by their real approach to the inventions of the great scientist. In 1929, Tsiolkovsky submitted to the military a technical proposal for the development of a jet aircraft. After long delays the military bureaucrats replied that such an aircraft was "technically unfeasible." As a result, works on jet aircraft propulsion started in the USSR only after such planes appeared in England and Germany. Unaware that he was being used as a pawn in propagandistic games, Tsiolkovskiy honestly believed in the progressiveness of the Soviet regime. Flattered by unusual public attention, he wrote: "The USSR government has done a lot of good. But what I like most of all is the dissemination among the public of the precise and technical sciences. As a result, I have the opportunity to speak everywhere and be understood in the most remote areas of my country."[23] While that was an obvious exaggeration, exploitation of Tsiolkovsky's name by the Soviet government did unintentionally help to form a scientifically correct public perception of future space exploration.

THOSE WHO READ TO BUILD

The works of the space pioneers and a subsequent "Space Boom" (roughly 1924–1935) prepared public awareness and acceptance of the possibility of a space flight. The collapse of the old imperial social structure led to a greater exposure of formerly underprivileged layers of society to education. Thus, space flight theory as created by space pioneers had much greater public exposure than before the Revolution. At the same time, unlike pre-Revolutionary times, many professional scientists dedicated themselves to the promotion of those ideas. This led to a younger generation of scientists and engineers who undertook practical research and development.

Real experiments with rocket technology appeared almost simultaneously with the "Space Boom." But unlike lectures and public presentations, such technology required enormous financial investments. In the Soviet Union, investments could only come from the government, which it turned out was not interested. The only state institution that pursued rocketry experimentation was the military. Space enthusiasts in the USSR faced the same problem as their colleagues in other countries, to start space exploration one would have to create a weapon. To begin their long way to a distant goal of space travel, they would have first to abandon that goal.

Initially, public groups of rocketeers were created under the umbrella of OSOAVIAKHIM (Society for the Support of Aviation and Chemical Development). Although officially "public," that organization in fact was controlled by the government and financed mostly by the military. It was established in 1927 to promote and provide technical training for young people enrolled in mandatory military service. Starting in 1931, OSOAVIAKHIM Reactive Motion Research Groups (GIRD) appeared in up to 90 cities, but only two—in Moscow and Leningrad—survived long enough to produce practical results.[24] Lack of funds and lack of determination forced others into oblivion.

At first, GIRDs were voluntary rocket clubs. That caused some of the members to joke about the Russian abbreviation of GIRD as of the "Group of Engineers Who Work for Nothing." But in 1932, the military promised financial help upon achieving some practical results. The Moscow GIRD (MosGIRD) was transformed into a professional research organization under OSAVIAKHIM command. The founder and the first chairman of MosGIRD, Fredrik Tsander, was considered an "incurable space buff" by the military and he was quickly replaced by a more "down-to-Earth" young aviation engineer named Sergey P. Korolev (1907–1966). This same Korolev became the father of Soviet rocketry and led the nation's space effort two decades later.

The task of MosGIRD was to develop for the military technology demonstrators in three major areas:

— Unguided liquid-propellant rockets;
— Rocket gliders with liquid-propellant engine;
— Ramjet engines.

Only the first of those directions produced practical results. The first Soviet liquid-propellant rocket, called 09 (sometimes referred to as GIRD-09), was launched on August 17, 1933. It was designed at one of the MosGIRD's sections under the military engineer Mikhail K. Tikhonravov (1900–1974). Years later, he would lead the development of the Sputnik satellite that reached orbit in 1957. The second rocket, GIRD-X, took-off in November 1933 after an untimely death of its principal designer and space pioneer Fredrik Tsander.

It should be noted that limited development of rocket-propelled weapons had been underway in the USSR since 1921. In 1928, a military rocket laboratory was moved from Moscow to Leningrad and received the designation GDL (Gas Dynamics Laboratory). By the early 1930s, it was researching several major rocket applications:

— Unguided solid-propellant ballistic missiles for salvo launches;
— Solid-propellant engines for take-off acceleration of heavy aircraft;
— Liquid-propellant rocket engines.[25]

Both MosGIRD and GDL were supervised and subsidized by the Chief of the Red Army Armaments Office, Mikhail N. Tukhachevskiy (1893–1937), who proposed to unify them into a single research and development organization. He established the Reactive Scientific Research Institute (RNII) at the end of 1933 in Moscow. RNII, which changed names a number of times and is known today as the M.V. Keldysh Research Center, played a major role in the development of rocket technology in the Soviet Union. Although the RNII's existence was known outside the military establishment, the details of its programs were secret. Thus, a "lion's share" of all research in rocket technology in the USSR was carried out for the military.

By the mid-1930s only two active public rocket groups remained in the USSR. They could be called "public" only with a certain degree of tolerance: they both worked under the OSOAVIAKHIM umbrella and were partially subsidized by the military. The first was the GIRD in the city of Leningrad (LenGIRD). Unlike the smaller but more technically oriented MosGIRD, the Leningrad group continued active space flight propaganda and boasted a large membership (more than 400 participants). For promotional purposes it even organized launches of rocket models. In 1932, LenGIRD started public courses on jet propulsion. Such inclinations toward public relations could be explained by the active cooperation of LenGIRD with Nikolay Rynin and other space pioneers. And the "granddaddy" of Russian popular science literature—Yakov Perelman—was a Vice-Chairman of LenGIRD. At the same time, LenGIRD was designing several liquid-propellant rockets for practical applications under the leadership of its chairman, Vladimir V. Razumov (1890–1967). One of the projects envisaged a gyroscope-controlled rocket with cameras as a payload. Another rocket, *LRD-D-1* (better known as the *"Razumov-Shtern Rocket"*) was planned for probing the atmosphere. Unfortunately none were completed. Only *LRD-D-1* was launched in 1934, but its liquid-propellant engine had been replaced with a solid one. In 1934, LenGIRD was transformed into a smaller Rocket Section of the regional OSOAVIAKHIM. Its activity (which continued until 1941) was limited to research of acceleration's influence on living organisms and tests of solid-propellant rocket engines.[26]

The second research group was the "Rocket Section" of the OSOAVIAKHIM Stratosphere Committee in Moscow. It was formed in 1934 by a group of former MosGIRD engineers who left RNII due to the internal conflicts. Under the leadership of Aleksandr I.

Polyarniy, the Rocket Section designed and tested an atmosphere-probing rocket called *R-06* (1937). It is still unclear what happened to this group later, but it was probably transferred back to the RNII or under military command of some sort. According to official Soviet sources, the group ceased to exist in 1938. The same sources, however, indicate that members of the group were still active a year later. On May 19, 1939, they tested the first Soviet two-stage rocket, designed by Igor A. Merkulov, a solid-propellant booster with a ramjet engine on the second stage.[27]

All these events of the 1930s coincided with a visible and rapid decline of public interest in space travel. One reason for this is clear: the time of Soviet mass euphoria was over. Research and development of rocket technology required not the public clubs but professional scientific research organizations. Existence of such organizations was justifiable from an economic standpoint only for the development of military technology. Thus, the technological foundation which would make space flight possible inevitably put a wall of secrecy between the technology and the public.

And there were also other reasons. At the end of the 1920s, under the direct influence of Josef Stalin, who was quickly gaining absolute power, the Bolshevik leadership introduced a new political course of direct Communist party control over the economy (Collectivization and Industrialization). Along with the abandonment of the NEP and all associated freedoms, the Communist party tightened its ideological grip over all spheres of social life. The predominance of state interests (which Stalin obviously associated with his own interests) was imposed as the top priority for everybody.

Apart from "sheer craziness" and "lack of practical usefulness," the idea of space flight was seen by party bureaucrats as a challenge to the "ideological unity" of the Soviet society. Facing global issues such as space travel, people would inevitably develop allegiances to the whole planet Earth—not to a select country and its dictator. Space pioneers understood the international nature of future space activities from the outset. In one of his early fiction books *Une Zemli* (*Beyond the Planet Earth*, 1896), Tsiolkovskiy described how a spaceship was built by a joint effort of the representatives of six industrial countries: Russia, the United States, England, France, Germany, and Italy. As soon as the Society for the Study of

Interplanetary Communications was formed in 1924, it attempted to establish contacts with space proponents in other countries. The organizers of the 1927 space exhibition in Moscow went even further by proclaiming themselves *cosmopolitans*—citizens of the whole planet.[28] Such ideas, "politically incorrect" in the mid-1920s, were considered high treason ten years later.

The official propaganda ridiculed "space crackpots" for their lack of realism and because of their "luring [of the] public into useless day-dreaming." The most that was tolerated publicly involved possible practical applications of rocket technology such as atmospheric probing. In their turn, rocket specialists clearly understood the difficulties in perfecting the new technology. They realized that under the circumstances, excessive optimism about future space flight would do more harm than help. The popular science books published at the time by leading RNII engineers, Sergey Korolev's *Raketniy polet v stratosfere* (*Rocket Flight in Stratosphere*, 1934) and Georgiy Langemak and Valentin Glushko's *Rakety: ikh ustroistvo i primeneniye* (*Rockets: Their Designs and Applications*, 1935), did not reflect the recent "Space Boom." Instead of Fredrik Tsander's slogan "Forward to Mars!" popular just several years earlier, Korolev in his book cautiously remarked that "Theoretically, a rocket does not have a flight ceiling."[29] The "All-Union Conference on Stratosphere Exploration" (Leningrad, 1934) and a meeting on the "Use of Reactive Devices for Stratospheric Studies" (Moscow, 1935) were the last publicly open forums in the USSR where rocket technology was discussed.

Strengthened ideological oppression by the ruling Communist party was supplemented by repressions against real and imagined political opponents. False accusations of millions of honest people were widely used by the NKVD (People's Commissariat of Internal Affairs) to justify the "Party line" that "the saboteurs penetrated all spheres of Soviet society."

The first victim of repressions among space enthusiasts was Yuriy Kondratyuk. After publishing his book in 1929, Kondratyuk met with Tsiolkovskiy and presented him his signed photograph. But soon after that, in 1930, Kondratyuk was arrested on false charges of industrial sabotage. The NKVD "investigation" was so sloppy that it saved Kondratyuk's life. Had the investigators uncovered his real past, he would have faced a death penalty. Instead, Kondratyuk spent two years in a special NKVD design bureau for convicted engineers, designing coal-mining facilities.[30] At least twice, in 1932 and 1933, Yakov Perelman and Sergey Korolev tried to invite him to join GIRD. Kondratyuk had to decline those invitations because he had just come out of jail. For him to work on defense-related matters would have required a thorough background check by the NKVD—something he could not afford. Yuriy Kondratyuk was lost to the cause of rocket technology forever.[31]

It is not clear if the curtailing of LenGIRD's activities were also due to Stalin's repression. That was quite possible, since the mysterious assassination of the Leningrad Communist party leader Sergey M. Kirov in 1934 caused a radical reshuffle among all city authorities. That assassination marked a beginning of a new, even more severe campaign of repression known as the "Great Purges." The worst was yet to come.

In 1937, at the pinnacle of the "Great Purges," Marshall Mikhail Tukhachevskiy was falsely accused of espionage and executed. Everything associated with his name, including the development of rocket technology, was put under suspicion. Soon the leadership of Tukhachevskiy's creation—the rocket research institute NII-3 (formerly the RNII)—was executed as well. Next year, two of its leading designers, Valentin Glushko and Sergey Korolev, were announced as "public enemies" and sentenced to long prison terms. The OSOAVI-AKHIM Chairman, Corps Commander Robert P. Eideman (1895–1937), was also executed with Tukhachevskiy. A severe reshuffling of the OSOVIAKHIM followed, and this effectively eliminated any remains of public participation and awareness in the development of rocket technology in the Soviet Union.

A SPACE FLIGHT RENAISSANCE

The German invasion of the Soviet Union during World War II required all the efforts of the nation for victory. Research in new technology, including rockets, was limited and found little practical application in the USSR at that time. Several attempts to develop liquid-propellant rocket interceptors and rocket accelerators for aircraft did not go beyond experiments. The pre-war development of long-range guided cruise missiles was abandoned after the arrest of their main designers—Korolev and Glushko. The only mass-produced rocket weapon was a salvo launcher with solid-propellant missiles, unofficially called *Katyusha*. Similar missiles were also successfully used by the Soviet Air Force.

There is no data on any public activity in the USSR related to rocket technology and space exploration during the war. Many space enthusiasts perished at that time. Nikolay Rynin and Yakov Perelman were among hundreds of thousands who starved to death in Leningrad, besieged by the Nazi invaders. Yuriy Kondratyuk volunteered as a private in the Red Army and was declared missing in action in early 1942. There were rumors after the war that Kondratyuk was captured by Germans and was put to work at the rocket center in Peenemünde. The story apparently circulated when a fragment of Kondratyuk's hand-written notebook was discovered in a German archives. The notebook was presumably found by German troops on the battlefield.[32]

There was at least one other event during the war that later influenced the growth of public interest in space exploration. In 1941, Soviet optician Dmitriy D. Maksutov (1896–1964) invented a new type of astronomical telescope—the Maksutov-Cassegrain (catadioptric) system—which is widely used today around the world. Among other advantages, Maksutov's invention allowed the folding of a light path three times, making the telescope very compact. After the war that telescope was produced in large numbers for public schools, boosting massive interest in astronomy among students.

Following the capture of German rocket technology at the end of the war, the Soviet government launched a large-scale rocket development program in May 1946. It far exceeded anything that had been done in that area beforehand. Sergey Korolev (who, along with Valentin Glushko, gained freedom in 1944) was put in charge of the recreation of the German V-2 ballistic missile. His small design office (about 90 engineers and technicians) years later became a large design bureau with thousands of employees. Today it is known as the S.P.Korolev Rocket-Space Corporation "Energia." Valentin Glushko became head of a separate design bureau for rocket engines, which today bears his name—the V. P. Glushko Scientific-Production Corporation "Energomash." A large number of other research organizations and production facilities were reoriented to rocket technology. Broad cooperation from different branches of the economy to produce rocket components was established. All of this was controlled by the Communist Party Central Committee through a complex bureaucratic structure. That massive effort fostered experiments with V-2 missiles in 1947 and the launch of a Soviet-built copy, R-1, a year later. Development of rocket technology in the USSR progressed rapidly thereafter and by 1953 it reached a point where intercontinental ballistic missiles (ICBM) became technically feasible.

All those developments were concealed from the public by a thick wall of secrecy. Nevertheless, public interest in rocket technology and space flight was constantly growing. There were several contributing factors. One was the booming progress in jet aviation. Unlike early missiles, the Soviet leadership publicly demonstrated new aircraft as part of a campaign to deceive the West about the real power of Soviet military. In both the public mind and popular science literature, jet planes and rockets were often associated as something of a similar nature.

Dissemination of space flight ideas ironically received a boost from the least expected source. In the late 1940s, Soviet leadership started another campaign of ideological brainwashing. Spearheaded mostly against Jews, the so-called "struggle against cosmopolitans" was also intended to "shield" the Soviet people from the "harmful influence from the West." Soviet historians were ordered to discover new facts to support Communist party allegations that all major advances in science and technology in the world in the recent 300 years were made in Russia and the USSR. Obviously chauvinistic by nature, that campaign nevertheless had an unusual positive side effect. Following the "Party line," official propaganda actively promoted the ideas of great Russian scientists of the past, including Tsiolkovskiy, Tsander, and Kondratyuk. Their books were reprinted in 1947 in large quantities.[33] A complete collection of Tsiolkovskiy's works was also published in 1954. After Stalin's death in 1953, the "anti-cosmopolitan" campaign ended but promotion of the alleged Soviet preeminence in scientific discoveries continued for "patriotic purposes." This propaganda was intended to demonstrate that Soviet technology was the same or higher when compared with that of the West. Thus, the Soviet people were not surprised when Soviet authorities announced the future launch of an artificial satellite. The public perception was:

"If the Americans can do this, then our scientists can do this also and probably better."

Despite the "fight against Western influence," the military did not disregard foreign experience and managed to translate and publish in the press several of the works of Oberth, Esnault-Pelterie, and other pioneers of rocketry. Moreover, the ideas of foreign authors very often were presented in the widely-circulated popular science literature, although without references to the original sources. It would not be surprising, for example, to see in a Russian book the drawing of the notorious Antipodal Bomber of Eugen Sänger with Soviet red stars on wings and a caption: "A project of a rocket plane of the nearest future."[34]

The popularity of astronautics was also based on a constantly-growing public interest in astronomy. Professional astronomers such as Vladimir P. Tsesevich, Boris A. Vorontsov-Velyaminov, and others published popular science books on basic astronomy and on astronomical observations. Later, those books were reprinted several times in thousands of copies. In the 1950s, Vorontsov-Velyaminov prepared an astronomy textbook for high school which, along with the Maksutov telescopes, became major instruments of mandatory astronomical public education. That textbook was probably the most published astronomy book in the world, given that in more than thirty years its updated editions have been reprinted in millions of copies (for each tenth grade student each year).

The Soviet public was also informed by the press about the "Flying Saucer" phenomenon abroad. Although official censorship did not allow the "UFO paranoia" to appear in the USSR, the subject of possible alien life in space attracted a lot of attention. That interest was also unintentionally fueled by Gavriil A. Tikhov (1875–1960), an astronomer who analyzed spectrum of the seasonal changes on Mars. That analysis led him to the conclusion that flora existed on Mars similar to what could be found on Earth on high mountains. Official Soviet propaganda immediately praised Dr. Tikhov as the founder of a new science—astrobiology—the study of life forms on another planets. Publicity around Tikhov's works sparked numerous public discussions and presentations. Popular lectures with the inevitable topic "Is there a life on Mars?" became standard entertainment in clubs and recreation facilities in the early 1950s.

The science fiction writer Aleksandr P. Kazantsev (1906-?) went even further. He was one of the first in the Soviet Union to gather evidence of alleged alien visits to our planet. One of his allegations of the mid-1950s sparked tremendous public interest. He stated that the notorious "Tunguska Meteorite" of 1908 was nothing but a nuclear explosion of an alien spaceship. By his logic, the object could not have been an ordinary large meteorite, since meteorites by their physical nature could not explode in midair before the impact. Numerous expeditions to the epicenter of the Tunguska phenomenon, as well as subsequent experiments, proved that Kazantsev was right in his main argument—the unknown celestial body really exploded in mid-air leaving no craters but hundreds of square kilometers of fallen dead forest destroyed by a vertical shock wave. That gigantic explosion (estimated power—10 megatons of TNT) above the Siberian wilderness remains a scientific mystery to this day.

Communist party ideologists never liked science fiction literature. They considered it "a deviation from socialist realism"—a potential threat to official ideology. Such literature was tolerated but never allowed to flourish. When ideological oppression substantially eased after Stalin's death, however, science fiction literature experienced a renaissance in the Soviet Union. If space flight was a rare topic in Soviet science fiction in the 1930s, it became a dominant theme in the mid-1950s. Interestingly, some Soviet writers actively exploited the idea of international cooperation in space. In his novel *Lunnaya doroga* (*Lunar Road*), Aleksandr Kazantsev showed how an American astronaut landed on the Moon first, suffered technical setbacks, and was rescued by a Soviet robotic rover remotely controlled from the Earth. We know now that the Soviet Union launched robotic spacecraft, including rovers, to the Moon during the time of the *Apollo* program. Similar situations were depicted in another novel *220 Dney na zvezdolete* (*220 Days in a Starship*) by Georgiy Martynov. The Americans were the first on Mars but the Russians came to their help in distress. It should be noted that such ideas of international cooperation and mutual help were expressed at the peak of the Cold War with all its mutual distrust and fear of nuclear attack.

In the years before Sputnik, Soviet popular science literature on rocket/space technology was published on a far grander scale than before World War II. The writings of Dr. Karl A. Gilzin were especially popular among high school students. His books provided a broad picture of contemporary space projects along with careful explanations of all associated physical phenomena. In the 1950s they became one of the main sources for the Soviet people to learn what had been published in *Aviation Week, Spaceflight, Flight International* and other aerospace magazines abroad (foreign magazines in the USSR were not publicly available). In a sense, Gilzin replaced Yakov Perelman in the popular literature on space in the Soviet Union.

Unlike the United States, the identities of Soviet rocket designers were kept secret and they were deprived of the opportunity of appealing to the public directly. Yet, some of them managed to break through that wall of secrecy and greatly influence the public perception of space flight. Ari A. Shternfeld (1905–1980) was born in Poland and educated in France. He became famous in the early 1930s for his work on space flight dynamics which won several international awards. In those years, a number of foreign rocket technology specialists were invited to work in the USSR. Shternfeld was among them and he was employed by the RNII in 1935. After the war, Shternfeld was not directly involved in the development of ballistic missiles and he was allowed to "go public." Beginning in the early 1950s he actively promoted ideas of space flight in a series of popular science articles and books. Those publications described a vast variety of future spacecraft and orbital space stations, as well as major directions of human colonization of space. Unlike Karl Gilzin, who summarized the foreign experience, Schternfeld based his writings mostly on his own and other Soviet research. His books became a channel through which Soviet rocket scientists communicated their ideas to the public. In one such book *Iskustvennye sputniki zemli (Artificial Satellites, 1956)*, Schternfeld publicized and explained all aspects of Sputnik a year before its launch.

THE BATTLE FOR SPUTNIK

Accepting the possibility of a satellite launch in the nearest future, the Soviet public did not know that such a possibility depended more on political decisions than on technical capability. To develop space technology, rocket scientists would have to convince first their political, military, and administrative leaders. And that task was probably more difficult than designing the satellite itself. Government bureaucrats saw space science and technology only as an impermissible deviation from the main task of rocketry—ICBM production. Mikhail Tikhonravov was the person who put his reputation and career on the line to break through that ignorance.

By the late 1940s, this "father" of the first Soviet liquid-propellant rocket held the military rank of Colonel-Engineer and worked as Deputy Commander of the secret military rocket scientific institute, the NII-4. Very little is known of his research on missile applications; even today most of it is still classified. What is known is that it was Tikhonravov who proposed the "rocket packet" concept based on Tsiolkovskiy's ideas—a multistage missile with parallel stages. Research on that concept eventually resulted in the development of the first Soviet ICBM and the Sputnik launcher from Sergey Korolev's design bureau.[35]

At the same time, Tikhonravov never forgot the dream of his youth—space flight. In 1945–1946 he initiated project VR-190: a rocket to carry two pilots into space on a suborbital trajectory. Similar projects were proposed at that time by German writer Willy Ley in the United States and by the British Interplanetary Society (using a modified V-2 in both cases). Although the VR-190 project received Stalin's approval in 1946, military bureaucrats managed to put it under wraps. Despite the secrecy, Karl Gilzin published a description of that project, "synthesized" with the foreign proposals, in his widely popular book *Putechestviye k dalekim miram* (*Journey to the Distant Worlds*, 1956).[36]

Starting in 1948, Tikhonravov and his researchers worked hand-in-hand with Korolev's design bureau on possible applications for artificial satellites. When Tikhonravov made an official presentation on the subject at a secret scientific conference in 1950, he incurred the wrath his Soviet overseers and the group was disbanded and he was demoted from his position.[37] But the battle for space was far from over. On October 2, 1951, Tikhonravov "went public" with an article in a newspaper for schoolchildren, *Pionerskaya Pravda*. Written in a simplistic propaganda style, his "Flight to the Moon" article nevertheless contained important indications. It explicitly stated that Soviet technology had already reached a level making possible space flight.[38] That statement was immediately noticed by American intelligence and congressional analysts, but the American public was not informed.[39] On the other hand, Tikhonravov's article was an important step in the creation of a public perception in the Soviet Union that space flight was not a dream anymore, but a realizable goal. Apparently, he chose the format of an article for children deliberately. It was in many ways a precedent to study the reaction of the authorities. Tikhonravov did not disclose any secrets but it was still rather risky to express "heretical" ideas publicly during Stalin's reign. It was tolerated if such "heresy" was expressed by fiction writers, but for secret rocket designers it was dangerous. Official censorship eventually allowed that article to appear probably because it was consistent with the "Party line" of promoting Soviet science.

Tikhonravov was not alone in his struggle. Apart from his former colleagues from the RNII—Korolev and Glushko—he had powerful allies in the USSR Academy of Sciences. Among them was Dr. Anatoliy A. Blagonravov (1894–1975), a prominent military scientist who, from 1949 onwards, supervised the use of Korolev's missiles in scientific experiments. Another one was Dr. Mstislav V. Keldysh (1911–1978)—the mathematical genius behind Soviet nuclear and missile programs who simultaneously headed the NII-1 research institute of the aviation industry (the former RNII) and the Section on Applied Mathematics in the Steklov Mathematics Institute of the USSR Academy of Sciences. Both were helpful in creating a "pro-space lobby" inside the Academy's leadership. The newly-elected President of the USSR Academy of Sciences, Dr. Aleksandr N. Nesmeyanov (1899–1980) apparently supported these space proponents. At the Congress of the World Peace Council in Vienna in 1953, he made an unprecedented statement that modern science was capable of launching an artificial satellite and a "stratoplane" to the Moon.[40] Nesmeyanov not only confirmed Tikhonravov's words, but became the first Soviet high-ranking official to publicly support space exploration.

Meanwhile, far from the public eye, Tikhonravov, Korolev and other space proponents prepared the final assault. In February 1954, Korolev managed to convince his boss—the Minister of Defense Industries Dmitriy F. Ustinov (1908–1984)—to examine proposals for an artificial satellite.[41] After a discussion of possible satellite applications at a special meeting chaired by Dr. Keldysh at the USSR

Academy of Sciences on March 16, proposals were prepared by a group of researchers from NII-4 under Tikhonravov's leadership. The final document was called a "Memo on an Artificial Satellite." It was edited by both Korolev and Keldysh, and approved by Nesmeyanov.[42] The timing was right, since the configuration of the first Soviet ICBM was completed. Sergey Korolev considered it very important that the development of an ICBM parallel the development of space experiments that might be carried by it. On May 20, 1954, a government decree authorized development of the R-7 ICBM. A week later—on May 27, 1954—the "Memo on an Artificial Satellite" reached Ustinov, who apparently made it known to higher authorities. As a result, the government authorized detailed research into artificial satellites in August 1954.[43] It was not a final victory: a decision on satellite construction had not been made. At the same time preparatory work for the decision had been laid.

The Soviet public was unaware of these developments in the secret world of the military-industrial complex. Yet their publicly visible consequences became apparent very soon. In September 1954, an Interdepartmental Commission on Interplanetary Communications (MKMS) was established at the USSR Academy of Sciences. Chairman of the MKMS was Dr. Leonid I. Sedov (1907—), a specialist in hydrodynamics, while Mikhail Tikhonravov was appointed as his deputy.[44] The Commission included many prominent scientists, but only a few of them were directly involved in rocket technology. The role of that Commission in Sputnik's development is still not quite clear. On one hand, it was a "figurehead" body created mostly for public show that helped to conceal the identities of the real space designers. For example, Leonid Sedov and Anatoliy Blagonravov were for years associated with the launch of the Sputnik although they had never been directly involved. On the other hand, the MKMS obviously influenced the choice and development of scientific equipment for the future satellite. It was a sort of a scientific advisory committee for Soviet political leaders and the military-industrial complex. For the Soviet public, the creation of the Commission was an indication of the first practical step toward a real space flight.

That same year, in 1954, the newly published *Bolshaya Sovetskaya entsiklopediya* (*Great Soviet Encyclopedia*) contained an article "Interplanetary Communications," written by Tikhonravov. That article included scientific and technical aspects of what is known today as "Astronautics in the West and Cosmonautics in Russia." Simultaneously with the creation of MKMS, the USSR Academy of Sciences issued a commemorative gold medal named after Konstantin Tsiolkovskiy. The medal was to be awarded to scientists and engineers for "outstanding achievements in interplanetary communications." Later, the first recipient of this medal was Sergey Korolev for the launch of Sputnik. Since the publication of the encyclopedia and the activities of the Academy of Sciences were directly supervised by the Communist party's Central Committee, Tikhonravov's article and Tsiolkovskiy's medal indicate that the idea of space exploration had found a favorable response from the authorities.

It is known that the preparatory committee of the International Geophysical Year (IGY) issued an invitation in October 1954 to build an artificial satellite for the comprehensive study of the Earth. It was a challenge meant for only two countries: the United States and the Soviet Union.[45] The United States accepted that challenge when President Dwight Eisenhower announced the *Vanguard* program on July 29, 1955. An immediate Soviet reaction—Leonid Sedov made a statement in Copenhagen on August 2, 1955—that the USSR would launch a satellite as well—created a long-lasting impression among historians that the Soviet satellite program was primarily motivated by competition with the United States. The following facts cast doubt on that conventional wisdom:

- The "Memo on an Artificial Satellite," sent to Ustinov in May 1954, did not contain any reference to either IGY or to the competition with the Americans.[46]
- The principal decision on satellite research was adopted by the Soviet government two months before the resolution of the IGY preparatory committee on satellites.
- The first reference to the political importance of a future satellite appeared in the memo on logistical questions of satellite development, prepared at the Korolev design bureau in June 1955.[47]

The centralized power structure of the Soviet Union was very intolerant toward any kind of unauthorized statements. Thus, it is doubtful that Sedov made his announcement of August 2, 1955 on his own without prior approval from the authorities, especially since the Soviet satellite program did not officially begin until January 30, 1956.

Apparently, the government decree of January 30, 1956, was not a beginning but a conclusion of two years of preparatory work conducted by the rocket industry and the USSR Academy of Sciences. The fact that developing satellites, lunar probes, and launch vehicles was discussed on August 30, 1955 at a meeting of the Military-Industrial Commission—one of the most powerful branches of Soviet government—indicates how comprehensive and thorough that preparatory work was.[48] In this respect, IGY and the American *Vanguard* program were not the prime reasons but additional stimuli to design and develop a Soviet satellite.

The secret decree of January 30, 1956 authorized development of *Object D*—a heavy satellite with numerous scientific instruments to conduct research in the interest of the IGY. Eventually the *Object D* was launched as *Sputnik 3* in 1958. When design of that satellite lagged behind schedule, Sergey Korolev initiated construction of the *PS* (Simple Satellite) which became *Sputnik 1*. Tikhonravov joined the Korolev design bureau as head of a special department to develop satellites and in the process the "father" of the first Soviet liquid-propellant rocket became the "father" of Sputnik.[49]

The development of the *PS* was a purely political act designed to deprive the Americans from being first. That was clearly indicated by Korolev in his letter to Soviet officials in January 1957.[50] Yet that does not mean that political prestige was the driving force behind Sputnik. The Soviet satellite program was more comprehensive than the public knew: along with *PS* and *Object D*, there was parallel development of a photo-reconnaissance satellite (*Object OD-1*) and of a prototype of a human spacecraft (*Object OD-2*).[51]

While the Soviet satellite program secretly took shape, the USSR Academy of Sciences engaged in a highly publicized public relations offensive on the issue of a future satellite. The Soviet Union joined the International Astronautical Federation in 1956. A conference of astronomers to discuss the physics of the Moon, Mars, and other planets took place in Leningrad in February 1956. In September of that year, the USSR Academy of Sciences submitted an outline of satellite experiments to the IGY preparatory committee (the announced experiments were planned for *Object D*).[52] Analysts of the United States Congress detected at least a dozen Soviet press announcements on the future satellite in 1956–1957.[53] They included such data as the approximate size, shape, and weight of Sputnik, its expected orbital parameters, estimated conditions of visual observations, and frequency of radio transmissions. All that data later appeared to be very close to the reality. The final warning came on September 17, 1957. During a public meeting to commemorate Tsiolkovskiy's centenary, Sergey Korolev made a report in which he clearly indicated that an artificial satellite would be launched "in the nearest future."[54] That warning, as well as all others, was ignored overseas.

The psychological shock that the American public experienced after Sputnik should be attributed not to the secrecy surrounding the Soviet satellite but to the arrogance of American mass media and to Cold War paranoia. On the other side, the Soviet public accepted the launch of *Sputnik 1* in a "matter-of-fact" manner. The educational level of the Soviet people by 1957 was much higher than in the 1930s. Thus, far greater numbers were ready to accept the space era as an inevitable stage in the progress of human civilization. At the same time, official propaganda about Soviet scientific and technological advances prepared the public to believe that it was natural that the USSR should be the first in space. Sputnik substantially boosted public interest in astronomy, rocket technology, and related areas. But most of all people were proud of their country's accomplishments.

CONCLUSIONS

The evolution of public perception about space flight in Russia and the Soviet Union had a long history. Public perception evolved under the influence of science fiction and popular science literature, which disseminated among the population new discoveries in science and technology. The works of Russian and foreign space pioneers were available for public consumption in the USSR and became a major contributing factor to the public understanding of space flight as a reality.

The correct public understanding of major problems related to space flight was formed in the USSR during the 1930s. That fostered practical experiments in rocket technology. By the 1950s, the Soviet public was ready to understand and accept the beginning of the space age as a natural "next step" in the development of human civilization.

Until the mid-1950s Soviet authorities did not support the idea of space flight, considering military applications of rockets as the only legitimate application. At the same time, the names of Russian space pioneers were actively exploited by officials for political purposes. Public perception of space flight in the USSR was often affected in a negative way by the ideological doctrine of the Communist party.

The Soviet satellite program was initially unrelated to the International Geophysical Year and competition with the United States. It was only later that these two factors contributed to changes in the initial program and stimulated its speedy development.

Soviet satellite development combined tight secrecy with controlled "leaks" of information, stemming from the fact that space technology in the USSR was a spinoff of the development of strategic weapons. The "leaked" information prepared the Soviet public for the launch of *Sputnik 1*, while the foreign public was not properly informed due to the politically motivated ignorance of the Western mass media.

The author would like to express his appreciation to Asif A. Siddiqi, the NASA Contract Historian and a researcher of Russian space technology, for providing additional materials for the paper and for useful discussions on the topic.

1.   Valentin P. Glushko, ed., *Kosmonavtika—Little Encyclopedia* (Moscow, USSR: Soviet Encyclopedia, 1970), p. 342.
2.   Ron Miller, *The Dream Machines* (Malabar, FL: Krieger Publishing Co., 1993), p. 54.
3.   A. V. Datsenko, *I Will Fly Out There …* (Moscow, USSR: Prapor, 1989); Glushko, ed., *Kosmonavtika—Little Encyclopedia*, p. 219.
4.   G. P. Svishchev, ed., *Aviatsiya entsiklopediya* (Moscow, Russia: Bolshaya Rossiyskaya entsiklopediya, 1994), p. 555.
5.   Miller, *Dream Machines*, p. 68.
6.   Yuri Mozzhorin, ed., *Kosmonavtika SSSR* (Moscow, USSR: Mashinostroeniye, 1986), p. 22.
7.   V. P. Glushko, ed., *Kosmonavtika entsiklopediya* (Moscow: Sovetskaya entsiklopediya, 1985), p. 422.
8.   *Ibid*, , p. 436.
9.   G. S. Vetrov, *S. P. Korolev i kosmonavtika: pervye shagi* (Moscow: Nauka, 1994), p. 19. See also A. Yu. Ishlinskiy, ed., *Akademik S. P. Korolev: ucheniy, inzhener, chelovek* (Moscow: Nauka, 1986).
10. Datsenko, *I Will Fly Out There …*, p. 92.

11. *Ibid.*, pp. 104–105.
12. Glushko, ed., *Kosmonavtika—Little Encyclopedia*, p. 293.
13. Vetrov, S. P. *Korolev i kosmonavtika*, p. 7.
14. *Ibid.*, p. 9.
15. Boris V. Raushenbakh, *Hermann Oberth* (Moscow, Russia: Nauka Press, 1993), p. 126.
16. V. I. Fedosiev and G. B. Sinyarev, *Introduction to Rocket Technology* (Moscow, USSR: OboronGIZ, 1961), p. 12.
17. Vetrov, S. P. *Korolev i kosmonavtika*, p. 9.
18. Datsenko, *I Will Fly Out There ...*, pp. 161–62.
19. Miller, *Dream Machines*, p. 175.
20. Glushko, ed., *Kosmonavtika entsiklopediya*, p. 395.
21. Vetrov, S. P. *Korolev i kosmonavtika*, p. 15.
22. Yaroslav Golovanov, *Korolev: fakty i mify* (Moscow, Russia: Nauka Press, 1994), p. 116; Glushko, ed., *Kosmonavtika—Little Encyclopedia*, p. 290; Miller, *Dream Machines*, p. 156.
23. Mozzhorin, ed., *Kosmonavtika SSSR*, p. 24.
24. Glushko, ed., *Kosmonavtika entsiklopediya*, p. 332.
25. Valentin P. Glushko, *Rocket Engines of GDL-OKB* (Moscow, USSR: APN, 1975), p. 9.
26. Glushko, ed., *Kosmonavtika entsiklopediya*, p. 332; Glushko, ed., *Kosmonavtika—Little Encyclopedia*, p. 372.
27. Glushko, ed., *Kosmonavtika entsiklopediya*, p. 332; Mozzhorin, ed., *Kosmonavtika SSSR*, p. 34.
28. Golovanov, *Korolev*, p. 116.
29. *Ibid.*, p. 201.
30. Datsenko, *I Will Fly Out There ...*, p. 187.
31. *Ibid.*, p. 191; Golovanov, *Korolev*, p. 149.
32. Valery Rodikov, "Who Are You, Engineer Kondratyuk?" (in Russian), in V. Shcherbakov, ed., *Zagadki zvezdnykh ostrovov: kniga pyataya* (Moscow: Molodaya gvardiya, 1989), p. 52.
33. K. A. Gilzin, *Travel to Distant Worlds* (Moscow, USSR: DetGIZ, 1956), p. 278.
34. These basic ideas, and their linkages to Soviet efforts, are discussed in Michael Stoiko, *Soviet Rocketry* (New York: Holt, Rinehart and Winston, 1970).
35. B. N. Kantemirov, "History of the Design Selection for the First ICBM, R-7" (in Russian), audiotape, paper at 10th International Symposium on History of Astronautics and Aeronautics, Moscow, Russia, June 20–27, 1995, translated from Russian by the author.
36. B. N. Kantemirov, "Flight—His Dream and His Work" (in Russian), *Zemlya i vselennaya* (November-December 1991), pp. 54–56; Anatoly Shiryaev and Valery Baberdin, "Prior to the First Leap into Space" (in Russian), *Krasnaya zvezda*, April 27, 1996, p. 5; Gilzin, *Travel to Distant Worlds*, pp. 230–31.
37. Kantemirov, "Flight—His Dream and His Work," p. 56.
38. M. K. Tikhonravov, "Flight to the Moon" (in Russian), *Pionerskaya pravda*, October 2, 1951, p. 4.
39. Martin Caidin, *Man Into Space* (New York: Pyramid Books, 1961), p. 171.
40. *Ibid.*
41. B. V. Raushenbakh, ed., *Materialy po istorii kosmicheskogo korabl 'vostok'* (Moscow, USSR: Nauka Press, 1991), p. 209.
42. Yu. P. Semenov, ed., *Raketno-Kosmicheskaya Korporatsiya "Energiya" imeni S. P. Koroleva* (Korolev: RKK Energiya named after S. P. Korolev, 1996), p. 86.
43. *Ibid.*, p. 86.
44. F. G. Krieger, *Behind the Sputniks* (Washington, DC: Public Affairs Press, 1958), p. 7.
45. Nicholas Daniloff, *The Kremlin and the Cosmos* (New York: Alfred A. Knopf, 1972), p. 55.

PETER A. GORIN

46. Raushenbakh, ed., *Materialy po istorii kosmicheskogo korabl 'vostok'*, pp. 5–15.
47. Semenov, ed., *Raketno-Kosmicheskaya Korporatsiya "Energiya" imeni S. P. Koroleva*, p. 87.
48. *Ibid.*
49. *Ibid.*, p. 88; "Great Russian Scientist" (in Russian), *Pravda*, September 18, 1957.
50. M. V. Keldysh, ed., *Tvorcheskoye naslediye Akademika Sergeya Pavlovicha Koroleva: izbrannyye trudy i dokumenty* (Moscow, USSR: Nauka Press, 1980), p. 369.
51. *Ibid.*, p. 373; Raushenbakh, ed., *Materialy po istorii kosmicheskogo korabl 'vostok,'* p. 210; Semenov, ed., *Raketno-Kosmicheskaya Korporatsiya "Energiya" imeni S. P. Koroleva*, p. 98.
52. Krieger, *Behind the Sputniks*, p. 9.
53. Caidin, *Man Into Space*, pp. 171–72. See also, V. Petrov, *Earth's Artificial Satellite* (Moscow, USSR: Voenizdat, 1958).
54. Keldysh, ed., *Tvorcheskoye naslediye Akademika Sergeya Pavlovicha Koroleva*, p. 376; A. Shternfeld, *Artificial Satellites* (Moscow, USSR: Technical Literature, 1958).

CHAPTER 2

Korolev, Sputnik, and the International Geophysical Year

Asif A. Siddiqi

From the perspective of historical inquiry, the institutional and political machinations behind the genesis of Sputnik have remained a largely ignored area of scholarship. Embellished by speculation and fueled by Soviet secrecy, the story behind Sputnik has assumed the form of a parable, cobbled together from rumors and mythology, and colored by an eagerness to fill in the blanks of what we did not know. Thus, while the post-mortem effects of Sputnik have been the subject of much scholarly debate, the origins and motivations that led to the launch of the first artificial satellite have remained, to a large degree, in the realm of conjecture. In recent years, with the dissolution of the Soviet state in 1991, mythology came into confrontation with reality. Declassified primary documents have provided a rich resource and incentive to look back again at an event which had such a profound impact on the course of events in the latter part of the twentieth century.

MYTHS

In ... *The Heavens and the Earth*, a seminal contribution to the understanding of the political history of the space era, author Walter A. McDougall creates a masterful tapestry of the political dynamics between the two superpowers both before and after the launch of Sputnik. But the event itself is consigned to a three-page account constructed from Western sources which themselves were based on hearsay and speculation.[1] Thus, even the mythology of Sputnik was at best a skeleton of a story. We knew when it was launched, what it looked like, and possibly who built it, but not much else. Thus, almost by default, in the historiography of space exploration, the Soviet space program, and indeed the so-called "space race," was said to have begun on October 4, 1957. One of the most entrenched paradigms of this history was that Sputnik was a political tool to demonstrate Soviet superiority in a new domain. James E. Oberg, one of the more ubiquitous followers of the former Soviet space program, wrote in his landmark 1981 book *Red Star In Orbit*:

> In the light of ... domestic and international problems [the] proposal for a Soviet artificial satellite ... suddenly became much more attractive to Khrushchev. First, it would signal to dissident political forces within the Soviet Union that Khrushchev was really leading the country to a glorious future; second, it would overawe the traditionalist "artillery generals" in the Red Army and allow a reorganization of the armed forces, including a reduction in obsolete ground forces ...; last, it would demonstrate in an unequivocal manner the existence of the long-range missile system, which was intended to discourage potential attack from the United States. Under these circumstances, what had first appeared to be a pointless diversion of technical resources suddenly became—as far as Khrushchev personally was concerned—a powerful idea ... "in the summer of 1957, the Central Committee ... finally endorsed the project." It should be noted that neither science nor world opinion seems to have entered into consideration.[2]

With the collapse of the Soviet Union and the opening of the archives, there has been an undue eagerness to engage in revisionism, especially vis-á-vis the view from the Soviet side. But in looking at the genesis of Sputnik, the "old" paradigms require a second look. Was Khrushchev personally involved in seeing the project through completion? If so, what were his motives? Did the International Geophysical Year play a role in the launch of Sputnik? On a larger level, did the space race really begin on October 4, 1957? Why did the Soviets launch Sputnik when they did? In looking at Sputnik's birth, was mythology the same as history?

CONCEPTION

In the pantheon of Soviet space historiography, one man's name stands out in both Western and Soviet scholarship, that of Sergey Pavlovich Korolev. Often called the founder of the Soviet space program, his contributions to the emergence of the Soviet Union as a space-faring nation have been amply chronicled in recent years, culminating in an 800-page *magnum opus* published in Russia in 1994.[3] Korolev had become absorbed in dreams of space exploration during his short tenure as a member and eventual leader of an amateur Soviet rocketry group in the early 1930s.[4] Lesser well-known is the name of Mikhail Klavdiyevich Tikhonravov, another former glider pilot who worked with Korolev in the 1930s building the first Soviet liquid-propellant rockets. Their paths diverged during World War II and in its aftermath they were working in different institutions, both contributing to the new long-range ballistic missile effort. Korolev had the auspicious title of "Chief Designer," by dint of his official role as head of the Department No. 3 of the Specialized Design Bureau at the Scientific Research Institute No. 88 ("NII-88" in its Russian abbreviation).[5] Stalin had established the NII-88 (pronounced *nee-88*) in 1946 to serve as the leading engineering organization in Soviet industry to develop long-range missiles.

During the following decade, Korolev's department, which eventually became an independent organization, the Experimental Design Bureau No. 1 (OKB-1), focused efforts on developing a series of ballistic missiles for the Soviet armed forces. Since the primary thematic thrust of Korolev's group was military missiles, there was negligible work on projects which had purely scientific utility. Dedicated wholly to the grand ideals of space exploration, Korolev did make a few spurious efforts to interest the leadership in artificial satellites in the late 1940s, but none of these ever proved to have any results until he combined his lobbying with Tikhonravov's independent work at the NII-4 ("nee-4"), an unrelated military institution dedicated to research on applications of ballistic missiles. After authoring several important R&D reports on the possibility of space launch vehicles and artificial satellites in the late 1940s and early 1950s, Tikhonravov emerged in 1954 with a detailed technical exposition entitled "Report on an Artificial Satellite of the Earth."[6]

It was at the same time, on May 20, 1954, that the Soviet government formally tasked Korolev's Design Bureau to develop the first Soviet intercontinental ballistic missile (ICBM), the R-7. Korolev did not waste time. Just seven days later, he sent Tikhonravov's satellite report to the Soviet government with an attached cover letter stating:

> I draw your attention to the memorandum of Comrade M. K. Tikhonravov, "Report on an Artificial Satellite of the Earth," and also to the forwarded materials from the U.S.A. on work being carried out in this field. The current

development of a new product [the R-7 ICBM] makes it possible for us to speak of the possibility of developing in the near future an artificial satellite ... It seems to me that in the present time there is the opportunity ... for carrying out the initial exploratory work on a satellite and more detailed work on complex problems involved with this goal. We await your decision.[7]

If Korolev's goal was to elicit a formal decree for his proposal, his appeal was not very successful. However, his request appears to have been passed on through various levels of the government and reached the office of missile and nuclear industry chief Vyecheslav A. Malyshev, officially the Minister of Medium Machine Building. Prompted by Korolev's persuasive arguments, Malyshev, along with three other top defense industry officials, submitted a proposal to Soviet leader Georgiy M. Malenkov asking permission to carry out "work on the scientific-theoretical questions associated with space flight."[8] No doubt interested in the military applications of Tikhonravov's satellite, Malenkov approved the idea. Armed with a modicum of support, Korolev commenced a modest research project at his Design Bureau coordinated with Tikhonravov's own work at the NII-4. Incredibly, as this research was ongoing, the satellite issue remained divorced from further governmental involvement, as Korolev was diverted to more important matters relating to work on military missiles such as the R-7 ICBM. It was, however, the very first intervention by the Soviet government on an issue related to space exploration.

Korolev's satellite work may have continued at a leisurely pace through the mid-1950s with lukewarm governmental support were it not for some surprising and well-publicized events outside of the USSR. In the spring of 1950, a group of American scientists led by James van Allen met in Silver Spring, Maryland, to discuss the possibility of an international scientific program to study the upper atmosphere and outer space via sounding rockets, balloons, and ground observations. Strong support from Western European scientists allowed the idea to expand into a worldwide program timed to coincide with a period of intense solar activity, July 1, 1957 to December 31, 1958. The participants named this period the International Geophysical Year (IGY) and created the *Comité speciale de l'année géophysique internationale* (the "Special

Committee for the International Geophysical Year" or "CSAGI") to establish an agenda for the program. Soviet representatives, including Academy of Sciences Vice-President Academician Ivan P. Bardin, served on the Committee, but do not appear to have had any significant contribution to its proceedings. In fact, the May 1954 deadline for submissions for participation in the IGY passed without any word from Soviet authorities. At a subsequent meeting in Rome on October 4, 1954, Soviet scientists silently witnessed the approval of a historic U.S.-sponsored plan to orbit artificial satellites during the IGY.[9] The satellite proposal clearly surprised the Soviet delegation, and perhaps had repercussions within the USSR Academy of Sciences. In the fall of 1954, the Academy established the Interdepartmental Commission for the Coordination and Control of Work in the Field of Organization and Accomplishment of Interplanetary Communications, a typically longwinded title which obscured its primary role, a forum for Soviet scientists to discuss space exploration in abstract terms, both in secret and in public.[10]

The existence of the Commission was announced on April 16, 1955, in an article in a Moscow evening newspaper; Academician Leonid I. Sedov, a relatively well-known gas dynamics expert was listed as the Chairman of the Commission.[11] Unlike the title of the body, the primary duty of the Commission was stated with unusual explicitness: "One of the immediate tasks of the Commission is to organize work concerning building an automatic laboratory for scientific research in space."[12] In hindsight, it is clear that the Commission, a part of the Astronomy Council in the Academy, had very little input or influence over *de facto* decision-making in the Soviet space program, although one of its functions was to collect proposals from various scientists on possible scientific experiments which could be mounted on future satellites. Sedov himself played a major role as Chairman by appearing at numerous international conferences talking in very general terms on the future of space exploration. None of its members had any direct connection or contact with the missile and space program, although they were clearly aware of the broad nature of Korolev's work. The latter appears to have had little to do with the formation or work of the Commission. He evidently attended one meeting in 1954 to inquire about the group's work.[13]

While this Commission had little real authority, its Chairman Sedov may have played a crucial role in connecting Korolev's satellite efforts with the International Geophysical Year. The chain of events was set off on July 29, 1955, by U.S. President Dwight D. Eisenhower's Press Secretary James C. Hagerty who announced at the White House that the United States would launch "small Earth-circling satellites" as part of its participation in the IGY.[14] It was at this same time that the International Astronautical Federation was holding its Sixth International Astronautical Congress at Copenhagen, Denmark. Heading the Soviet delegation was Sedov and Kirill F. Ogorodnikov, the editor of a respected astronomy journal in the USSR. The two were called into action by an announcement on August 2 by Fred C. Durant III, the President of the Congress, who reported the Eisenhower administration's intentions of launching a satellite during the IGY. Not to be outdone, Sedov convened a press conference the same day at the Soviet embassy in Copenhagen for about 50 journalists during which he announced that "In my opinion, it will be possible to launch an artificial Earth satellite within the next two years." He added that "The realization of the Soviet project can be expected in the near future."[15]

It is quite unlikely that Sedov was speaking on his own authority and possibly had taken cues from highly-placed Party officials who were aware of the government's approval in August 1954 of exploratory research on space issues. Perhaps a Party or Academy of Sciences official back in Moscow had decreed that Durant's statement warranted a response from Sedov. Certainly, there had been much discussion on the possibility of Soviet satellites by that time, although no single project had received approval. What is known is that the two pronouncements, one by the Eisenhower administration, and the one by Sedov, were the subject of relatively intense scrutiny by the press all over the world. This response appears to have been critical for Korolev.

By coincidence, it was on July 16, 1955, that Tikhonravov, along with OKB-1 engineer Ilya V. Lavrov as coauthor, finished his latest study on artificial satellites.[16] Based on work originating from the May 1954 document, the two suggested a reduced mass of 1,000–1,400 kilograms for an automated satellite. They also proposed the formation of a group of 70–80 people to carry out the task of designing and building the satellite and to work on future piloted spacecraft. (Korolev wrote in the margins: "Too many, 30–35 people.") The Chief Designer, more attuned to the political reality of such a project, also added that "the creation of [a satellite] would have enormous political significance as evidence of the high development level of our country's technology."[17] In a move symptomatic of Korolev's relentless perseverance of the space issue since the early 1950s, Korolev also had one of his sector chiefs at the OKB-1 prepare a technical report on the possibility of sending a probe to the Moon using modified versions of the R-7 ICBM.

The activity on the space front reached its zenith on August 30, 1955 when Korolev attended two different meetings, one with the defense community and one with the scientific community, to discuss the new satellite report. The former was at the offices of the powerful Military-Industrial Commission, the coordinating mechanism for management of the entire Soviet defense industry. Presiding over the meeting was the Commission's new Chairman Vasiliy M. Ryabikov. Also in attendance were Academician Mstislav V. Keldysh, a noted scientist involved in research and development on several high profile military programs, and Col.-Engineer Aleksandr G. Mrykin, a senior artillery officer responsible for overseeing the procurement of new ballistic missiles for the Soviet armed forces.[18] At the meeting Korolev spoke of both his satellites and lunar probes. Notorious for his legendary short temper and larger-than-life personality, Mrykin was not receptive to Korolev's old arguments of the possibly great political importance of a Soviet satellite. The artillery officer told Korolev in no uncertain terms that only when the R-7 had completed its flight testing would they consider a satellite. Fortunately for Korolev, he had Keldysh's support, and that may have tipped the scales. While details of the deliberations remain extremely sketchy, it appears that Ryabikov approved the use of an R-7 ICBM for a modest satellite program. Lunar probes were not considered. There were probably two factors working in Korolev's favor: the possible use of a satellite for military purposes; and the demonstration of Soviet science and technology during the IGY.

Armed with Ryabikov's approval, Korolev attended a second secret meeting the same day at the offices of the "chief scholarly secretary" of the Academy of Sciences Gennadiy V. Topchiyev. Many other scientists and designers including Keldysh, Tikhonravov, and rocket engine specialist Valentin P. Glushko were present. Korolev reported to the distinguished assemblage that the Council of Chief Designers at a recent meeting had conducted a detailed examination on modifying the original R-7 into a vehicle capable of launching a satellite into orbit. No doubt, he also spoke of the government's interest on the matter. At the end of his speech he formally proposed to build and launch a series of satellites into space, including one with animals, and for the Academy to establish a formal commission to carry out this goal. The Chief Designer had a specific timetable in mind. He told his audience, "As for the booster rocket, we hope to begin the first launches in April-July 1957...before the start of the International Geophysical Year."[19] If earlier, Korolev's satellite plans had been timed for the indefinite future, the Eisenhower administration's announcement in July 1955 completely changed the direction of Korolev's attack. Not only did it imbue Korolev's satellite proposal with a new sense of urgency, but it also gave him a specific timetable to aim for. If the United States was planning to launch during the IGY, then the Soviets would launch one a few months *before* the beginning of the International Geophysical Year, guaranteeing a first place finish. The attending scientists at the meeting accepted the new proposal, and at Korolev's recommendation Keldysh was designated the Chairman of the commission. Korolev and Tikhonravov would serve as his deputies.

The following day, on August 31, a smaller group, including Korolev, Tikhonravov, and Keldysh met to discuss some of the proposals for satellite instruments which many scientists had submitted to Sedov's Commission in the past year. A few days later Tikhonravov and Keldysh convened with some prominent Soviet scientific scholars to explain details of the satellite design and how their instruments were being considered. Korolev himself approved a preliminary scientific program in September 1955, a program which included the study of the ionosphere, cosmic rays, the Earth's magnetic fields, luminescence in the upper atmosphere, the Sun, and its influence on the Earth, and other natural phenomena. The detailed development of a scientific program was left in the hands of the two existing commissions of the Academy headed by Anatoliy A. Blagonravov and Leonid I. Sedov.[20]

The approval by the Academy to conduct a purely scientific research program accelerated matters considerably. In the ensuing months, several important meetings were held, both by Keldysh's commission and by the Council of Chief Designers, which elaborated the details of the project. Between December 1955 and March 1956, Keldysh consulted a huge number of distinguished scholars to refine the scientific experiments package. They included numerous famous Soviet scientists, many of whose names were public knowledge unlike those who were actually developing the spacecraft.[21] It was a large-scale operation with a single coordinating mechanism which, because of its "civilian" nature, had little precedent. Korolev himself was very conscious of the fact that an official decree on the project had yet to be issued, which meant that a rocket was still not officially available for the project. The magnitude of the immediate tasks, however, obscured that important issue for the time being. There were continuous problems with the program, especially since many who were cooperating did not share Korolev's enthusiasm for the project.

It took about four months for Ryabikov's spoken approval in August 1955 to translate into a formal decree of the Soviet government. As a purely scientific project managed by the Academy of Sciences, it was not considered a top priority. In fact, the Soviet government probably viewed the satellite project in much the same manner as they viewed the continuing series of scientific rocket flights into the upper atmosphere which also used military missiles for "civilian" purposes. They were relatively inexpensive, unobtrusive, and ignored by the political leadership. The evidence suggests that First Secretary Nikita S. Khrushchev was not even consulted on the matter.

The satellite project became a reality on January 30, 1956, when the USSR Council of Ministers issued decree number 149–88ss. The document approved the launch in 1957 of an unoriented artificial satellite, designated the "Object D," in time for the International Geophysical Year. As per Tikhonravov's previous computations, the mass of the satellite was limited to 1,000 to 1,400 kilograms of which 200 to 300 kilograms would be scientific instruments. Apart from the Academy of Sciences, five industrial ministries would be involved in the project. The responsibility for preparing a Draft Plan for the Object D fell on the shoulders of Sergey S. Kryukov, at the time a Department Chief at the OKB-1. Tikhonravov served as the "chief scientific consultant."[22]

The decree in support of the Object D was not enough for Korolev, who did not want to consign his dreams of space exploration to a single decree, one among possibly hundreds signed by the Council of Ministers the same month. He wanted a direct verbal promise from the Soviet leadership on the satellite project, in particular from Khrushchev himself. His chance came in February 1956, during a high level state visit to the OKB-1. Khrushchev, escorted by the top Presidium members Aleksey I. Adzhubey, Nikolay A. Bulganin, Vyecheslav M. Molotov, and Mikhail G. Pervukhin, as well as Korolev's boss Minister of Defense Industries Dmitriy F. Ustinov, were on hand to congratulate the OKB-1 on their recent success with a new missile and also to review the progress on the R-7 ICBM project.[23]

The visit was important for Khrushchev since it was his first direct exposure to the top-secret ballistic missile program, an effort which had essentially been run by a number of industrial bureaucrats since Stalin's death out of view from Party leaders like Khrushchev.[24] During the visit, the delegation were shown around the premises of the institute by Korolev on a tour which culminated with a presentation of a full-scale non-functional model of the R-7 intercontinental ballistic missile. The guests were apparently stunned into silence by the size of the vehicle. Like a good performer, Korolev waited a few seconds for the sight to sink in, before giving a brief presentation on the vehicle. Glushko then began an elaborate presentation, much different from Korolev's, filled with extraneous technical details "like he was talking to first course students at the neighboring forestry institute ... rather than the higher leadership."[25] Recognizing the pointlessness of a technical treatise, Korolev cut Glushko short, before summarizing with a succinct conclusion. After a short discussion on the R-7's capabilities, Korolev innocuously added that: "Nikita Sergeyevich [Khrushchev], we want to introduce you to an application of our rockets for research into the higher layers of the atmosphere, and for experiments outside the atmosphere."[26] The Soviet leader expressed polite interest, although it was clear by this time that most of the guests were becoming tired and bored with the proceedings. Undeterred, Korolev first showed them huge photographs of suborbital missiles that were used for biological and geophysical investigations. Detecting that his guests were in a hurry to leave, the Chief Designer quickly moved ahead and pointed everyone's attention to a display in a corner of the room of a model of the Object D satellite. Invoking the name of a legendary Soviet scientist, Korolev hurriedly explained that it would now be possible to realize the dreams of Tsiolkovskiy with the use of the R-7 missile. Perceiving that the audience was not much impressed by the speech, Korolev pointed out that the United States had stepped up their satellite program, but that compared to the "skinny" U.S. launch vehicle, the Soviet R-7 could significantly outdo that project in terms of the mass of the satellite. In closing, he added that the costs for such a project would be meager, since the basic expense for the launcher was already allocated for in the R-7 booster. Khrushchev began to exhibit some interest and asked Korolev if such a plan might not harm the R-7 weapons research program given that was the primary focus of work at Korolev's Design Bureau. Clearly oversimplifying the difficulties involved, Korolev shot back that unlike the United States, which was spending millions of dollars to develop a special rocket to launch a satellite, all the Soviets would have to do would be to replace the warhead with a satellite on the R-7. Khrushchev hesitated for a second, perhaps suspicious of Korolev's intentions, but answered back: "If the main task doesn't suffer, do it."[27]

After over two years of explicit lobbying the artificial satellite project was a reality. And it owed its approval more than anyone to Korolev. Tikhonravov had provided the technical expertise, Keldysh had helped with his political clout, but it was finally Korolev's repeated requests, letters, meetings, reports, and entreaties which finally forced the decision. Korolev also had a climate conducive to his needs. His standing among the military and industrial community had evolved over the years from maverick engineer to genius manager. His successes with a series of ballistic missiles pleased both the military and the industry. And it could not hurt that both of these sectors, by 1956, were populated by individuals who were sympathetic to the Chief Designers unquenchable thirst for space exploration. Clearly, Korolev alone could not have done it. Events outside his control, such as the Eisenhower administration's announcement, Sedov's press conference, the fall of the Beriya group in the nuclear weapons industry, and Khrushchev's rise to power were pivotal events in the road to approval. But hindsight suggests that the Soviet space program was born on January 30, 1956, and without Korolev it would never have been conceived.

LABOR

The Object D (or D-1) was so named since it would be the fifth type of payload to be carried on an R-7, Objects A, B, V, and G being designations for different nuclear warhead containers.[28] The satellite was a complex scientific laboratory, far more sophisticated than any other IGY proposal from the period. While Kryukov's engineers depended a great deal on Tikhonravov's early work on satellites, much of the actual design was a journey into uncharted territory for the OKB-1. There was little precedent for creating pressurized containers and instrumentation for work in Earth orbit, while long-range communications systems had to be designed without the benefit of prior experience. The engineers were aware of the trajectory tracking and support capabilities for the R-7 missile, and this provided a context for determining the levels of contact with the vehicle. The fact that the object would be out of contact with the ground for long periods of time (unlike sounding rockets) meant that new self-switching automated systems would have to be used. The selection of metals to construct the satellite also presented problems to the engineers, since the effects of continuous exposure to the space environment was still in the realm of conjecture. The experiments and experience from sounding rocket tests provided a database for the final selection.

Technical work on the vehicle officially began on February 25, 1956, with contracts handed out on March 5. Tikhonravov's group at the NII-4 and Korolev's Design Bureau at the NII-88 were the two most active participants in this process, but numerous other organizations provided various elements of the complete satellite. By June 14, Korolev finalized the necessary changes to the basic version of the R-7 ICBM in order to use it for a satellite launch. The new booster would incorporate a number of major changes including the use of uprated main engines, deletion of the central radio package on the booster, and a new payload fairing replacing the old one used for a nuclear warhead.[29] A month later, on July 24, 1956, Korolev formally approved the initial

Draft Plan for the Object D. The document was co-signed by his senior associates Tikhonravov, Konstantin D. Bushuyev, Sergey O. Okhapkin, and Leonid A. Voskresenskiy.[30]

By mid-1956 the Object D project was beginning to fall significantly behind schedule. Some subcontractors were particularly lackadaisical in their assignments, and parts were often delivered which did not fit the original specifications. On September 14, Keldysh made a personal plea at a meeting of the Academy of Sciences Presidium for speeding up work, invoking a threat all would understand: "We all want our satellite to fly earlier than the Americans."[31] Events in the satellite program took an abrupt turn in the waning months of 1956. Actual test models of the Object D, expected to be ready by October, remained unfinished. By the end of November, Korolev began to suffer from great anxiety, no doubt compounded by his extraordinarily busy plans, traveling from Kaliningrad to Kapustin Yar to Tyura-Tam to Molotovsk and back several times to oversee various projects.[32] Part of this anxiety was due to serious concerns that his project would be suddenly preempted with a satellite launch from the United States. He had been informed of a September 1956 launch of a missile from Patrick Air Force Base at Cape Canaveral, Florida, which, according to his erroneous information, was a failed attempt to launch a satellite into orbit.[33] A second concern were the results of static testing of the R-7 engines on the ground. Instead of the projected specific impulse of 309–310 seconds, the R-7 engines would not produce more than 304 seconds, too low for the heavy Object D satellite. He realized that perhaps he was making this effort too complicated. Why not attempt to launch something simpler on the first orbital attempt instead of a sophisticated one-and-a-half-ton scientific observatory?

At the end of November Tikhonravov was perceptive enough to detect Korolev's anxiety and verbalized it: "What if we make the satellite a little lighter? Thirty kilograms or so, or even lighter?"[34] Keldysh was at first opposed to the idea, but eventually ceded to the strong-willed Korolev. This time Korolev would not depend on dozens of other subcontractors; he made sure that the smaller satellite would be designed and manufactured completely in his own Design Bureau with the help of only two outside organizations: the Scientific-Research Institute of Current Sources under Nikolay S. Lidorenko for the design of the onboard batteries and the NII-885 under Chief Designer Mikhail

S. Ryazanskiy for the radio-transmitters. On January 5, 1957, Korolev sent off a letter to the government which described his revised plan. He asked that permission be given to launch two small satellites, each with a mass of 40–50 kilograms, in the period April-June 1957 immediately *prior* to the beginning of the IGY. This plan would be contingent upon the timetable for the R-7 program which Korolev admitted was behind schedule; the first launch of the missile was set for March 1957 at the earliest. Each satellite would orbit the Earth at altitudes of 225 X 500 kilometers and contain a simple shortwave transmitter with a power source sufficient for 10 days operation. Korolev did not obscure the reasons for the abrupt change in plans:

> ... the United States is conducting very intensive plans for launching an artificial Earth satellite. The most well-known project under the name "Vanguard" uses a three-stage missile ... the satellite proposed is a spherical container of 50 centimeters diameter and a mass of approximately 10 kilograms. In September 1956, the U.S.A. attempted to launch a three-stage missile with a satellite from Patrick Base [sic] in the state of Florida which was kept secret. The Americans failed to launch the satellite ... and the payload flew about 3,000 miles o\r approximately 4,800 kilometers. This flight was then publicized in the press as a national record. They emphasized that U.S. rockets can fly higher and farther than all the rockets in the world, including Soviet rockets. From separate printed reports, it is known that the U.S.A. is preparing in the nearest months a new attempt to launch an artificial Earth satellite and is willing to pay any price to achieve this priority.[35]

While Korolev's information on U.S. plans may have been in error, his instincts were not that far off. The United States could have launched a satellite by early 1957, but various institutional and political obstacles precluded such an attempt.

By January 25, 1957, the Chief Designer approved the initial design details of the satellite, now officially designated Simple Satellite No. 1 (PS-1).[36] Although there was some token resistance to Korolev's revised plan, primarily from Keldysh, his letter appeared to have adequately invoked the specter of U.S. eminence in the field of military technology. On February 15, the USSR Council of Ministers formally issued a new decree (no. 171–83ss) entitled "On Measures to Carry Out in the International Geophysical Year," approving the new proposal.[37] The two new satellites, PS-1 and PS-2, weighing approximately 100 kilograms each, would be launched in April–May 1957 after one or two fully successful R-7 launches. Eisenhower's plan to launch an American satellite during IGY was the deciding factor on a launch date. The Object D launch, meanwhile was pushed back to April 1958. Focused on a more modest objective, Korolev wasted little time. He quickly sent out technical specifications for the initial satellite PS-1 to the two subcontractors. By this time there was an impressive sight at the Tyura-Tam launch base in Soviet Central Asia: the first flight article of the magnificent R-7 was on the launch pad.

The first three launches of the R-7 ICBM in May-July 1957 were all failures, completely disrupting Korolev's schedule to launch a satellite before the beginning of the IGY. The days following the last failure were the lowest point for Korolev and his associates. Suddenly everything they had labored for over three years had been put into doubt. There was severe criticism from higher officials and even talk of curtailing the entire program. For Korolev, the headaches were compounded by the cumulative delays of his Simple Satellite project. It was now a month *into* the IGY and the R-7 itself had not flown a successful mission. His dreams, his position, his status were all in jeopardy, and this began to affect his temperament. In mid-June he had written to his wife from the launch site, "Things are not going very well again," adding with a note of optimism, "… right here and now, we must strive for the solution we need!" By July things began to deteriorate. On the 8th he wrote "We are working very hard," but after the second launch failure, he wrote on the 23rd "Things are very, very bad."[38] Korolev's biographer wrote in 1987, "In all the postwar years, no days were more painful, difficult, or tense for Sergey Pavlovich Korolev than those of that hot summer of 1957."[39]

Apart from competition from the United States, Korolev had to unexpectedly deal with a different kind of threat at the time, one from within the USSR in the person of Chief Designer Mikhail K. Yangel of the Experimental Design Bureau No. 586 (OKB-586). In the first quarter of 1957, Yangel's Design Bureau at Dnepropetrovsk in Ukraine, on orders from ministerial boss Dmitriy F. Ustinov, had begun to explore the possibility of modifying their R-12 intermediate range ballistic missile into a satellite launch vehicle.[40] The missile itself, fueled by storable hypergolic propellants unlike the R-7, was the subject of a five year long development program, at first under Korolev's tutelage, but later transferred to Dnepropetrovsk. Prodded by the unending delays in the R-7 program, Yangel evaluated "the possibility of the *immediate* launch of a similar satellite [as Korolev's] using the simplest of booster rockets based on the strategic R-12 missile."[41] Although analysis proved that a hastily modified two-stage R-12 could be used for this goal, it did not seem likely that a first launch could be carried out prior to either the R-7 or the Americans. To Korolev's relief, the plan was shelved.

Back at the launch range of Tyura-Tam, the fourth R-7 launch on August 21, 1957 was successful. The missile and its payload flew 6,500 kilometers, the warhead finally entering the atmosphere over the target point at Kamchatka. Korolev was so subsumed by euphoria that he stayed awake until three in the morning speaking to his deputies and aides about the great possibilities that had opened up, the future, and mostly about his artificial satellite.[42] It was extremely unusual for the Soviets to publicize successes in any military field, so it was all the more odd when six days after the R-7 launch, the official news agency TASS released a brief communiqué:

A few days ago a super-long-range, intercontinental multistage ballistic missile was launched. The tests of the missile were successful; they fully confirmed the correctness of the calculations and the selected design. The flight of the missile took place at a very great, hitherto unattained, altitude. Covering an enormous distance in a short time, the missile hit the assigned region. The results obtained show that there is the possibility of launching missiles into any region of the terrestrial globe. The solution of the problem of creating intercontinental ballistic missiles will make it possible to reach remote regions without resorting to strategic aviation, which at the present time is vulnerable to modern means of antiaircraft defense.[43]

Clearly it did not have the intended effect on the U.S. public or media, since for the most part, little attention was given it. Those that did pay lip service to the announcement spoke only to dismiss the claim, a stance justified partly by the black hole of information on Soviet ballistic missiles. It would take 38 more days before the entire world would take notice that a new age had arrived, heralded by that same intercontinental ballistic missile.

BIRTH

Work on the "simple satellite" PS-1 had continued at an uneven pace since development of the object began in January 1957. Between March and August, engineers carried out computations to select and refine the trajectory of the launch vehicle and the satellite during launch. These enormously complicated computations for the R-7 program were initially done by hand using electrical arithrometers and six-digit trigonometric tables. When more complex calculations were required, engineers at the OKB-1 were offered the use of a "real" computer recently installed at the premises of the Academy of Sciences at Keldysh's request. The gigantic machine filled up a huge room at the department and may have been the fastest computer in the USSR in the late 1950s: it could perform ten thousand operations per second, a high-end capability for Soviet computing machines of the time.[44]

There were many debates on the shape of the first satellite, with most senior OKB-1 designers preferring a conical form since it fit well with the nose cone of the rocket. At a meeting early in the year, Korolev had a change of heart and suggested a metal sphere at least one meter in diameter.[45] There were six major guidelines followed in the construction of PS-1:

— the satellite would have to be of maximum simplicity and reliability while keeping in mind that methods used for the spacecraft would be used in future projects;
— the body of the satellite was to be spherical in order to determine atmospheric density in its path;
— the satellite was to be equipped with radio equipment working on at least two wavelengths of sufficient power to be tracked by amateurs and to obtain data on the propagation of radio waves through the atmosphere;
— the antennae were to be designed so as to not affect the intensity of the radio signals due to spinning;
— the power sources were to be onboard batteries ensuring work for two to three weeks; and
— the attachment of the satellite to the core stage would be such that there would be no failure to separate.

The five primary scientific objectives of the mission were:

— to test the method of placing an artificial satellite into Earth orbit;
— to provide information on the density of the atmosphere by calculating its lifetime in orbit;
— to test radio and optical methods of orbital tracking;
— to determine the effects of radio wave propagation through the atmosphere; and
— to check principles of pressurization used on the satellite.[46]

The satellite as it eventually emerged was a pressurized sphere, 58 centimeters in diameter made of an aluminum alloy. The sphere was constructed by combining two hemispherical casings together. The pressurized internal volume of the sphere was filled with nitrogen at 1.3 atmospheres which maintained an electro-chemical source of power (three silver-zinc batteries), two radio-transmitters, a thermo-regulation system, a ventilation system, a communications system, temperature and pressure transmitters, and associated wiring. The two radio transmitters operated at frequencies of 20.005 and 40.002 megacycles at wavelengths of 1.5 and 7.5 meters. The signals on both the frequencies were spurts lasting 0.2 to 0.6 seconds, providing the famous "beep-beep" sound to the transmissions. The antennae system comprised four rods, two with a length of 2.4 meters each and the remaining two with a length of 2.9 meters each. Tests of this radio system were completed as early as May 5, 1957, using a helicopter and a ground station.[47] The total mass of the satellite was 83.6 kilograms of which 51.0 kilograms was simply the power source. The lead designer for PS-1 was Mikhail S. Khomyakov. Oleg G. Ivanovskiy was his deputy.[48]

Korolev, of course, kept close tabs on the development of PS-1 and continuously saw to it that the spherical satellite was kept spotlessly clean and shiny not only for its reflective qualities, but perhaps also for its overall aesthetic beauty. On one occasion he flew into a rage at a junior assembly shop worker for doing a poor job on the outer surface of a *mockup* of the satellite. "This ball will be exhibited in museums!" he shouted.[49] An aide from Moscow telephoned Korolev at Tyura-Tam on June 24 to inform him that he had just signed the document specifying the final configuration of the satellite. The launch vehicle earmarked for the satellite was a slightly uprated version of the basic R-7 ICBM variant. The modifications included the omission of a 300 kilogram radio-package from the top of the core booster, the changing of burn times of the main engines; the removal of a vibration measurement system, the use of a special nozzle system to separate the booster from the satellite installed at the top of the core stage, and the installation of a completely new payload shroud and container replacing the warhead configuration.[50] The length of the booster with the new shroud was 29.167 meters, almost four meters shorter than the ICBM version.

The Council of Ministers had formally approved the simple satellite program in February 1957. With one R-7 success under his belt, Korolev now needed final permission from the State Commission to proceed with a satellite launch. It was supposed to be merely a formality, since the Soviet government had already approved the satellite attempt, but the process appears to have been fraught with difficulty, suggesting that even at this late stage, there were individuals on the Commission who were not interested in a satellite. At a Commission meeting in late August, Korolev formally asked for permission to launch a satellite if a second R-7 ICBM successfully flew in early September. Convincing the Commission proved to be much harder than expected and the meeting ended in fierce arguments and recriminations. Not easily turned away, Korolev tried again at a second session soon after, this time using a political ploy: "I propose let us put the question of national priority in launching the world's first artificial Earth satellite to the Presidium of the Central Committee of the Communist Party. Let them settle it."[51] It worked. None of the members wanted to take the blame for a potential miscalculation, and Korolev got what he wanted. A final document for launch, "The Program for Carrying Out a Test Launch of a Simple Unoriented ISZ (the Object PS) Using the Product 8K71PS," was later signed by:

— Vasiliy F. Ryabikov (Military-Industrial Commission);
— Mitrofan I. Nedelin (Ministry of Defense);
— Dmitriy F. Ustinov (Ministry of Defense Industries);

— Valeriy D. Kalmykov (Ministry of Radio-Technical Industry); and

— Aleksandr N. Nesmeyanov (Academy of Sciences).[52]

The subsequent launch of the R-7 on September 7 was as successful as the one in August, and the R-7 ICBM flew across the Soviet Union before depositing its dummy warhead over the Kamchatka peninsula.[53] In the summer, Korolev and the other Chief Designers began to informally target the satellite launch for the one hundredth anniversary of spaceflight visionary Tsiolkovskiy's birth on September 17, but achieving this date proved increasingly unrealistic. Instead of being at Tyura-Tam for a space launch on that day, Korolev and R-7 rocket engine designer Glushko were both in attendance at the Pillard Hall of the Palace of Unions in Moscow for a special celebration of the great visionary's birthday. In a long speech to the distinguished audience, Korolev, whose real job was not revealed, prophesized that, "in the nearest future the first test launches of artificial satellites of the Earth with scientific goals will take place in the USSR and the USA."[54] The audience had little idea of the accuracy of the prediction.

On September 20 Korolev was at Moscow for a meeting of the State Commission for the PS-1 launch.[55] Chairman Ryabikov, Korolev, Keldysh, and Marshall Nedelin were the principal participants and established October 6 as the target date of the launch based on the pace of preparations. At the same meeting, the Commission decided to publicly announce the launch of PS-1 only after completion of the first orbit. A communiqué to this effect was written up by Ryabikov himself on September 23.[56] The frequencies for tracking by amateurs had already been announced earlier in the year in the issues of the journal *Radio* although details of the program had obviously been omitted. Korolev himself flew into Tyura-Tam on September 29 staying in a small house close to the primary activity area near site two.

The preparations for launching were for the most part uneventful save for the last minute replacement of one of the batteries on the flight version of PS-1. Still apprehensive over a last minute U.S. launch, Korolev abruptly proposed to the State Commission that the launch be brought forward two days. His concerns were apparently prompted by plans for a conference in Washington, D.C. in early October as part of IGY proceedings. On the 6th, the day of PS-1's scheduled launch, a paper entitled "Satellite Over the Planet" was to be presented by the American delegation. He believed that the presentation was to be timed to coincide with a hitherto unannounced launch of a U.S. satellite.[57] KGB representatives assured Korolev that this was not so, but Korolev was convinced that a launch of Army Jupiter C might be attempted. In the end, the schedule for PS-1's launch was moved forward two days to the 4th; Korolev signed the final order for launch at four in the afternoon on the 2nd and sent it to Moscow for approval.[58]

The R-7 was transported and installed on the launch pad in the early morning of October 3 escorted on foot by Korolev, Ryabikov, and other members of the State Commission. Fueling began early the following morning at 5:45 a.m. local time.[59] Korolev, under a great amount of pressure, remained cautious throughout the proceedings. He told his engineers, "Nobody will hurry us. If you have even the tiniest doubt, we will stop the testing and make the corrections on the satellite. There is still time ... ."[60] Most of the engineers, understandably enough, did not have time to ponder over the historical value or importance of the upcoming event. PS-1's deputy designer Ivanovskiy recalled "... Nobody back then was thinking about the magnitude of what was going on: everyone did his own job, living through its disappointments and joys."[61]

On the night of the 4th, huge flood lights illuminated the launch-pad as the engineers in their blockhouse checked off all the systems. In the command bunker accompanying Korolev were some of the senior members of the State Commission. All launch operations were handled by two men, a civilian and a military officer. Representing the civilians was Korolev's deputy Leonid A. Voskresenskiy, one of the most colorful characters in the history of the Soviet space program. A daredevil motorcyclist with a legendary penchant for taking risks, he had been with the program since the early days in 1945 when the Soviets had scoured Germany for the remains of the A-4 missile. Lt.-Col. Aleksandr I. Nosov represented the military. Both men were 44 years old at the time. The actual command for launch was entrusted to the hands of Boris S. Chekunov, a young artillery forces lieutenant. He later recalled the final moments as the clock ticked past midnight local time: "When only a few minutes remained until lift-off, Korolev nodded to his deputy Voskresenskiy. The operators froze, awaiting the final order. Nosov, the chief of the launch control team, stood at the periscope. He could see the whole pad. 'One minute to go!' he called."[62] Another senior engineer in the bunker recalled:

> With the exception of the operators, everybody was standing. Only N. A. Pilyugin and S. P. Korolev were allowed to sit down. The launch director [Nosov] began issuing commands. I kept an eye on S. P. Korolev. He seemed nervous although he tried to conceal it. He was carefully examining the readings of the various instruments without missing any nuance of our body language and tone of voice. If anybody raised their voice or showed signs of nervousness, Korolev was instantly on the alert to see what was going on.[63]

The seconds counted down to zero and Nosov shouted the command for lift-off. Chekunov immediately pressed the lift-off button. At exactly 2228 hours 34 seconds Moscow Time on October 4, the engines ignited and the 272,830 kilogram booster lifted off the pad in a blaze of light and smoke. The five engines of the R-7 generated about 398 tons of thrust at launch. Although the rocket lifted off gracefully, there were problems. Delays in the firing of several engines almost resulted in a launch abort. Additionally, at T+16 seconds, the System for the Synchronous Emptying of the Tanks (SOBIS) failed, which resulted in higher than normal kerosene consumption. A turbine failure due to this resulted in main engine cut-off one second prior to the planned moment.[64] Separation from the core stage, however, occurred successfully at T+324.5 seconds, and the 83.6 kilogram PS-1 successfully flew into a free-fall elliptical trajectory. The first human-made object entered orbit around the Earth inaugurating a new era in exploration.

With most State Commission members still in the bunker, engineers at Tyura-Tam awaited confirmation of orbit insertion from the satellite in a van set up about 800 meters from the launch pad. As a huge crowd waited outside the van, radio operator Vyecheslav I. Lappo from the NII-885, who had personally designed the onboard transmitters, sat expectantly for the first signal. The Kamchatka station picked up signals from the satellite and there was cheering but Korolev cut everybody off: "Hold off on the celebrations. The station people could be mistaken. Let's judge the signals for ourselves when the satellite comes back after its first orbit around the Earth."[65] Eventually the distinct "beep-beep-beep" of the craft came in clearly over the radio waves and the crowd began to celebrate. Chief Designer Ryazanskiy who was at the van immediately telephoned Korolev in the bunker. The ballistics experts at the Coordination-Computation Center in Moscow had determined that the satellite was in an orbit with a perigee of 228 kilometers and an apogee of 947 kilometers, the latter about 80 kilometers lower than planned due to the early engine cut-off. Inclination of the orbit to the Earth's equator was 65.6 degrees while orbital period was 96.17 minutes.[66] Experts at the Moscow Center also ascertained that the satellite was slowly losing altitude, but State Commission Chairman Ryabikov waited until the second orbit was over prior to telephoning Soviet leader Nikita S. Khrushchev.

According to conventional wisdom, Khrushchev's reaction to the launch was unusually subdued for an event of such magnitude, indicating that he, like many others, did not immediately grasp the true propaganda effect of such a historic moment. He told the press later that:

When the satellite was launched, they phoned me that the rocket had taken the right course and that the satellite was already revolving around the earth. I congratulated the entire group of engineers and technicians on this outstanding achievement and calmly went to bed.[67]

Khrushchev's son, however, recalls his father's reaction was a little more enthused. The older Khrushchev at the time was on visit to Kiev to discuss economic issues with the Ukrainian Party leadership. Around 11 p.m., these negotiations were interrupted by a telephone call. Khrushchev quietly took the call, then returned back to his discussions, without saying anything. Eventually, as his son recalled, the news was too difficult to keep under wraps:

He finally couldn't resist saying [to the Ukrainian officials]: "I can tell you some very pleasant and important news. Korolev just called (at this point he acquired a secretive look). He's one of our missile designers.

> Remember not to mention his name—it's classified. So, Korolev has just reported that today, a little while ago, an artificial satellite of the earth was launched."[68]

The Soviet leader was evidently animated the rest of the evening, speaking in glowing terms about the new era of missiles which could "demonstrate the advantages of socialism in actual practice" to the Americans.

The official Soviet news agency TASS released the communiqué Ryabikov had authored on the morning of October 5. Published in the morning edition of *Pravda*, it was exceptionally low-key and was not the headline of the day:

> For several years scientific research and experimental design work have been conducted in the Soviet Union on the creation of artificial satellites. As has already been reported in the press, the first launching of the satellites in the USSR were planned for realization in accordance with the scientific research program of the International Geophysical Year. As a result of very intensive work by scientific research institutes and design bureaus the first artificial satellite in the world has been created. On October 4, 1957, this first satellite was successfully launched in the USSR. According to preliminary data, the carrier rocket has imparted to the satellite the required orbital velocity of about 8,000 meters per second. At the present time the satellite is describing elliptical trajectories around the Earth, and its flight can be observed in the rays of the rising and setting Sun with the aid of very simple optical instruments (binoculars, telescopes, etc.).[69]

The Soviet media did not ascribe a specific name for the satellite, generally referring to it as "Sputnik," the Russian word for "satellite," often also loosely translated as "fellow traveler."

As the media tumult over Sputnik began to mount in the West, the Soviet leadership began to capitalize on the utter pandemonium pervading the discourse on the satellite in the United States. On October 9, *Pravda* published a long report anonymously authored by Korolev and other designers detailing the construction and design of the satellite.[70] The parties responsible for this great deed were, of course, not named. Having been involved in the defense industry, the real job titles of the members of the Council of Chief Designers had always remained secret, although Tikhonravov and others had freely published under their own names through the 1950s on topics of general interest. This suddenly changed as their names disappeared from official histories. Beginning with the launch of Sputnik, of the four major contributors to its success, Korolev, Glushko, and Keldysh were referred in the open press as the Chief Designer of Rocket-Space Systems, the Chief Designer of Rocket Engines, and the Chief Theoretician of Cosmonautics respectively. The fourth, Tikhonravov, did not even have a pseudonym for himself.

The titles not only hid their identities, but also added an element of attraction and enigma to the men behind the world's first space program. New editions of histories of Soviet rocketry published prior to 1957 ceased to carry Korolev's name, and Soviet encyclopedias now merely listed him as heading a laboratory in an unspecified "machine building" institute in the USSR. Glushko meanwhile was now said to be laboratory chief at the Moscow Institute of Mineral Fuels.[71] Korolev, certainly in recognition of the key role he played, was allowed to write in no less an important newspaper as *Pravda*, but under the pseudonym "Professor K. Sergeyev." His first article titled "Research into Cosmic Space" was published on December 12, 1957. Khrushchev claimed at the time that as the years went by "the photographs and names of these illustrious people will be made public," but that for the moment "in order to ensure the country's security and the lives of these scientists, engineers, technicians, and other specialists, we cannot yet make known their names or publish their photographs."[72]

CONCLUSION

Khrushchev was clearly cognizant of the satellite project, but he seems to have been remarkably uninterested in it. Certainly, he used Sputnik to advance his political agenda *post-facto*, but the launch itself was never intended as anything more than a response to Korolev's formidable powers of persuasion. As such, the timing of the Sputnik launch was motivated by a single reasoning: Korolev's drive to preempt a U.S. satellite launch attempt during the International Geophysical Year. At first, it was a competition with Vanguard. Spurred by the July 1955 announcement of U.S. satellite plans for the IGY, Korolev, joined by Tikhonravov and Keldysh, convinced both the government and the Academy of Sciences within a month to proffer support for a complex Soviet satellite project timed for launch before the IGY. A second jolt came as a result of miscommunication about a U.S. Army missile launch in September 1956. Putting the heavy scientific satellite on the backburner, Korolev's engineers put together a much simpler satellite to beat any American attempt. Once again, they timed it for launch before the start of the IGY. This 84 kilogram ball, although delayed several months, lifted off into orbit on October 4, 1957, and opened a new era.

The political and cultural shock bequeathed by Sputnik set events in motion that eventually gave rise to Apollo, perhaps the central artifact of the so-called "space race" of the Cold War. Conventional wisdom suggests that the race began on October 4, 1957, and ended on July 20, 1969, with the Moon landing. But as we begin to dig deeper into the origins of the space race, it is clear that the race began not with the launch of Sputnik, but in fact with the Eisenhower administration's announcement in July 1955, more than two years before Sputnik. And perhaps fortunately for the Soviet Union, it was a race in which one of the participants, the United States, did not even know it was running until it was too late.

1. Walter McDougall, ... The Heavens and the Earth: A Political History of the Space Age (New York: Basic Books, Inc., 1985), pp. 60–62.
2. James E. Oberg, Red Star in Orbit (New York: Random House, 1981), pp. 29–30.
3. See Yaroslav Golovanov, Korolev: fakty i mify (Moscow, Ruussia: Nauka Press, 1994).
4. For a detailed look at Korolev's scientific activities in the 1930s see G. S. Vetrov, S. P. Korolev i kosmonavtika: pervye shagi (Moscow, Russia: Nauka Press, 1994).
5. Yu. P. Semenov, ed., Raketno-Kosmicheskaya Korporatsiya "Energiya" imeni S. P. Koroleva (Korolev, Russia: RKK Energiya named after S. P. Korolev, 1996), p. 22.
6. For a detailed english language summary of the details of Tikhonravov's research during the 1940s and early 1950s as well as the famous 1954 report itself, see Asif A. Siddiqi, "Before Sputnik: Early Satellite Studies in the Soviet Union, 1947–1957," Spaceflight (October 1997): 334–337; Asif A. Siddiqi, "Before Sputnik: Early Satellite Studies in the Soviet Union, 1947–1957—Part 2," Spaceflight (November 1997): 389–392. Tikhonravov's document has been reproduced as M. Tikhonravov, "Report on an Artificial Satellite of the Earth" (in Russian) in B. V. Raushenbakh, ed., Materialy po istorii kosmicheskogo korabl "vostok" (Moscow, USSR: Nauka Press, 1991), pp. 5–15.
7. The text of this letter in a censored version has been published as S. P. Korolev, "On the Possibility of Work on an Artificial Satellite of the Earth" (in Russian), M. V. Keldysh, ed., Tvorcheskoye naslediye Akademika Sergeya Pavlovicha Koroleva: izbrannyye trudy i dokumenty (Moscow, USSR: Nauka Press, 1980), p. 343.
8. Semenov, Raketno-Kosmicheskaya Korporatsiya "Energiya" imeni S. P. Koroleva, p. 86. The coauthors of the proposal were: B. L. Vannikov (First Deputy Minister of Medium

Machine Building), M. V. Khrunichev (First Deputy Minister of Medium Machine Building), and K. N. Rudnev (Deputy Minister of Defense Industries). Of interest is the fact that Malyshev, Vannikov, and Khrunichev were all high officials in the *nuclear* weapons industry. Rudnev was the only one from the missile industry.

9.   Edward Clinton Ezell and Linda Neumann Ezell, *The Partnership: A History of the Apollo-Soyuz Test Project* (Washington, D.C.: NASA SP-4209, 1978), p. 16; Nicholas Daniloff, *The Kremlin and the Cosmos* (New York: Alfred A. Knopf, 1972), p. 54.

10.   A. Yu. Ishlinskiy, ed., *Akademik S. P. Korolev: ucheniy, inzhener, chelovek* (Moscow USSR: Nauka Press, 1986), p. 453; Boris Konovalov, "The Genealogy of Sputnik" (in Russian), in V. Shcherbakov, ed., *Zagadki zvezdnykh ostrovov* (Moscow, USSR: Molodaya gvardiya, 1989), p. 115.

11.   "Commission on Interplanetary Communications" (in Russian), *Vechernaya moskva*, April 16, 1955, p. 1. An English translation of the announcement is included in F. J. Krieger, *Behind The Sputniks: A Survey of Soviet Space Science* (Washington, D.C.: Public Affairs Press, 1958), pp. 328–30. The names of only four other members were announced at the time: V. A. Ambartsumyan, P. L. Kapitsa, B. V. Kukarin, and P. P. Parenago. A larger 27 member list was submitted to the International Astronautical Federation in October 1957.

12.   "Commission on Interplanetary Communications," 1955.

13.   Of the 27 Commission members listed in 1957, only two individuals, A. A. Blagonravov and D. Ye. Okhotsimskiy, were directly involved in the ballistic missile and space programs. The former headed the Commission for Upper Atmosphere Research of the Academy of Sciences which oversaw all scientific suborbital launches, while the latter was one of the leading mathematicians at the Department of Applied Mathematics of the V. A. Steklov Mathematics Institute of the Academy of Sciences (OPM MIAN) who was involved in the early design of the R-7 ICBM. See also Ishlinskiy, *Akademik S. P. Korolev: ucheniy, inzhener, chelovek*, p. 453.

14.   Ezell and Ezell, *The Partnership*, p. 18.

15.   Robert W. Buchheim and the Staff of the Rand Corporation, *Space Handbook: Astronautics and its Applications* (New York: Random House, 1959), p. 277; "We'll Launch 1st Moon, and Bigger, Says Russ," *Los Angeles Examiner*, August 3, 1955; John Hillary, "Soviets Planning Early Satellite," *The New York Times*, August 3, 1955.

16.   Ishlinskiy, *Akademik S. P. Korolev: ucheniy, inzhener, chelovek*, p. 445; Yaroslav Golovanov, "The Beginning of the Space Era" (in Russian), *Pravda*, October 4, 1987, p. 3. Note that in Semenov, *Raketno-Kosmicheskaya Korporatsiya "Energiya" imeni S. P. Koroleva*, p. 86, it is stated that the report was authored only by Lavrov and it was completed on June 16, 1955, not July 16, 1955. Tikhonravov himself has, however, claimed they *both* authored the report.

17.   Semenov, *Raketno-Kosmicheskaya Korporatsiya "Energiya" imeni S. P. Koroleva*, p. 87.

18.   *Ibid.* Keldysh's official posts were: Director of the NII-1 and Chief of the OPM MIAN. Mrykin's official post was First Deputy Commander of the Directorate of the Chief of Reactive Armaments (UNRV). The UNRV was subordinate to the Chief Artillery Directorate (GAU) of the General Staff of the Ministry of Defense.

19.   Ishlinskiy, *Akademik S. P. Korolev: ucheniy, inzhener, chelovek*, p. 455; Golovanov, *Korolev: fakty i mify*, pp. 523–24; Golovanov, "The Beginning of the Space Era." Others present at this meeting were M. A. Lavrentiyev and G. A. Skuridin.

20.   Ishlinskiy, *Akademik S. P. Korolev: ucheniy, inzhener, chelovek*, pp. 455–56; Christian Lardier, *L'Astronautique Soviétique* (Paris: Armand Colin, 1992), p. 107; Golovanov, "The Beginning of the Space Era." Blagonravov's commission was at the time directing the scientific investigations on board suborbital rockets, while Sedov's commission had recently been established as a public forum for Soviet scientists to discuss space exploration.

21. These included: atmospheric specialists V. I. Krasovskiy, L. V. Kurnosovaya, and S. N. Vernov; the young mathematicians from the OPM MIAN T. M. Eneyev, M. L. Lidov, D. Ye. Okhotsimskiy, and V. A. Yegorov; solar battery expert N. S. Lidorenko; and the more famous Academicians L. A. Artsimovich, V. L. Ginsburg, A. F. Ioffe, P. L. Kapitsa, B. P. Konstantinov, and V. A. Kotelnikov. See Ishlinskiy, *Akademik S. P. Korolev: ucheniy, inzhener, chelovek*, p. 456; Golovanov, "The Beginning of the Space Era."

22. Keldysh, *Tvorcheskoye naslediye Akademika Sergeya Pavlovicha Koroleva*, p. 362; Ishlinskiy, *Akademik S. P. Korolev: ucheniy, inzhener, chelovek*, p. 445; Konovalov, "The Genealogy of Sputnik," pp. 116–117; Golovanov, *Korolev: fakty i mify*, p. 529; Semenov, *Raketno-Kosmicheskaya Korporatsiya "Energiya" imeni S. P. Koroleva*, p. 87. B. Konovalov, "Dash to the Stars" (in Russian), *Izvestiya*, October 1, 1987, p. 3. The five industrial ministries were: the Ministry of Defense Industries, the Ministry of Radiotechnical Industry, the Ministry of Ship Building Industry, the Ministry of Machine Building, and the Ministry of Defense.

23. Sergey Khrushchev, *Nikita Khrushchev: krizisy i rakety: vzglyad iznutri: tom 1* (Moscow, Russia: Novosti, 1994), p. 97.

24. William P. Barry, *The Missile Design Bureaux and Soviet Piloted Space Policy, 1953–1974*, Draft of University of Oxford D.Phil., Dissertation, 1995.

25. Khrushchev, *Nikita Khrushchev: krizisy i rakety: vzglyad iznutri: tom 1*, p. 106.

26. Khrushchev, *Nikita Khrushchev: krizisy i rakety: vzglyad iznutri: tom 1*, p. 109.

27. Khrushchev, *Nikita Khrushchev: krizisy i rakety: vzglyad iznutri: tom 1*, pp. 110–11.

28. Raushenbakh, *Materialy po istorii kosmicheskogo korabl "vostok"*, p. 209. A, B, V, G, and D are the first five letters of the Russian cyrillic alphabet.

29. Ishlinskiy, *Akademik S. P. Korolev: ucheniy, inzhener, chelovek*, p. 446; Golovanov, *Korolev: fakty i mify*, p. 530; Timothy Varfolomeyev, "Soviet Rocketry that Conquered Space: Part 1: From First ICBM to Sputnik Launcher," *Spaceflight* (August 1995): 260–63.

30. Golovanov, *Korolev: fakty i mify*, p. 530; Lardier, *L'Astronautique Soviétique*, p. 107. Tikhonravov was officially an employee of the NII-4 but was temporarily working as the Chief Consultant to the NII-88 OKB-1.

31. An edited version of Keldysh's speech has been published as M. V. Keldysh, "On Artificial Satellites of the Earth" (in Russian), V. S. Avduyevskiy and T. M. Eneyev, eds. *M. V. Keldysh: izbrannyye trudy: raketnaya tekhnika i kosmonavtika* (Moscow, USSR: Nauka Press, 1988), 235–240; See also Golovanov, *Korolev: fakty i mify*, p. 530.

32. Kaliningrad was the location of the OKB-1, while sea trials of the R-11FM were carried out near Molotovsk. Kapustin Yar and Tyura-Tam were the two missile launch ranges.

33. This was a Jupiter C missile (no. RTV-1) which flew a distance of 5,300 kilometers on September 20, 1956, during a re-entry test. A live third stage could have put a small payload into orbit, but this was not the intended goal.

34. Golovanov, "The Beginning of the Space Era"; Golovanov, *Korolev: fakty i mify*, p. 532.

35. The complete text of Korolev's letter is reproduced as S. P. Korolev, "Proposal on the First Launch of an Artificial Satellite of the Earth Before the Start of the International Geophysical Year" (in Russian) in Keldysh, *Tvorcheskoye naslediye Akademika Sergeya Pavlovicha Koroleva*, pp. 369–70.

36. Ishlinskiy, *Akademik S. P. Korolev: ucheniy, inzhener, chelovek*, p. 447.

37. Semenov, *Raketno-Kosmicheskaya Korporatsiya "Energiya" imeni S. P. Koroleva*, pp. 88, 632.

38. Golovanov, "The Beginning of the Space Era."

39. *Ibid.*

40. V. Pappo-Korystin, V. Platonov, and V. Pashchenko, *Dneprovskiy raketno-kosmicheskiy tsentr* (Dnepropetrovsk, Ukraine: PO YuMZ/KBYu, 1994), p. 60; S. N. Konyukhov and V. A. Pashchenko, "History of Space Launch Vehicles Development," presented at the 46th International Astronautical Congress, October 2–6, 1995, Oslo, Norway, IAA-95-IAA 2.2.09. The range of the missile was about 2,000 kilometers.

41.   Yu. Biryukov, "From the History of Space Science: The Price of Decision—First Place (The First Satellites)" (in Russian), *Aviatsiya i kosmonavtika* (October 1991): 37–39. Author's emphasis.

42.   Golovanov, "The Beginning of the Space Era"; Golovanov, *Korolev: fakty i mify*, p. 514; Council of Veterans of the Baykonur Cosmodrome, *Proryv v kosmos: ocherki ob ispitatelyakh spetsialistakh i stroitelyakh kosmodroma Baykonur* (Moscow, Russia: TOO Veles, 1994), pp. 25, 174.

43.   "Report on Intercontinental Ballistic Missile" (in Russian), *Pravda*, August 27, 1957. A complete English translation of the press release is included in Krieger, *Behind The Sputniks*, pp. 233–34.

44.   Ishlinskiy, *Akademik S. P. Korolev: ucheniy, inzhener, chelovek*, p. 447; V. Lysenko, ed., *Three Paces Beyond the Horizon* (Moscow, USSR: Mir Publishers, 1989), p. 58.

45.   I. Minyuk and G. Vetrov, "Fantasy and Reality" (in Russian), *Aviatsiya i kosmonavtika* (September 1987): 46–47.

46.   M. K. Tikhonravov, "The Creation of the First Artificial Earth Satellite: Some Historical Details," *Journal of the British Interplanetary Society* 47 (May 1994): 191–194.

47.   Ibid.; G. A. Kustova, *Ot pervogo Sputnika do "Energii"—"Burana" i "Mir"* (Kaliningrad, Russia: RKK Energiya, 1994), p. 37; Jacques Villain, ed., *Baïkonour: la porte des étoiles* (Paris: Armand Colin, 1994), p. 26; Golovanov, *Korolev: fakty i mify*, p. 537.

48.   O. G. Ivanovskiy, *Naperekor zemnomy prityazhenyu* (Moscow, USSR: Politicheskoy literatury, 1988), pp. 167–169.

49.   Mikhail Florianskiy, "October 4—For the First Time in the World," *Moscow News Supplement* (1987), no. 40.

50.   Varfolomeyev, "Soviet Rocketry that Conquered Space"; Keldysh, *Tvorcheskoye naslediye Akademika Sergeya Pavlovicha Koroleva*, p. 365; Yu. A. Mozzhorin et al., eds., *Nachalo kosmicheskoy ery: vospominaniya veteranov raketno-kosmicheskoy tekhniki i kosmonavtiki: vypusk vtoroy* (Moscow, Russia: RNITsKD, 1994), pp. 60–61.

51.   Council of Veterans of the Baykonur Cosmodrome, *Proryv v kosmos*, pp. 29–30.

52.   Semenov, *Raketno-Kosmicheskaya Korporatsiya "Energiya" imeni S. P. Koroleva*, p. 90. "ISZ" is the Russian abbreviation for Artificial Satellite of the Earth. The 8K71PS was the industrial designation for the modified version of the R-7 used for the satellite launch.

53.   Yu. V. Biryukov, "Materials from the Biographical Chronicles of Sergey Pavlovich Korolev" (in Russian), in B. V. Raushenbakh, ed., *Iz istorii Sovetskoy kosmonavtiki* (Moscow, USSR: Nauka Press, 1983), p. 238; Lardier, *L'Astronautique Soviétique*, p. 93; Golovanov, *Korolev: fakty i mify*, p. 517. According to some Western sources, Soviet leader Khrushchev is said to have been present at this launch, but this is unconfirmed by Soviet or Russian sources.

54.   S. P. Korolev, "On the Practical Significance of K. E. Tsiolkovskiy's Proposals in the Field of Rocket Technology" (in Russian), in B. V. Raushenbakh, ed., *Issledovaniya po istorii i teorii razvitiya aviatsionnoy i raketno-kosmicheskoy nauki i tekhniki* (Moscow, Russia: Nauka Press, 1981), p. 40. This is a complete version of his speech. An abridged English translation has been reproduced in Institute of the History of Natural Sciences and Technology, *History of the USSR: New Research. 5: Yuri Gagarin: To Mark the 25th Anniversary of the First Manned Space Flight* (Moscow, Russia: Social Sciences Today, 1986), pp. 48–63. Note that the latter does not include the above quote.

55.   The State Commission for the Launch of the Object PS-1 comprised the following members: Chairman V. M. Ryabikov (Chairman of the VPK); V. P. Barmin (Chief Designer of GKSB SpetsMash); I. T. Bulychev (Deputy Chief of Military Communications of the Ministry of Defense General Staff); V. P. Glushko (Chief Designer of OKB-456); S. P. Korolev (Chief Designer of OKB-1); V. I. Kuznetsov (Chief Designer of NII-944); A. A. Maksimov (from the UNRV); A. G. Mrykin (First Deputy Chief of the UNRV); M. I. Nedelin (Deputy Minister of Defense for Reactive Armaments); A. I. Nesterenko (Commander of the

NIIP-5); G. N. Pashkov (Deputy Chairman of the VPK); N. A. Pilyugin (Chief Designer of NII-885); M. S. Ryazanskiy (Chief Designer and Director of NII-885); S. P. Shishkin (Chief Designer at KB-11). Others involved were: A. F. Bogolomov (Chief Designer of the OKB-MEI); M. V. Keldysh (Director of NII-1 and Chief of the OPM MIAN); I. T. Peresypkin (Minister of Communications); K. N. Rudnev (Deputy Minister of Defense Industries); G. R. Udarov (Deputy Chairman of the State Committee for Defense Technology); and S. M. Vladimirskiy (Deputy Chairman of the State Committee for Radio Electronics). See Yu. A. Skopinskiy, "State Acceptance of the Space Program: Thirty Years of Work" (in Russian), *Zemlya i vselennaya* (September–October 1988): 73–79; Lardier, *L'Astronautique Soviétique*, p. 285.

56. Ishlinskiy, *Akademik S. P. Korolev: ucheniy, inzhener, chelovek*, p. 447; Lardier, *L'Astronautique Soviétique*, pp. 108–9; Konovalov, "The Genealogy of Sputnik," pp. 122–123.

57. Golovanov, *Korolev: fakty i mify*, p. 537–38.

58. This document was not actually signed until the morning of the launch. See Ishlinskiy, *Akademik S. P. Korolev: ucheniy, inzhener, chelovek*, p. 448.

59. Mozzhorin, *Nachalo kosmicheskoy ery*, p. 63.

60. Golovanov, "The Beginning of the Space Era."

61. *Ibid.*

62. Ivan Borisenko and Alexander Romanov, *Where All Roads Lead to Space Begin* (Moscow, USSR: Progress Publishers, 1982), p. 66.

63. Mozzhorin, *Nachalo kosmicheskoy ery*, p. 63. The author of this quote was (at the time) OKB-1 engineer Ye. V. Shabarov.

64. Ishlinskiy, *Akademik S. P. Korolev: ucheniy, inzhener, chelovek*, p. 448, 464; B. Ye. Chertok, *Rakety i lyudi: Fili Podlipki Tyuratam* (Moscow, Russia: Mashinostroyeniye, 1996), p. 197.

65. Mozzhorin, *Nachalo kosmicheskoy ery*, p. 64.

66. Ishlinskiy, *Akademik S. P. Korolev: ucheniy, inzhener, chelovek*, p. 464.

67. Daniloff, *The Kremlin and the Cosmos*, pp. 65–66.

68. Khrushchev, *Nikita Khrushchev: krizisy i rakety: vzglyad iznutri: tom 1*, pp. 337–38.

69. "Announcement of the First Satellite" (in Russian), *Pravda*, October 5, 1957. A complete English translation of this announcement is included in Krieger, *Behind The Sputniks*, pp. 311–12.

70. "Report on the First Satellite" (in Russian), *Pravda*, October 9, 1957. A complete English translation of this article in included in Krieger, *Behind The Sputniks*, pp. 313–25.

71. *Soviet Space Programs, 1962–65; Goals and Purposes, Achievements, Plans, and International Implications*, Prepared for the Committee on Aeronautical and Space Sciences, U.S. Senate, 89th Cong., 2nd Sess. (Washington, D.C.: U.S. Government Printing Office, December 1966), pp. 149–50.

72. *Soviet Space Programs, 1962–65*, pp. 71–72.

CHAPTER 3

Korolev's Triple Play: *Sputniks 1, 2 and 3*

James J. Harford

Soviet politics, planning, and technology spanning the period 1946–1958 made possible the launching of an artificial satellite to the surprise of the West. The strategy used by Sergey Pavlovich Korolev, with the support of Mstislav Keldysh, in bringing the satellite from conceptualization by Mikhail Tikhonravov to actuality was nothing short of inspiring. In this essay I will explore the early work on *Sputnik 3*, which was planned to be *Sputnik 1*; the hurried development of *Sputnik 1* when *Sputnik 3* was not ready; the even more hurried development of *Sputnik 2* (the Layka carrier) at Premier Nikita Khrushchev's behest; the actual launches and the casual reaction, at first, by Kremlin officialdom to *Sputnik 1*'s success; and then the quick switch to braggadocio when the world impact was realized.[1]

INITIAL SOVIET REACTION TO THE *SPUTNIK 1* LAUNCH

While it jolted the rest of the world, the successful launch of *Sputnik 1* on October 4, 1957, received casual treatment, at first, in Moscow. Korolev's former colleague, Academician Boris Raushenbakh, told me, some 35 years later, "Look up the pages of *Pravda* for the first day after the launch. It got only a few paragraphs. Then look at the next day's issue, when the Kremlin realized what the world impact was."[2]

The article in the October 5 *Pravda* was, indeed, tersely phrased. Positioned modestly in a right hand column part way down on the first page, it did not even mention the satellite in its head. Titled routinely, "TASS Report," it gave the facts of the launch clinically.

The article gave basic information—size, weight, orbital inclination, radio frequency on which the beep could be heard—and it credited the great Tsiolkovskiy with having established the feasibility of artificial Earth satellites decades earlier.[3]

THE NEXT DAY'S TURNABOUT

The next day's *Pravda* was something else.[4] "World's First Artificial Satellite of Earth Created in Soviet Nation" stretched across the top of page one, which was devoted almost entirely to the achievement. But the lead story in the right hand column did not recount the feat itself, with first hand reports from the Soviet protagonists—their names were top secret, after all, so they could not even be contacted. Instead, the column was datelined New York, and it quoted in detail the congratulations of Russia's fiercest Cold War rival, the USA. The words, generous in their praise, were from Joseph Kaplan, chairman of the U.S. National Committee for the International Geophysical Year. Both the Soviet and American satellite programs were carried out under IGY auspices, with the results available to the world's scientists. Below the Kaplan item were congratulatory bulletins from A. C. B. Lovell, the British astronomer, and from a member of the USSR's political family, Pavel Novatskiy of the Polish Academy of Sciences—the latter headed "Big Victory."

Big victory it certainly was. Poems lyricized the event, like "Leap into the Future" and "Scouting the Celestial Deep." An ephemeris, showing the times when the carrier rocket would be visible over cities in the USSR, as well as Detroit and Washington, was printed like a train timetable. It was the moment to cash in on the performance of a feat that Nikita Khrushchev could never have dreamed would have so powerful an effect. He, after all, had been apathetic about "just another Korolev rocket launch," as Raushenbakh described the attitude of the Soviet premier and his claque on first hearing the news.

In the days to come, though, *Pravda* was delighted to print the praises of friends and enemies. Reactions from Peking and Shanghai (that friendship would dissolve only a few years hence), Warsaw, Paris, Vienna, Rome, London, and an especially long one from New York, ran under a big headline that ran across the page, "Russians Won the Competition."[5]

THE U.S. MUFFED THE CHANCE TO BE FIRST

It was a competition which the Americans should have won hands down. The concept of putting up a satellite had been known to the world's space enthusiasts for many years. In America serious proposals to launch a spacecraft into Earth orbit had been discussed since the mid-1940s. Robert P. Haviland recalls that when he was working in the Navy's Bureau of Aeronautics in 1945, motivated by a report on space rockets in a document captured from the Peenemünde Germans, he "wrote a 4–5 page memo proposing a Navy satellite development but it was scorned."[6]

In February 1946, the U.S. Army Air Corps asked the major air frame companies to submit secret proposals for the design of an "earth orbiting satellite." Douglas Aircraft was notified on July 1 that its design was judged the winner. Today's aerospace companies will find it hard to believe that the Air Corps evaluation was completed in less than four months! What's more, Douglas was funded for the study at the "unheard of amount in those days" of $1 million but their contract was switched to the newly-formed Project Rand (for Research and Development) in Santa Monica, California.[7] Rand's eventual report on "Preliminary Design of an Experimental World-Circling Spaceship" predicted, with keen perception, that "The achievement of a satellite craft by the United States would inflame the imagination of mankind, and would probably produce repercussions in the world comparable to the explosion of the atomic bomb."[8] Alas, nothing followed the study, and so it would not be a U.S. satellite that would generate those repercussions.

TSIOLKOVSKIY SHOWED IN 1903 THAT A SATELLITE COULD BE ORBITED

In Russia, as mentioned earlier, Konstantin Tsiolkovskiy had shown mathematically in 1903 how a device launched at a certain velocity would achieve Earth orbit. Then in 1948, forty-five years later, the visionary Mikhail Tikhonravov had made the case to Korolev for developing just such a device. At first he was unable to get support for the concept. His presentation to a meeting of the Academy of Artillery Sciences was treated skeptically. Golovanov quotes the remarks of the Academy president, Anatoliy Blagonravov, at the meeting: "The topic is interesting. But we cannot include your report. Nobody would understand why ... They would accuse us of getting involved in things we do not need to get involved in … ." However, what Blagonravov said officially was not what he thought instinctively. This courteous, mild-mannered, chain-smoking, white-haired former general, who would become in later years one of the chief spokesmen for the Soviet Union in the United Nations Committee on the Peaceful Uses of Outer Space, was bothered by the wary reception of Tikhonravov's ideas. "There was no way he could escape the thought that this ridiculous report was in fact not very ridiculous at all." Blagonravov, risking the derision of his colleagues, put the report back on the agenda, thereby giving Tikhonravov—and Korolev—license to study possible satellite designs.[9]

THE R-7 ICBM CARRIER MADE A SPUTNIK LAUNCH FEASIBLE

Some five years later, towards the end of 1953, having redesigned the R-7 rocket to carry a heavier payload, Korolev had drafted a proposed decree for the Central Committee of the Communist Party which included the possibility of using the vehicle to launch a satellite. However, while the draft "was making its way to the top" mention of the satellite was struck out.[10] Not until May 26, 1954 did Korolev formally propose the satellite launch to Dmitriy Ustinov, Minister of Armaments.[11] By then the R-7 was capable of propelling an H-bomb warhead of 5 tons—the actual size had not yet been determined—over an intercontinental ballistic trajectory. It could easily orbit a satellite of some 1.5 tons. According to Korolev's deputy, Vasiliy Mishin, Korolev had to propose a Sputnik launch as part of the test program of the ICBM program.[12] In any case, Korolev's proposal to Ustinov is so delicately phrased that R-7 is not mentioned at all, but merely referred to as the

> … new article which permits speaking about the possibility of designing an artificial Earth satellite within the next few years. By a certain reduction of the weight of the payload it will be possible for the satellite to achieve the necessary velocity of 8,000 m/sec.[13]

As Roald Sagdeev, longtime head of the USSR Space Research Institute, put it, "Korolev and his colleagues could have only had a vague idea of how heavy the final reentry vehicle (for the ICBM warhead) should be. Sakharov [Andrey] was still far from knowing how to make this deadly weapon relatively compact and easily portable. Since rapid progress was required, rocket designers adopted a worst-case strategy and started to develop an ICBM that, as it was discovered later, had a substantial excess of launch capability, or throw weight."[14]

U.S. *VANGUARD* CHOSEN OVER PROJECT ORBITER

Meanwhile more substantive thinking on possible satellite designs had been resumed in the U.S. The most expedient design approach came from von Braun's team at the Army's Redstone Arsenal in Huntsville, Alabama. Starting with a meeting in early 1954 with George Hoover of the Office of Naval Research, von Braun and his colleagues eventually came up with Project Orbiter, which would have been an Army-Navy-Air Force design using already-developed Army Ordnance weapons technology to put a small satellite into orbit.[15] But the Eisenhower administration had reasons to choose a different approach. Eisenhower wanted to keep the military out of the IGY program, which was dedicated to scientific purposes. That was the surface reason. The other one, more telling, was—ironically—based on military strategy. He wanted to be consistent with his well publicized "Open Skies" stance at a time when U-2's and spy satellites were being developed to begin reconnoitering the USSR. Eisenhower reasoned that a satellite put up as part of the IGY program would strengthen the freedom of the skies policy and would be less likely to disturb Nikita Khrushchev's sensibilities about overflight than one sponsored by three military services. Also, the Soviets were likely to launch an IGY satellite themselves. And so a project named *Vanguard*—which proved to be an embarrassing choice of names—was chosen to carry the U.S. banner into the space age. Based on a sounding rocket developed by the Naval Research Laboratory, but under the auspices of the National Science Foundation, it was going to be a riskier venture than Orbiter because it called for substantially modified first and second stages, a new third stage, and new rocket engines and guidance technology.

SEDOV REVEALS SOVIET SATELLITE PLAN AT IAF

On July 29, 1955, the Eisenhower administration announced that the U.S. would launch *Vanguard* for scientific purposes during the 1957–58 IGY. A few days later, at the Sixth Congress of the

International Astronautical Federation in Copenhagen, a delegation of Soviet scientists, appearing at IAF for the first time, revealed at a press conference that the USSR, too, might be in the game. Leonid Sedov, head of the Soviet delegation and the newly appointed chairman of an Academy of Sciences Commission on Interplanetary Communications,[16] choosing his words carefully, said:

> From a technical point of view, it is possible to create a satellite of larger dimensions than that reported in the newspapers which we had the opportunity of scanning today. The realization of the Soviet project can be expected in the comparatively near future. I won't take it upon myself to name the date more precisely.[17]

But, Sedov, it seems, was speculating, since no official decision had yet been made that there would be a Soviet satellite in IGY. In fact, not until January 30, 1956, would the Council of Ministers issue a decree authorizing its development.

TIKHONRAVOV TRANSFERRED TO KOROLEV BUREAU

It was at this time that Korolev was able to arrange for Mikhail Tikhonravov and his team, which had been working on a satellite concept at Special Design Bureau #385 in the Ural Mountains, to join him. One of the members of that team, Konstantin Feoktistov, recalled:

> We wanted to build a satellite but Korolev had that responsibility. Tikhonravov was transferred to Korolev's bureau in early 1956 after the Party and the Goverment had authorized Korolev to proceed with the development of a satellite. But the rest of U.S. in the group had to apply to Korolev individually. However, Korolev relied on the advice of Tikhnoravov, his old friend and, by the end of 1957 I was chosen, although it was difficult to leave #385.[18]

SPUTNIK PLANS LAG, KELDYSH CRITICIZES INDUSTRY

Even the January decision, however, was not followed by sufficiently aggressive support for the satellite development in the opinion of Korolev and his close ally, Mstislav Keldysh. Time to beat the Americans had flown by and the Soviet establishment was not yet revved up. Nine months later, on September 14, in what must have been exasperation, and probably with Korolev's prodding, Keldysh appeared before the Presidium of the Soviet Academy of Sciences to state his case.[19] He first reviewed patiently, like a good teacher, the physics of placing a satellite in orbit. Then he covered the pioneering scientific measurements which the Soviet satellite would make—the Earth's magnetic fields, the "ionic composition of the upper layers of the atmosphere," the "corpuscular radiation of the Sun," cosmic radiation, and possible micrometeorites.

He must have caused eyes to widen a bit when he said "... we are considering placing a live organism in the satellite—a dog. It turns out that the perception of a dog is the most similar to the perception of a man—biologists so consider it. The dog will live there in the absence of a gravitational field, in conditions of irradiation, of cosmic radiation. The dog will be undergoing all sorts of dangers, because if the satellite is hit by a large meteoric particle, it will broach the satellite ... ."

By then he surely had their full attention, but he went on. "We, of course, can't stop at the task of creating an Earth satellite. We, naturally, are thinking of further tasks—of space flight. The first project along these lines, I believe, will be to fly around the Moon and photograph it from the side which is always hidden to us." Then came hard criticism: "We have come up short in a whole series of tasks in the Academy of Sciences, and are lagging now. Back in August we were to turn in the dimensions of the equipment and their mode of attachment to the rocket ... We delayed this work, which resulted in a notable delay in the planning and development of the satellite itself."

He called out specific industry laggards: "In general, the radiotechnological industry isn't helping us enough ... They are sluggishly regarding the creation of this satellite ... We have already committed one breach in delaying the delivery of size specifications and other information to the Korolev Design Bureau." Finally, the ultimate motivation came out baldly: "... it would be good if the Presidium were to turn the serious attention of all its institutions to the necessity of doing this work on time ... we all want our satellite to fly earlier than the Americans'."[20]

Keldysh and Korolev had not only the Academy of Sciences and industry to motivate, they had to deal with the objections of the military generals, who feared that the satellite project would slow down the development of the R-7 ICBM. The fear was under-standable since the R-7's first five launch attempts had failed.[21] But foot-dragging by the support institutions was only one of Korolev's problems. In spite of Keldysh's speech, the satellite which was supposed to have the honor of being first continued to fall behind schedule. In fact, it would lag so badly that it would become *Sputnik 3*—a huge, sophisticated satellite, eventually launched on May 15, 1958. *Sputnik 3's* 1,327 kg payload included virtually all of the instruments called out in Keldysh's speech.

"SIMPLEST" SPUTNIK MOVES AHEAD OF BIG SCIENCE SATELLITE

"But," said Georgiy Grechko, one of the engineers who worked at the Korolev design bureau during those days, and who later became a cosmonaut, "these devices were not reliable enough so the scientists who created them asked us to delay the launch month by month. We thought that if we postponed and postponed we would be second to the U.S. in the space race so we made the simplest satellite, called just that—*Prostreishiy Sputnik*, or 'PS.' We made it in one month, with only one reason, to be first in space."[22]

It was at this time, on August 21, 1957, that the R-7, in its sixth attempt, propelled a dummy H-bomb warhead all the way to Kamchatka, some 6,000 kilometers. With that success the confident Korolev made his move to beat the Americans with his PS. His next obstacle was a skeptical State Commission for the R-7 ICBM. Discussion of Korolev's satellite proposal before that body, according to a 1992 report by a journalist in a Moscow magazine, was "sharp, the opponents arguing primarily about the tight timing," and "complete agreement was not achieved." Korolev had to go back to the Commission a second time. This time he tried a different ploy. Why not, he challenged the members, put the question of authorization to the Presidium of the Central Committee of the Communist Party—in the context of whether or not the USSR should try to be the first country in the world to launch a satellite? The Commission blanched. "Nobody wanted to be scapegoat." And so the project proceeded.[23]

With the Commission's nervous acquiescence, Korolev bulldozed the development, in a little more than a month, of a plain, polished 83.6 kg sphere containing only a radio transmitter, batteries, and temperature measuring instruments, with the intent to place it in orbit on a rocket which had failed in five of its first six launch attempts. It was a very hectic month, and while the satellite was simple, the attention given to its manufacture was unsparing, especially by Korolev, himself.

"I coordinated the production, testing, launch preparations and the launch itself," recalled Oleg Ivanovskiy, looking back 36 years during a 1993 interview. Ivanovskiy had been deputy to Mikhail Khomyakov, *Sputnik 1*'s principal designer. He recalled that there were problems:

> For example, there were two peculiarities which satellites had that missiles did not. One was that the satellite required precise thermal control and the other was that vacuum sealing was used to assure reliable performance. We had to find new techniques of manufacturing the surfaces in order to achieve the necessary optical and thermal qualities. We had no experience in this work. We needed vacuum chambers. I recall one episode when we had to persuade the production shop that the satellite was a new item, not a missile. Korolev, with his iron character, was able to influence the attitude of people. The Party directed that new paint be put on the factory walls. Korolev put the satellite on a special stand, draped in velvet, in order that the workers would show reverence towards it. He supervised the carrying out of the production schedule every day personally.[24]

One of the metalworkers who was assigned to the *Sputnik 1* manufacture was Gennadiy Strekalov, later to become a cosmonaut. "My teacher in metalworking did the finishing," he told me. "Two half spheres were stamped, then machined, then the masters did the finishing."[25] Strekalov, who, in 1995, flew to the *Mir* space station with American astronaut Norman Thagard and then participated in the first Mir-Shuttle rendezvous, is as proud of his work on *Sputnik 1* as he is of his four orbital flights.[26]

"Korolev came over to the shop and insisted that both halves of the sputnik's metallic sphere be polished until they shone, that they be spotlessly clean," recalled Konstantin Feoktistov, who would be the first engineer-cosmonaut to go into orbit in the three-man *Voskhod* seven years later. "The people who developed the radio equipment were actually the ones demanding this. They were afraid of the system overheating, and they wanted the orbiting sphere to reflect as many rays of the Sun as possible."[27]

An idea of how intense were the preparations for the launch of the "simplest satellite," and how sensitive Korolev was to the event's historical importance, comes from the recollections, on the 30th anniversary of the launch, of one of the design team, Mikhail Florianskiy:

> The jettisoning of the nose cone and the process of separation of the sputnik from its carrier was being tested at the assembly shop late in August 1957. It is not a complex procedure, but fraught with possible surprises. Everything was going on normally—or so it seemed—when Korolev all of a sudden subjected the plant's chief engineer to a terrible dressing down. Korolev was berating him—what for!—for the poor quality of the surface of the *mockup* of the sputnik! The quality of the surface is really important in flight because the heat conditions of the sputnik depend on it, but why the dressing down now, when quite a different process was being tested?

Korolev said angrily: "This ball will be exhibited in museums!"[28]

> It was Korolev's aesthetic as well as engineering sense that had led him to insist on the ball shape for *Sputnik 1*, although one of the early designs proposed was for a cone-shaped structure. "Today, after decades have passed," recalled colleague Mark Gallay in a 1980s interview, "we simply cannot imagine the first sputnik to be anything other than what it was: an elegant ball ... with an antenna thrown back like a galloping horse."[29] There were two flight-ready spheres built, one for the launch and another one for ground testing, developing the welding and other fabrication techniques. The second one would later be launched too, with the carrier for the dog Layka on *Sputnik 2*.

REBROV RECALLS THE LAUNCH PREPARATIONS

A recollection of the preparations for *Sputnik 1,* and the launch itself from Baikonur, comes from Colonel Mikhail Rebrov:

> People in the "space room" worked in white smocks, performing each operation with the greatest thoroughness. The rocket was assembled in the big hangar. Silence fell when the Chief Designer appeared. At the time Sergey Korolev was exacting and more strict than ever ... .

Only two days were left. The carrier rocket was rolled out to the launch pad in the early morning of October 2, 1957. Korolev walked in front, together with all the other chief designers. They walked in silence the entire 1.5 km long way from the assembly-testing building to the pad. No one will ever know what was going through Sergey Korolev's mind at the time. Later on, when the sputnik was installed in orbit, and its call sign was heard over the globe, he said: "I've been waiting all my life for this day!" The moment of the blast-off has been described many times. Then the rocket got out of the radio zone. The communication with the sputnik ended. The small room where the radio receivers were was overcrowded. Time dragged on slowly. Waiting built up the stress. Everyone stopped talking. There was absolute silence. All that could be heard was the breathing of the people and the quiet static in the loudspeaker ... And then from very far-off there appeared, at first very quietly and then louder and louder, those "bleep-bleeps" which confirmed that it was in orbit and in operation. Once again everyone rejoiced. There were kisses, hugs and cries of "Hurrah!" The austere men, who were greeted out of space by the messenger they had made, had tears in their eyes.[30]

AMERICANS SHOCKED BUT DETERMINED TO RESPOND

Hardly tearful, more like rueful, was the reaction of the American space community. Even though the possibility that the Soviets had been making plans for launching a satellite was known in the U.S., and not only to government insiders, the fact of the launch was a deep jolt to the space professionals.[31]

At that time, I was Executive Secretary of the professional society for space engineers—the American Rocket Society (ARS). *Astronautics* magazine, the main ARS publication, could not resist reminding its readers that, "A little over two years ago, when the government's guided missile policy committee decided against the von Braun-Medaris[32] satellite proposal in favor of Project *Vanguard*, there were four dissenters who voted to send the idea on to the National Security Council for further consideration. One of them was the 'Lone Eagle,' Charles A. Lindbergh, the last one to make aviation headlines of the same magnitude as 'Sputnik.'"[33]

Grim determination characterized the American rocketeers. "They [the Russians] must not be allowed to win this game—a game with far-reaching political, social and economic consequences," *Astronautics* editorialized.[34]

In the same issue, a very insightful interpretation of the techno-logical significance of the Russian feat by Martin Summerfield of Princeton University had an upbeat aspect, but was somber in its reflection on the advanced state of Soviet space technology:

> The success of the Russian "Sputnik" was convincing and dramatic proof to people around the world of the real prospects of space travel in the not distant future. The fact that a 23-in. sphere weighing 184 lb. has been placed in an almost precise circular orbit indicates that a number of important technological problems such as high thrust rocket engines, lightweight missile structures, accurate guidance, stable autopilot control, and large scale launching methods have been solved, at least to the degree required for a satellite project.[35]

I got a phone call at my home in Princeton about 7:00 p.m. on Friday evening, October 4, from *The New York Times* aeronautics reporter, Richard Witkin. Had I heard? What is the reaction of the U.S. rocket community? My response is not even in my memory. But the impact of the launch on the U.S., as well as on my own career, would be powerful, indeed. The ARS at the time had a membership of about 5,000 engineers and scientists, most of them working on missile programs, although a few dozen were on Project *Vanguard*. By 1962, just seven years later, the membership quadrupled to 20,000, a growth so rapid that industry and govern-ment pressures caused a merger of ARS with the Institute of Aeronautical Sciences, the society for aeronautical engineers, into what is today the 40,000 member American Institute of Aeronautics and Astronautics (AIAA).

WORLD REACTION

The news of the launch in the world's leading newspapers got Second Coming treatment. *The New York Times*, receiving the story in the late afternoon of Friday, October 4, printed the next morning a rarely used three-line head in half-inch capital letters, running full length across the front page:

SOVIET FIRES EARTH SATELLITE INTO SPACE;
IT IS CIRCLING THE GLOBE AT 18,000 M.P.H.;
SPHERE TRACKED IN 4 CROSSINGS OVER US

Other world newspapers gave the event similar play. Then the interpretation began. *The Manchester Guardian* needed only a couple of days to begin to speculate apocalyptically on what the Russians might now do. An October 7 editorial titled "Next Stop Mars?" read, "The achievement is immense. It demands a psychological adjustment on our part towards Soviet society, Soviet military capabilities and—perhaps most of all—to the relationship of the world with what is beyond." Some of *The Guardian*'s speculation was downright clairvoyant. "We must be prepared to be told what the other side of the Moon looks like [Lunik 3 produced the photos only two years later], or how thick the cloud on Venus may be [revealed by the Soviet Venera and U.S. Mariner series starting in the '60s]." Accurately, *The Guardian* pointed out that, "The Russians can now build ballistic missiles capable of hitting any chosen target anywhere in the world." That, certainly, was true, but it would be some years before improvements in guidance technology made the capability an actuality. It certainly did not follow, as *The Guardian* stated a few sentences later, that "Clearly they have established a great lead in missile technology." This was one of the earliest inaccurate predictions of a missile gap.

French reaction was equally ebullient. "Myth has become reality: Earth's gravity conquered," bannered *Le Figaro*, and went on to report the "disillusion and bitter reflections" of "The Americans (who) have had little experience with humiliation in the technical domain."[36] For three weeks the world could hear the beeping of *Sputnik 1* before the radio died out, and it orbited more than 1,400 times before burning up in the atmosphere after three months in space.

Much has been written about the effect of the *Sputnik 1* launch on the world scene. Many American space enthusiasts, stricken with gloom at the time, now reflect that it might have been the best thing that could have happened to awaken the need for an aggressive space program. Only four years later President Kennedy would call for what became the Apollo program. An enormous infrastructure of space research and development centers, test and launch facilities, and supporting industry and university programs would come into being.

SOME NEGATIVE WESTERN REACTION TO SPACE BUILDUP

There were those who reacted negatively to the surge in space technology and its consequent spurring of the growth of high tech weapons. As former Ambassador to the Soviet Union George Kennan put it in his memoirs, "It [*Sputnik*] caused Western alarmists, such as my friend Joe Alsop, to demand the immediate subordination of all other national interests to the launching of immensely expensive crash programs to outdo the Russians in this competition. It gave effective arguments to the various enthusiasts for nuclear armament in the American military-industrial complex. That the dangerousness and expensiveness of this competition should be raised to a new and higher order just at the time when the prospects for negotiation in this field were being worsened by the introduction of nuclear weapons into the armed forces of the Continental NATO powers was a development that brought alarm and dismay to many people besides myself."[37]

SPUTNIK 2, PUSHED BY KHRUSHCHEV, READIED IN ONE MONTH

While all this introspection was taking place in the West, the creators of *Sputnik* were unable to be interviewed, take bows, be photographed, get medals. Sergey Korolev was literally back at the office, because Khrushchev, realizing what a hot property he had, gave him new orders. Do something bold, Sergey Pavlovich, to celebrate the upcoming 40th anniversary of the Revolution! It's only a month away.

Cosmonaut Grechko told the story:

> I heard this from Korolev himself with my own ears. After *Sputnik 1* Korolev went to the Kremlin and Khrushchev said to him,
> "We never thought that you would launch a Sputnik before the Americans. But you did it. Now please launch something new in space for the next anniversary of our revolution."
> The anniversary would be in one month! I'll bet that even with today's computers nobody would launch something into space in one month. It was, I think, the happiest month of his [Korolev's] life. He told his staff, and his workers, that there would be no special drawings, no quality check, everyone would have to be guided by his own conscience. The engineers would make drawings and give them directly to the workers. And we launched on November 3, 1957, in time for the celebration of the Revolution.[38]

The payload of *Sputnik 2* weighed 508 kg, more than six times the weight of *Sputnik 1*. A shroud housed a carrier containing the world's first space passenger, the mongrel dog Layka, plus a duplicate of the *Sputnik 1* sphere. Layka is reported to have barked and eaten food during his lonely sojourn but, alas, he died when the capsule overheated after failing to separate from its booster, thereby rendering the thermal control system inoperative. Animal groups protested, but the Soviets made Layka into a martyr for a noble cause. Veterans of the *Sputnik 2* project regard it as an even more significant achievement than *Sputnik 1*. Not a single engineering task had been performed on it until after *Sputnik 1* went up.

With the new triumph Khrushchev could not resist escalating his needling of the Americans. In a speech at the 40th anniversary of the Revolution, on November 6, he said, "It appears that the name *Vanguard* reflected the confidence of the Americans that their satellite would be the first in the world. But ... it was the Soviet satellites which proved to be ahead, to be in the vanguard ... In orbiting our earth, the Soviet sputniks proclaim the heights of the development of science and technology and of the entire economy of the Soviet Union, whose people are building a new life under the banner of Marxism–Leninism."

VANGUARD, RUSHED TO LAUNCH, EXPLODES ON PAD

Those derisive comments proved even more galling to the Americans a month later when, on December 6, the first attempt to launch the *Vanguard* satellite was an ignominious failure before the world's television cameras. The three stage rocket was designated TV-3, for Test Vehicle 3, and it had been originally scheduled to be just that, a test. But under Sputnik pressure, the "test" was moved up several months, and made into a full-fledged attempt at a satellite launch. But the vehicle got only a few feet off the ground before sagging back, buckling, bursting into a huge conflagration and tossing its tiny 1.47 kg payload, still transmitting, some yards away.

Pravda delightedly reproduced the front page of the *London Daily Herald* which showed a photo of the *Vanguard* being readied on the launch pad next to one of the explosion. Superimposed above the immense Herald headline, which read, "OH, WHAT A FLOPNIK!" was *Pravda*'s comment, "Reklama and Deistvitelnost," or "Publicity and Reality."

U.S. EXPLORER AUTHORIZED, LAUNCHED, DISCOVERS VAN ALLEN BELTS

Following the *Sputnik 2* launch, a team composed of the von Braun group from the Army Ballistic Missile Agency—for the launch vehicle—and one from the Jet Propulsion Laboratory in California—for the space capsule—the latter headed by William H. Pickering, was given the nod to put up the first U.S. satellite. On its initial try the team launched the 14 kg *Explorer 1*, on January 31, 1958. Fortunately, Pickering and Ernst Stuhlinger, one of von Braun's staff chiefs, conscious of the odds against *Vanguard*'s success, had persuaded James van Allen of the University of Iowa to make the package of scientific instruments being readied for *Vanguard* compatible with *Explorer 1*'s Jupiter C launch vehicle.[39] This turned out to be particularly fortuitous because *Explorer 1*, with only one sixth the payload weight of *Sputnik 1*, scored a major scientific coup when its instruments sensed a pattern of radiation around the Earth leading to the discovery of the now famous van Allen Belts. It is an interesting footnote on scientific history that *Sputnik 1* could have made the discovery if it had installed simple instruments. *Sputnik 2*, in fact, did have the instruments, wrote van Allen: "a pair of Geiger-Mueller tubes ... which operated properly and yielded data for seven days"[40] but the Soviet scientists failed to develop the data necessary to interpret the discovery.

SPUTNIK-3 FINALLY ORBITED, NASA ESTABLISHED

Oleg Ivanovskiy, who worked on the *Sputniks 1, 2,* and *3,* speaks with deserved satisfaction of the Soviet achievements of those months in late 1957 and early 1958. But there was trouble too. "We had our first space failure," he told me, with the difficult *Sputnik 3.* "It was April 27, 1958, and it was caused by a rocket engine failure. The rocket went up about 12–15 km and the satellite fell separately. There was a search for the satellite. I remember that the pilots conducting the search were not allowed to know what they were looking for. 'Just search the area for anything unusual,' they were told, 'and don't attract the camels.' It was crazy secrecy. Finally one pilot came back and said he had seen something that sounded to us like the satellite. We sent out a rescue team in an armored vehicle. When we got it back some of the instruments could still operate." Ivanovskiy then showed me proudly a copper wire which he had recovered from one of the instruments which kept the satellite beeper from operating prematurely.

When the 1.5 ton *Sputnik 3* was launched successfully, on May 15, 1958, it caused even more anxiety in the West. Any doubt that the Russians would soon have the capability to send an ICBM to the United States was demolished. Lyndon Johnson, then Senate Majority Leader, had demanded a congressional investigation of the impact of *Sputnik 1* only a few days after its launch. On the Senate floor in January he had made recommendations originating in his Preparedness Subcommittee to "Start work at once on the development of a rocket motor with a million pound thrust," "Put more effort in the development of manned missiles (satellites)," and "Accelerate and expand research and development programs, provide funding on a long-term basis, and improve control and administration within the Department of Defense or through the establishment of an independent agency."

During the same months in early 1958 President Eisenhower, with his science adviser, James Killian, also concluded that a civilian space agency was needed, and directed Hugh Dryden, head of the National Advisory Committee for Aeronautics (NACA), to prepare legislation which would create such an agency on that relatively small organization's structure. The result was Public Law 85–568, signed on July 29, 1958, by the President, calling for the creation of the National Aeronautics and Space Administration (NASA). It was quite an about-face for a president whose staff members had belittled *Sputnik 1* as "a silly bauble" and "a neat scientific trick" and who, himself, had said that it had not bothered him "one iota."[41]

Sputnik 3's large load of scientific instruments was designed to measure micrometeorites, density of the upper atmosphere, cosmic rays, solar radiation, the presence and effect of high energy particles and the Earth's own radiation environment.[42] It could have performed a *tour de force* of scientific research in virgin territory. *The Manchester Guardian* reported that "This impressive list (of instruments) is a telling demonstration of the fact that the latest Russian sputnik has been launched for strictly scientific purposes."

SPUTNIK 3 MISSES CHANCE TO MAP VAN ALLEN BELTS

Unfortunately, to the great embarrassment of the Soviets, the huge vehicle missed one more chance at what would have been its most significant achievement—mapping the radiation belts. *Explorer 1's* instruments, which had revealed the existence of the radiation from the belts, had at first been overwhelmed by its intensity, and it took some time for the van Allen analysis team to understand what it had measured. *Sputnik 3* could have mapped the belts systematically, but failed to do so because of a defective tape recorder. The recorder, designed to store and transmit to Earth the information collected by the instruments when it was out of direct radio contact, had failed utterly to transmit the necessary data. Roald Sagdeev recounts what happened:

> A scientific team landed at the Baikonur Cosmodrome for the final integration and testing of hardware on *Sputnik 3*. Korolev invited everyone for the last briefing before the final okay was to be given and the countdown started ... . It was the first impressive collection of scientific instruments, each of which was reported to be functioning normally. However, trouble was soon discovered in some of the supporting hardware. The problem was with the more or less routine tape recorder, whose function was to accumulate data from the different experiments and to prepare messages for the ground station. The spacecraft, revolving around the globe, would only be in contact with the ground station during periods of "direct radio visibility." Simply speaking, the ground station would be unable to sense the signals from the spacecraft when it was behind the horizon ... . With such a crucial role, members of the scientific team were extremely worried about the troubled tape recorder and they recommended postponing the actual launch to give the technicians a chance to fix it. However, the tape recorder's ambitious engineer, Alexei Bogomolov ["I too often had to depend on his hardware," Sagdeev footnotes acidly] did not want to be considered a loser in the company of winners. He suggested that the testing failure was simply caused by electromagnetic interference from the multiplicity of different electrical circuits in the test room. He boldly proposed to launch *Sputnik 3* on time. To the great disappointment of the scientific team, Korolev accepted Bogomolov's suggestion ... . During the flight, however, it was confirmed that Bogomolov had been dead wrong. His tape recorder did not work. Consequently, the scientific information gathered was limited by the area of direct radio visibility ... . Each scientific group had results, but because of the recorder failure they had to guess whether the phenomena discovered were of local or planetary significance.[43]

The most disappointed scientist, says Sagdeev, was Sergey Vernov, a renowned physicist. The detectors on *Sputnik 3* sensed extremely high levels of radiation, but was it local or did it exist around the Earth? Some six weeks earlier, on March 26, *Explorer 3* had been launched (*Explorer 2* had failed to orbit), carrying the first tape recorder ever launched on a satellite. As Van Allen wrote, it "functioned beautifully in response to ground command and fulfilled our plan of providing complete orbital coverage of radiation intensity data."[44]

The Soviets, without tape recorded data, were hogtied. As Van Allen recalled, the team was at first puzzled that they "were encountering a mysterious physical effect of a real nature."[45] They "worked feverishly in analyzing the data from *Explorers 1* and *3* (by primitive hand reduction of pen-and-ink recordings) and organizing them on an altitude, latitude, and longitude basis."[46] In an interesting sidelight on the whole episode, at first there was suspicion that the intense radiation was coming from a Soviet nuclear test. Subsequent analysis, however, proved that it was "geomagnetically trapped corpuscular radiation" distributed in a "belt" around the Earth. At a conference in the summer of 1958, the name "Van Allen radiation belt" was applied for the first time.[47] More confirmation of the origin of the radiation, as well as the discovery that there was also an outer belt, came two months later—again from the Americans—when *Explorer 4* mapped the belts from July 26 to September 19, 1958.

"On a purely observational basis," wrote Van Allen, "the *Sputnik 3* data actually represented discovery of the earth's *outer* radiation belt inasmuch as they were acquired before those of *Explorer 4* and *Pioneer 3*."[48] However, Vernov and colleagues had not yet interpreted the observational finding, although in what seems to have been hindsight, Vernov published a photo of what is now known as the van Allen belts in *Pravda* on March 6, 1959, alleging that the data were based on findings that he had reported at an IGY conference in August, 1958. However, colleagues of van Allen who heard the paper maintain that the fragmentary data available at that time from *Sputniks 1, 2, and 3* could not have formed the basis for such a finding, and, what's more, no such finding was reported in the paper.[49]

Afterwards, a Russian joke circulated that the belts were to be called the van Allen-Vernov radiation belts. "What did Vernov do? He discovered the Van Allen belts."

THE SPUTNIK/*VANGUARD*/EXPLORER RACE 1957–58

October 4, 1957	USSR:	*Sputnik 1* (83.6 kg) launched.
November 3	USSR:	*Sputnik 2* (508.3 kg), with dog Layka as passenger, launched.
December 6	U.S.A.:	*Vanguard* TV-3 explodes on launch pad.
January 31, 1958	U.S.A.:	*Explorer 1* (14 kg), America's first satellite, discovers the van Allen radiation belts.
February 5	U.S.A.:	A second *Vanguard* try fails.
March 5	U.S.A.:	*Explorer 2* fails to orbit.
March 17	U.S.A.:	*Vanguard 1* (1.47 kg) successfully orbits, establishes the pear-shapedness of the Earth.
March 26	U.S.A.:	*Explorer 3* orbits, collects radiation and micrometeoroid data.
April 27	USSR:	First try to launch *Sputnik 3* fails.
April 28	U.S.A.:	Another *Vanguard* fails to orbit (third failure).
May 15	USSR:	*Sputnik 3* (1,327 kg) orbits, carrying large array of scientific instruments, but tape recorder fails, so it can't map Van Allen belts.
May 27	U.S.A.:	*Vanguard* fails for the fourth time.
June 26	U.S.A.:	*Vanguard* fails for fifth time.
July 26	U.S.A.:	*Explorer 4* orbits and maps Van Allen radiation belts for 2 1/2 months.
August 24	U.S.A.:	*Explorer 5* fails to orbit.
September 26	U.S.A.:	*Vanguard* fails for the sixth time.

1. Adapted from James J. Harford, *Korolev: How One Man Masterminded the Soviet Drive to Beat America to the Moon* (New York: John Wiley & Sons, 1997), pp. 121–37.
2. Raushenbakh comment to author at 41st International Astronautical Congress, Montreal, Oct. 11, 1991.
3. In *Rockets and Cosmic Space*, a monograph published by Tsiolkovskiy in 1903.
4. *Pravda*, October 6, 1957, pp. 1–2.
5. *Pravda*, October 8, 1957, p. 3.
6. Haviland telephone interview with author, May 22, 1995.
7. Yvonne Brill, one of the participants in the Rand study, letter to the author, June 15, 1995.

8. Project Rand Report SM-11827, May 2, 1946, from Rand 25th Anniversary Volume, Rand Corp., Santa Monica, CA, 1973, p. 3.

9. Yaroslav Golovanov, "The Beginning of the Space Era," *Pravda*, October 4, 1987, p. 3, translated into English in JPRS-USP-88-001, February 26, 1988, p. 48.

10. Georgiy Vetrov, private communication, June 13, 1995.

11. M. V. Keldysh, ed., *Tvorcheskoye naslediye Akademika Sergeya Pavlovicha Koroleva: izbrannyye trudy i dokumenty* (Moscow, USSR: Nauka Press, 1980), p. 343, and M. V. Tarasenko, *Voennyye aspekty Sovetskoi kosmonavtiki* (Moscow, Russia: Nikol, 1992), p. 16. Tarasenko reports that the recommendation called for "development of a 2–3 ton satellite, a recoverable satellite, a satellite for a long orbital stay of 1–2 people and an orbital station with regular Earth ferry communication."

12. Vasiliy Mishin interview with author, September 2, 1992.

13. Keldysh, *Tvorcheskoye naslediye Akademika Sergeya Pavlovicha Koroleva* ... .

14. Roald Sagdeev, *The Making of a Soviet Scientist* (New York: John Wiley & Sons, 1994), p. 155.

15. For a detailed account of the genesis of Project Orbiter see Ernst Stuhlinger and Frederick I. Ordway III, *Wernher von Braun: Crusader for Space* (Malabar, FL: Krieger Publishing Co., 1994), pp. 123–31.

16. The full name seems a satire on Soviet bureaucracy. It was the Interdepartmental Commission for the Coordination and Control of Scientific-Theoretical Work in the Field of Organization and Accomplishment of Interplanetary Communications of the Astronomical Council of the U.S.S.R. Academy of Sciences.

17. F. J. Krieger, *Behind the Sputniks* (Washington, DC: Public Affairs Press, 1958), p. 330.

18. Konstantin Feoktistov interview with author, Moscow, December 12, 1991.

19. V. S. Avduyevskiy and T. M. Eneyev, eds. *M. V. Keldysh: izbrannyye trudy: raketnaya tekhnika i kosmonavtika* (Moscow, USSR: Nauka Press, 1988), p. 235.

20. *Ibid.*, pp. 235–40.

21.

22. Georgiy Grechko interview with author, Moscow, May 16, 1993.

23. Council of Veterans of the Baykonur Cosmodrome, *Proryv v kosmos: ocherki ob ispitatelyakh spetsialistakh i stroitelyakh kosmodroma Baykonur* (Moscow: TOO Veles, 1994), pp. 77–78. The State Commission was chaired by Vasiliy Ryabikov. It included, besides Korolev himself, Marshal Mitrofan Nedelin, Georgiy Pashkov, Valentin Glushko, Nikolay Pilyugin, Viktor Kuznetsov, Mikhail Ryazanskiy, Vladimir Barmin, Aleksandr Mrykin, S. Shishkin, I. Bulychev, Aleksey Nesterenko, and A. Maksimov.

24. Oleg Ivanovskiy interview with author, Moscow, January 26, 1993.

25. Gennadiy Strekalov interview with author, Moscow, January 28, 1993.

26. The Mir-Shuttle rendezvous was Strekalov's fourth orbital flight, although his fifth space launch. He and Vladimir Titov were successfully ejected by the launch escape system when their *Soyuz T-10* spacecraft was engulfed by flames in 1983.

27. Nikolai Zheleznov, "Hello—(Bip-Bip) Scientist, Designer and Cosmonaut Speaking," *Soviet Life*, October 1982, p. 35.

28. Mikhail Floriansky, "October 4—For the First Time in the World," *Moscow News Supplement*, No. 40 (3288), 1987.

29. A. Yu. Ishlinskiy, ed., *Akademik S. P. Korolev: ucheniy, inzhener, chelovek* (Moscow: Nauka, 1986), p. 62.

30. Col. Mikhail Rebrov, "Sputnik No. 1," *Moscow News Supplement*, No. 40 (3288), 1987, p. 3.

31. US IGY officials had known of the Soviet plans since 1955 when *Pravda* had alerted its readers, and the delegates to the 1955 International Astronautical Congress in Copenhagen were informed as well. In June, 1957, an article appeared in the Soviet journal *Radio*, by one V. Vakhnin, alerting radio specialists on how to tune in on an orbiting satellite's signal.

32. Major General John B. Medaris, von Braun's boss at the Army Ballistic Missile Agency.

33. *Astronautics*, November 1957, p. 6.

34. *Ibid.*, p. 17.

35. Martin Summerfield, "Problems of Launching an Earth Satellite," *Astronautics*, November 1957, pp. 18–21, 86–88.

36. *Le Figaro* (Paris, France), October 7, 1957, pp. 4–5.

37. George F. Kennan, *George F. Kennan Memoirs 1950–1963* (New York: Pantheon Books, 1972), p. 140.

38. Georgiy Grechko interview.

39. James A. van Allen, *Origins of Magnetospheric Physics* (Washington, DC: Smithsonian Institution Press, 1983), p. 49. Stuhlinger and van Allen had begun discussions on the use of a satellite to investigate cosmic rays above the atmosphere more than three years earlier, in 1954, when van Allen was at Princeton University. Also, Pickering communication with author, March 18, 1996.

40. *Ibid.*, p. 93.

41. Glen P. Wilson, *Prologue: Quarterly of the National Archives*, Winter 1993, pp. 364–70.

42. These objectives are virtually the same as those developed at a conference at the University of Michigan on January 27, 1956, the results of which were published in *Scientific Uses of Earth Satellites* (Ann Arbor: University of Michigan Press, 1956).

43. Sagdeev, *Making of a Soviet Scientist,* pp. 156–57.

44. van Allen, *Origins of Magnetosphere Physics*, p. 82.

45. *Ibid.*, p. 66.

46. *Ibid.*

47. *Ibid.*, p. 72. van Allen reports that it was physicist Robert Jastrow who first used the term at a meeting of the International Atomic Energy Agency in Europe.

48. *Pioneers 1, 2, 3,* and *4* were attempted lunar probes launched in 1958 and early 1959.

49. van Allen, *Origins of Magnetosphere Physics*, pp. 129–30. Letter from van Allen to Robert Toth, *New York Herald Tribune*, March 13, 1959.

CHAPTER 4

Sputnik and the Creation of the Soviet Space Industry

William P. Barry[1]

The Sputnik project, and the tumultuous world reaction to it, led to the creation of a space industry unlike anything in the West. Here we are accustomed to an industry composed of familiar aerospace giants whose revenues are still primarily derived from the aircraft industry. However, the industry that outlived the Soviet space program appears to be an impenetrable bureaucratic maze of design bureaus, state research and production centers, science-production associations, and other "space corporations." One of the keys to making sense of this maze is to realize that the Russian Federation and the Ukraine do not have an "aerospace" industry.[2] Instead, the Soviet Union bequeathed its successors with a distinct "space" industry. This industry's unusual organization, habits of behavior, and high level of involvement in the space policy-making process are a result of long-standing traditions in Soviet research and development (R&D) combined with the unique events leading to Sputnik, and the reaction that followed it.

SOVIET RESEARCH AND DEVELOPMENT

Sputnik, and the industry it spawned, were products of the Soviet weapons R&D process. As in all Soviet industry, weapons research, development, and production were organized under specialized industrial ministries; e.g., the Ministry of Aviation Industry. All ministries followed the same highly compartmentalized approach to R&D. From the early 1930s onward, the Soviet R&D process was divided among three distinct types of organizations.[3] Basic research was carried out by Scientific-Research Institutes (Nauchno-

Issledovatel'skii Institut or NII). Series production of the final product was carried out by industrial plants or factories. The critical bridge between research and production was supplied by the Design Bureau (Konstruktorskoe Biuro or KB). The task of the design bureau was to develop new designs or applications of technology in light of basic research information (from the NII), the limitations and capabilities of industry, and the needs of the "customer."[4] Major KBs had the capability to produce experimental prototypes both for testing purposes and to prove the feasibility of production. (In some cases the design bureau was co-located with the factory that was assigned to produce its designs.) The most important design bureaus were distinguished by the title OKB (Opytno-Konstruktorskoe Biuro), meaning Experimental Design Bureau.

OKBs were led by an individual who was perhaps the closest thing to an entrepreneur that existed in the Soviet Union—the chief designer. These larger-than-life figures gave their organizations a distinct personality. OKBs were often simply referred to by the chief designer's name. In the aviation industry these include such famous organizations as the Tupolev and MiG (Mikoian and Gurevich) bureaus. The space industry has its own honor roll of famous chief designers; including, Korolev, Glushko, Yangel, and Chelomei. These chief designers not only had to lead the development of new technologies, but had to succeed in "selling" their new proposals to higher authorities. Failure on a project, or the inability to win new "contracts" could lead to dismissal of the chief designer and, in some cases, the elimination of his bureau.

Despite such pressure on chief designers, head-to-head competition among design bureaus was not the norm in Soviet industry.[5] However, the Soviet leadership did deliberately establish competition in parts of the weapons R&D complex. This expensive approach had its origins in aviation R&D of the 1930s.[6] As a result of the relatively poor performance of Soviet aircraft during the Spanish Civil War, Stalin purged and restructured the aviation industry. The leading Soviet aircraft designer, Andrei Tupolev, was imprisoned and many new design organizations were established under the control of promising young engineers.[7] Aircraft R&D was no longer entrusted to these experts, but became more tightly controlled by the political leadership. For the technically unschooled political leadership, competition between aircraft designers proved extremely useful. Competition, combined with intrusive oversight by state security organs, was highly effective in forcing the creation of the kind of Air Force the leadership wanted. Later, these control mechanisms were applied to some other high priority, high technology weapons programs, including missiles. The missile program chief designers were well aware of the high price that Tupolev, and others, had paid for failing to please the political leadership.[8] In such top priority, weapons programs chief designers lived a high-pressure life frequently characterized by ruthless political maneuvering.

STEPS TOWARD SPUTNIK

In the post-war era the Soviet leadership had no intention to waste resources on space projects. Yet, despite the tight controls, missile chief designers were able to turn their energies to starting the Soviet space program. Remarkably, early Soviet space enthusiasts had been rather successful in developing rockets in the 1930s, but this line of research was snuffed out in the pre-war purges. Virtually all of the leading rocket engineers were arrested and sentenced to forced-labor camps for various crimes. Some were lucky enough to work in prison design bureaus until they were released late in World War II. As the war ended, many of the earlier rocket enthusiasts were put to work analyzing German missile developments.[9] Thus, the Soviet post-war missile effort had a strong core of native talent with a theoretical background in rocketry, some practical experience, a great enthusiasm for space exploration, and considerable caution about expressing the latter.

In early 1946 the Soviet leadership formally established a ballistic missile development program.[10] There were a number of important parallels with the U.S. ballistic missile program of the time. In both cases, the programs began by exploiting German wartime developments and relied heavily on the talents of captured German rocket developers.[11] Both countries also pursued ballistic missiles as part of a three-pronged delivery system development effort. Long-range nuclear capable aircraft and winged (cruise) missiles were the other two lines of investigation. In the Soviet Union, as in the U.S., ballistic missiles were the least favored option, since success in building a reliable long-range ballistic missile seemed far in the future.[12]

However, there was a crucial difference between these U.S. and Soviet post-war weapons projects. In the U.S., R&D for the three delivery system projects was largely conducted by aircraft companies. In the USSR these projects were assigned to two separate industrial ministries. Aircraft and winged missile projects went to the Ministry of Aviation Industry. The liquid-fueled ballistic missile program was assigned to the Ministry of Armaments. The traditional explanation for this decision is that wingless missiles are ballistic projectiles—much more like artillery than aircraft. Thus, the artillery industry under the Ministry of Armaments was given control of long-range ballistic missiles. This was not unprecedented, for the Ministry of Armaments had earlier been involved in the development of solid-fuel projectiles including the "Katiusha" rockets. However, it is highly likely that there was more to the decision than was admitted at the time, for the Minister of Armaments was the highly ambitious and successful Dmitrii Ustinov.[13] Nevertheless, this was a fateful decision. Unlike the U.S., the Soviet ballistic missile program was an outgrowth of the artillery industry, rather than the aircraft industry.

Within a few years, the native Soviet space enthusiasts came to dominate the ballistic missile program. The German rocket specialists worked largely in isolation from, and apparently in competition with, their Soviet counterparts. However, the Soviet designers were able to take full advantage of German technical and missile manu-facturing experience. Within two years the Soviet missile team displaced the Germans by rapidly producing and testing more advanced missiles. By 1950 only about fifty of the original four hundred German rocket specialists remained in the Soviet Union.[14] The Soviet rocket team's rise was the result of a number of factors. Political maneuvering at the highest levels of government appears to have been a factor once again. However, more importantly, the Soviet rocket team proved extraordinarily fast and effective in their work.

One of the keys to the success of the Soviet rocket team was an organization known as the Council of Chief Designers. Although the bulk of the initial missile program was concentrated in a research institute under the Ministry of Armaments (NII-88), crucial subsystems were developed by other NIIs and KBs in other ministries. The leading missile designer in NII-88, Sergey Korolev, called a meeting of the other

TABLE 4.1:   THE COUNCIL OF CHIEF DESIGNERS (PRIOR TO 1953)

Chief Designer	Institution	Product	Ministry
S. P. Korolev	NII-88/OKB-1	Long-range Missiles	Armaments
V. P. Glushko	OKB456	Rocket Engines	Aviation Industry
V. I. Kuznetsov	NII-944	Gyro Instruments	Shipbuilding Industry
N. A. Piliugin	NII-885	Internal Control Systems	Means of Communication
M. S. Riazanskii	NII-885	Radio Control Systems	Means of Communication
V. P. Barmin	State Specialized Design Bureau (GSKB)	Ground Equipment	Machine and Instrument building

five major chief designers in November 1947 (see Table 4.1). The "Big Six" began to meet regularly in Korolev's office to coordinate their work, to discuss technical details, and to forge agreements about the overall course and direction of the missile program. This unofficial working-level organization allowed the leading missile engineers to present a "united front" to their overseers; this gave them greater control over key technical decisions.[15] At least two of the chief designers had an agenda that went well beyond ballistic missiles. Korolev, the driving force in the Council of Chief Designers, had been a leader among the pre-war rocket enthusiasts and had spent the war in a prison design bureau. Valentin Glushko, the chief designer of rocket engines, was an even earlier space enthusiast who had worked with Korolev in the 1930s and suffered the same fate. Thus, the Soviet missile program came to be led by a highly effective group of space enthusiasts who were able to coordinate their efforts outside the usual bureaucratic channels.

Under the tight supervision of Stalin's secret police, the missile designers were limited in the pursuit of their space ambitions; however, after Stalin's death (March 1953) the situation changed dramatically. The most important change came with the arrest of Lavrentii Beriia; the man who oversaw Stalin's secret police. Following Beriia's arrest in June 1953, police oversight of weapons

R&D largely evaporated. Ironically, weapons R&D became even more politicized as a result. Stalin's top lieutenants had been given responsibility for various weapons programs and the behind the scenes struggle for political succession appears to have involved arguments over the future of these weapons systems. Nikita Khrushchev's emergence as the champion of ballistic missiles was clearly a part of this struggle.[16] With Khrushchev's victory over his main political rivals in mid-1957, the missile program found itself with an extremely powerful ally.

However, the first two Soviet space projects had actually been started three years earlier. Although the evidence on the first project is still sketchy, it appears that a researcher in a Ministry of Defense research institute (NII-4) convinced his superiors to start a reconnaissance satellite effort in 1954. Interestingly, the researcher's argument for the project was built on evidence that the U.S. military was secretly pursuing a spy satellite project.[17] At about the same time, Soviet space enthusiasts (Korolev, in particular) began to lobby for a scientific satellite through the USSR Academy of Sciences.[18] Apparently, these scientific satellite proposals were prompted by a call for satellite projects from the International Geophysical Year (IGY) organizing committee.[19] Korolev's proposals were not enthusiastically received in the USSR. Nonetheless, the Soviet government did approve an IGY satellite proposal about a year later; in the summer of 1955. This was the same time that the U.S. announced that it would support an IGY satellite project.

Both Soviet satellite projects were rather low priority. The top priority for the missile chief designers was to develop an Intercontinental Ballistic Missile (ICBM). Since the end of World War II the Soviet leadership had focused on breaking the U.S. monopoly in nuclear weapons. Although the nuclear program had been remarkably successful, the Soviets had been unsuccessful in countering the U.S. delivery advantage. The U.S. could launch a nuclear strike from a ring of bases surrounding the USSR. However, the Soviets still needed a long-range delivery system in order to strike the U.S. Khrushchev came to favor the ICBM, and from 1953 onward the Soviet government made the R-7 (SS-6) missile a top priority. In fact, Korolev's organization, which became a design bureau known as OKB-1 in 1950, "spun-off" shorter range and naval missiles to other design bureaus (see Appendix 1). Among these new organizations was a design bureau and plant in Dnepropetrovsk, Ukraine, that later became the primary Soviet missile facility under the leadership of chief designer Mikhail Yangel. Unlike the satellite projects, the missile program had little problem acquiring the resources it needed.

The satellite projects might well have died on the vine if Khrushchev had not visited Korolev's design bureau in January 1956. Early that month, the Kremlin leadership was given a personal tour of OKB-1. Korolev seized the opportunity to sell Khrushchev on the IGY satellite project. Khrushchev appears to have approved the idea largely because Korolev promised him that he could quickly perfect the ICBM and then use it to beat the U.S. into space; and with a much larger satellite to boot.[20] A decree issued on January 30, 1956, grafted the Soviet IGY satellite project onto the top-priority ICBM program. Such brazen lobbying was a well-established tradition among the top chief designers, but in this case it had far-reaching consequences. Khrushchev was so taken by Korolev that he granted him direct access. For the next few years, Korolev was able to bypass the usual oversight mechanisms and lobby the leader of the Soviet Union directly.

Even with a powerful sponsor, the nascent Soviet space industry was still hampered by problems. Perfecting the R-7 ICBM was the pre-requisite to launching any satellite, and the chief designers' energies were focused on this until they succeeded in August 1957. Nonetheless, Korolev still found time to work on his pet satellite project. Unfortunately, by January 1957 it was clear that the IGY satellite would not be ready in time to beat the U.S. With characteristic boldness Korolev proposed substituting two "simple satellites" for the IGY satellite. This substitution was apparently approved in January 1957. Thus, the satellites we now know as *Sputnik 1* and *Sputnik 2* were born. (The Soviet IGY satellite was not successfully launched until May 1958; it was known as *Sputnik 3*.)

THE REACTION TO SPUTNIK

After Sputnik the Soviet government asserted that their space program was a well thought-out, long-term program of peaceful exploration. In fact, the Soviet space program started as an official afterthought—attached to the ICBM program only at the behest of its developer. The first satellite was "a result of the intensive work, by research institutions and design bureaus" not the product of a fully developed state industry.[21] Like the program it supported, the Soviet space industry was really born in the aftermath of Sputnik. It developed in fits and starts that reflected the preferences of the Soviet political leadership and long-standing feuds within the space industry.

Khrushchev appears to have been both surprised and delighted by the impact that Sputnik had on the world.[22] He had originally pinned his hopes for impressing the world on the creation of a Soviet ICBM. However, the unprecedented announcement of the first successful R-7 (SS-6) test in August 1957 went virtually unnoticed outside the Soviet Union. Only after Westerners could see *Sputnik 1* whizzing overhead did they take notice of the military potential of Soviet missiles. Khrushchev became an ardent proponent of space spectaculars as evidence of the superiority of Soviet communism. Korolev was happy to oblige by putting his long-suppressed ideas into practice with a series of ever-more astounding space launches.

The push for space spectaculars led to major changes at OKB-1 and in the missile industry. Korolev's OKB-1 redirected much of its attention away from missile work and toward the string of space missions. Not only was OKB-1 the home of the R-7 (SS-6) launcher, but it soon acquired the key personnel from other organizations that had been involved in satellite R&D. This included the bulk of the Ministry of Defense research institute (NII-4) that had been conducting research on reconnaissance satellites. OKB-1 was quickly metamorphosing from a missile KB into a "space" design bureau. Further evidence that the core mission of OKB-1 had changed is suggested by the fact that in 1959 two new subsidiaries of OKB-1 were created to handle missile design and production. The first of these, Filial #2 ("Branch #2") was set up in Krasnoiarsk to handle Korolev's newest ICBM design—the R-9.[23] Further development work, and production supervision for the R-7 was given to an organization, Filial #3 ("Branch #3") in Kuibyshev (now, Samara).[24] Korolev's space and missile empire had grown dramatically, but clearly his personal interests and energies were devoted to the space work at OKB-1.

Unfortunately for Korolev and the emerging space industry, Khrushchev's attention returned to ICBMs. Although the R-7 (SS-6) eventually met the design requirement of being able to deliver a nuclear warhead to the U.S., it turned out to be a very poor Cold War weapon. In response to Soviet missile rhetoric, the U.S. had rapidly escalated its nuclear posture—building forces capable of an immediate and massive strike. The Soviet military successfully argued that the R-7 was much too slow to respond to such a threat (it took hours to fuel and prepare for launch). The one weapon that could strike the U.S. was, therefore, too easily pre-empted by the hair-trigger U.S. forces. To the surprise of Western intelligence, the USSR deployed just a handful of R-7 (SS-6) ICBMs.[25] Soviet efforts turned instead to developing a second-generation of ICBMs. More interested in space than missiles, Korolev devoted relatively little attention to new ICBMs. However, one of the organizations that had spun-off from OKB-1 in the early 1950s was already hard at work on this problem. Mikhail Yangel, chief designer of OKB-586 in Dnepropetrovsk, was much more attuned to the demands of the military. He had already begun work on a storable fuel ICBM, the R-16 (SS-7), in 1956. It was ready for testing in 1960. Thus, Korolev's position as *the* chief designer in the missile and space fields began to erode.

Failure to take strategic demands seriously was not the only thing that led to a decline in Korolev's status. The "united front" provided by the Council of Chief Designers began to break up shortly after the success of *Sputnik 1*. Disagreements over technical details soon turned to bickering about control over space projects, resources, and who would get credit for the successes. In particular, the chief designer of rocket engines, Valentin Glushko, took umbrage at Korolev's dominance of the space program. In keeping with the traditional behavior of weapons chief designers, the feud between Korolev and Glushko grew so bitter and personal that each demanded the other's removal from the space and missile programs. Khrushchev's unsuccessful personal efforts to mediate this dispute contributed to his loss of enthusiasm for Korolev and the other missile chief designers.

By the Spring of 1960 Khrushchev was ready to take dramatic steps to shake up the space and missile industry. The precise reason why Khrushchev acted when he did is still unclear, but the problems he faced had been growing for several years. In addition to strategic nuclear problems and his frustration with the Council of Chief Designers, it appears that he also lost faith in the political leadership that was supposed to supervise the defense industry. Leonid Brezhnev, the Central Committee Secretary overseeing the defense industry (and one who had benefited greatly from his connections to Korolev), was "promoted" to the ceremonial post of President of the USSR. At around the same time Khrushchev decided to set up new competition for the space and missile industry.

As mentioned above, the use of competition between weapons design bureaus was not a new idea. However, Khrushchev put a new twist on it. In 1960 he not only established a new missile and space design bureau, but he set it up in a different ministry. Both Korolev and Yangel faced a new competitor, and so did their ministry. The competition was part of the aviation industry, not the armaments industry.[26]

Strangely enough, Khrushchev did not give this assignment to one of the major aircraft design bureaus. Instead, a winged (cruise) missile design organization was authorized to start work on both ICBM and space projects. OKB-52, under chief designer Vladimir Chelomei, had been created in 1955 to develop the P-5 (SS-N-3) cruise missile for the Soviet Navy. Chelomei was no stranger to political maneuvering. He had led the post-war effort to build an analog of the German V-1 "buzz bomb." However, Chelomei failed to produce a useable weapon. He was dismissed, and his design bureau was handed over to the MiG OKB.[27] Only by lobbying the Navy brass and Khrushchev himself, did Chelomei finally regain control of a design bureau.[28] Three years later, in 1958, Chelomei hired Khrushchev's only surviving son, Sergei. This connection helped Chelomei to gain access to the Soviet leader so that he could lobby for new projects.

OKB-52 clearly needed more resources to carry out the new space and missile projects that Khrushchev authorized in early 1960. The first major expansion came quickly, when the Miasishchev aircraft design bureau (OKB-23) and its associated production facility were subordinated to Chelomei[29] (see Appendix 2). In 1962, Chelomei also won control of OKB-301, the design bureau that had been headed by Semen Lavochkin until his death in 1960.[30] Chelomei's missile and space empire grew so quickly that he was reported to have claimed that he was "the most expensive man" in the Soviet Union.[31] Despite the wealth of resources devoted to Chelomei's projects, they did not bear fruit immediately.

Korolev's OKB-1 continued to run the manned space program since it was the only organization capable of building and launching such missions. However, Korolev's dominance over space and missile programs declined dramatically. His plans to build much larger booster rockets and new manned spacecraft were slowly starved of support. The R-9 missile was canceled and Filial #2 in Krasnoiarsk was separated from Korolev's organization. It was made an independent organization known as OKB-10 (see Appendix 2). By mid-1962 OKB-10 was tasked to develop a satellite project that had originated in Yangel's OKB-586.[32] In fact, Yangel had begun to take on the bulk of unmanned earth-orbital missions in 1961 with the advent of the Kosmos series of satellites.[33] Thus, Korolev's early dominance of the space industry dissipated rapidly with the rise of alternatives and competitors.

Unfortunately for Chelomei, his efforts were just beginning to pay off when his sponsor fell from power. A prototype of his UR-500 launcher (later known as the Proton) and the "Polet" satellite were first launched in November 1963. A second launch was carried out on Cosmonautics Day (April 12) the next year. Six months later, in October 1964, Khrushchev was ousted from power. For the space industry, the change of political leadership had huge consequences.

After Khrushchev was deposed there was a major reorganization and reprioritization in the space and missile industry. Rejecting the two-ministry approach, the new leadership created the Ministry of General Machinebuilding (MOM) and placed all of the space and missile industry under its control. Within two years the major KBs were confusingly renamed; OKB-1 became the Central Design Bureau of Experimental Machinebuilding (TsKBEM) and Chelomei's OKB-52 became the Central Design Bureau of Machinebuilding (TsKBM). More importantly, the new Soviet leadership, now headed by Korolev's old ally Leonid Brezhnev, began a relentless attack on Chelomei's empire. A number of Chelomei's projects were canceled, or turned over to other design bureaus.[34] Chelomei's Filial #2 was made independent, renamed KB Lavochkin, and given responsibility for developing planetary probes (a duty previously held by Korolev's organization) (see Appendix 2).

With the support of the new leadership, Korolev once again focused on manned space missions. Not only did Korolev have the support of Brezhnev, but his former minister, Dmitrii Ustinov, became the top Communist Party supervisor of the defense industry. Ustinov, who had his own reasons to dislike Chelomei, was the key figure in dissecting the empire that had been built around OKB-52. In short order, Korolev gained control over the Soviet manned lunar landing program, and the separate circumlunar program that Chelomei had controlled. Although Korolev died in January 1966, his organization carried the manned lunar programs forward under the control of his long-time deputy Vasilii Mishin.

Chelomei's organization survived largely due to support from the Soviet military. The Soviet General Staff favored a number of Chelomei's projects and was able to keep them going for quite some time. The UR-100 (SS-11) and UR-100N (SS-19) ICBMs were major Chelomei missile programs that survived. The ICBM version of the UR-500 was canceled, but it remained as a space launch vehicle (Proton). One of the reasons to keep the UR-500 alive was that it was to be the launcher for Chelomei's manned military space station. This was a program started as a counter to the U.S. Manned Orbiting Laboratory (MOL); which was also to be a manned reconnaissance platform. Although the U.S. canceled MOL in 1969, the USSR kept their military space station alive for nearly another decade. However, Chelomei lost considerable control over this program. Most importantly, after the U.S. won the race to the moon, Ustinov commandeered Chelomei's space station design (and much of the hardware) to turn it into a scientific space station. This project was handed over to Korolev's bureau; TsKBEM. The military version of the space station *did* fly three times under the cover names of Salyut 2, 3, and 5.[35] Yet, this project came to an end in 1977, when the last of these stations was de-orbited.

There were few major changes to the structure of the Soviet space industry from the late 1960s until the USSR collapsed in 1991. The changes that did occur reinforced the traditions and practices that had begun with Sputnik. The most significant alteration came in the mid-1970s when much of the Soviet R&D complex was re-organized into Science-Production Associations (NPOs) (see Appendix 3). A number of the space design bureaus and industrial plants were unified under design bureau control as NPOs. The designation of the design bureaus as leading organization in the NPOs reinforced the power and perquisites of the space chief designers. They were now formally responsible for production, as well as R&D.

The NPO unification was also used as an excuse to dispose of Korolev's successor as chief designer. Vasilii Mishin, who was still focused on putting a man on the moon, was "retired" when TsKBEM was merged with Glushko's engine design bureau to create NPO Energiia. Glushko took this opportunity to push his own agenda; part of which included re-writing space history so as to downplay Korolev's role and enhance his own.[36] The influence of Glushko and NPO Energiia were substantially enhanced in 1976 when Glushko was appointed to a position on the powerful Communist Party Central Committee. This occurred at the same time that Dmitrii Ustinov became a member of the ruling Politburo and Minister of Defense. Although Ustinov died in 1984, Glushko's influence continued until his death in 1989. After Glushko's death, the two great design bureaus went their separate ways. Glushko's engine design organization became known as NPO Energomash, while Korolev's former organization retained the name NPO Energiia.

Most of the other significant changes in the Soviet space industry relate to the continuing assault by Ustinov and Glushko on Chelomei's empire. Throughout this period, Ustinov continued to show favoritism for the OKBs that he had supervised as Minister of Armaments in the 1950s. Thus, Yangel's organization, renamed KB Iuzhmash in 1966, came to dominate the field of missile production while remaining involved with space work.[37] The other successors and spin-offs from Korolev's design bureau consolidated their control of the space industry. A landmark step in this process happened in 1981 when Chelomei's Filial #1 (the former Miasishchev design bureau) was handed over to NPO Energiia.[38] As can be seen in Appendix 3, Chelomei's former Filial #1 was subjected to frequent changes and reorganization that coincide remarkably with changes in the political leadership.[39] By the end of the Soviet period the design bureau (KB Saliut) had been separated from the production facility (the Khrunichev Machinebuilding Plant). Recently, these two organizations were reunited. In June 1993 the "Khrunichev State Rocket and Space Center" was created. In this case, however, the plant director (Anatolii Kiselev), and not the chief designer, was put in charge. [40]

CONCLUSIONS

From its very beginnings the Soviet space industry was an unusual hybrid. Space industries were born within the defense industrial complex and remained subordinated to it throughout the Soviet period. However, the space program was never merely a military program. It was, originally, a special state propaganda program; overseen and coordinated at the very highest levels of government. This military-political duality prevented the creation of a separate civil space organization on the model of NASA. Although the USSR tried to create the appearance of having a civil space agency (pretending first that the Academy of Sciences ran the space program, and later setting up Glavkosmos), the Soviet space industry remained subordinated to defense industrial ministries. Even in the post-Soviet period this did not change. Although the Russian Space Agency (RKA) was "spun-off" of the Ministry of General Machinebuilding in February 1992, control over the space industry remained in the hands of the Ministry of Defense Industry until it was abolished in Spring 1997.[41] Within a month (notably around April 12—Cosmonautics Day), it was announced that the Russian Space Agency (RKA) would assume oversight of the space industry.[42]

Although direct military influence over the Russian space industry has declined throughout the last thirty years, the habits of secrecy that characterized the defense industry did not. The tight Soviet secrecy about space programs, objectives, and industry was probably unavoidable due to the overlap between the missile and space programs. However, maintaining security also served other purposes. Most importantly, the habit of announcing intentions only after success had been achieved allowed the very small handful of leaders at the top to carry out their program with no real oversight. Thus, people who rose to positions of power could pursue their personal agendas with little fear of repercussions. This habit of behavior was evident throughout the

Soviet period and, only now, is it starting to change due to the glare of media and international scrutiny.

Members of the Politburo and the Central Committee were not the only ones who tried to treat the space program as their personal fiefdom. The pattern of behavior of Soviet space chief designers was also remarkable in its ruthless pursuit of power. Starting with Korolev, the most politically successful chief designers shamelessly exploited their connections to top political leaders to win approval of the all-important government decrees that drove major programs.[43] Moreover, the chief designers developed a reflexive aversion to competition. This may have been a reaction to their personal experiences under Stalin. Yet, whatever the cause, space chief designers were not merely content to win a "contract." They seemed compelled to try to destroy their rivals. This was evident in battles over programs, the struggles over control of R&D resources, and even in writing the history of the space program.

One of the most important distinctions between the former Soviet space industry and its counterpart in the West is that the Soviet industry was largely the outgrowth of design groups with no real connection to the aviation industry. The Korolev and Yangel OKBs were originally part of the artillery industry. The largely eclipsed Chelomei design bureau may have been subordinated to the aviation industry (until 1965), but its leader (Chelomei) was an outsider to that ministry. OKB-52 was established and expanded through Khrushchev's intervention. Moreover, Chelomei had been a cruise missile designer, not an aircraft designer. The non-aircraft roots of these organizations have had major implications for their approach to rocket and spacecraft design. This is especially evident in manned spacecraft, which tend to rely much less on pilot intervention than Western designs.[44]

For the former Soviet space industry the legacy of Sputnik has been long lasting. Although the structure of space industry is rooted in the traditions of Soviet weapons R&D, the approval process for Sputnik and the leadership handling of space as a political program established patterns of behavior that have had powerful influences. These influences are clearly evident in the structure of the former Soviet space industry, and in its current behavior. The restructuring of the Russian space industry announced in April 1997 may herald the opening of a more Western-style approach to organizing space R&D.

Yet, it is probably no accident that all of the organizations selected by the Russian Space Agency (RKA) to remain as the "powerful core" of the Russian space industry are the ones tied most closely to the original Korolev design bureau.[45] The echoes of Sputnik still ripple through the Russian space industry.

1. The views expressed in this chapter are those of the author. This chapter does not necessarily reflect the views of the Air Force Academy, the U.S. Air Force, or the U.S. Government.

2. Approximately 80 percent of the Soviet space industry was located in the Russian Federation, and inherited by it, after the dissolution of the USSR. The Ukraine remains home to several important scientific and industrial facilities. Although the other successor states inherited a number of facilities important to the operation of space systems (e.g., the Baikonur Cosmodrome in Kazakstan), only Russia and the Ukraine have retained a substantial space research and production capability.

3. Arthur Alexander, *R&D in Soviet Aviation*, R-589-PR, Santa Monica, CA: RAND, November 1970, p. 4.

4. The "customer" could be another ministry (e.g., the Ministry of Defense in the case of weapons), the producing ministry, or the political leadership.

5. In fact, traditional Marxist economic thought holds that such competition is a tremendous waste of resources. The elimination of such "wasteful" competition was seen as one of the great advances of socialism.

6. See: Mikhail Tsypkin, *The Origins of the Soviet Military Research and Development System (1917–1941)*, Ph.D. diss., Harvard University, 1985.

7. The most famous personal account of this era is now available in an excellent English translation. See: L. L. Kerber, *Stalin's Aviation Gulag: A Memoir of Andrei Tupolev and the Purge Era*, ed. and trans. by Von Hardesty (Washington DC: Smithsonian Institution Press, 1996).

8. Missile chief designer Korolev had studied under Tupolev, and wound up working for him in the "aviation gulag."

9. B. Konovalov, "From Germany—To Kapusti Yar" (in Russian), *Izvestiia*, April 6, 1991; and S. Averkov, "Missiles of the Third Reich" (in Russian), *Rabochaia Tribuna*, July 4, 1991.

10. This was accomplished through Council of Ministers Decree No. 1017–419ss, May 13, 1946. See, I. D. Sergeev, et al., *Khronika osnovnykh sobytii istorii raketnykh voisk strategicheskogo naznacheniia* (Moscow, Russia: TsIPK, 1994), pp. 227–34.

11. Although the "brains" of the German rocket program were snapped up by the West, the Soviets inherited most of the German rocket facilities. In October 1946, the Soviets rounded up several hundred of the remaining German rocket workers and shipped them to various locations in the Soviet Union.

12. In fact, Soviet nuclear weapons were built only for delivery by aircraft until 1953.

13. Political maneuvering is also suggested by the fact that responsibility for solid fuel missiles was assigned to yet another ministry (the Ministry of Agricultural Machinebuilding—formerly known as the Commissariat of Munitions). This ministry was controlled by another major industrial leader, Boris Vannikov, who later became a major figure in the nuclear weapons program.

14. US Central Intelligence Agency, *National Intelligence Estimate Number 11-6-54: Soviet Capabilities and Probable Programs in the Guided Missile Field*, (Top Secret) Washington,

DC, October 5, 1954, as declassified June 29, 1993, by CIA Historical Review Program, pp. 5–6. (Those that remained were guidance and control specialists, most of whom returned to Germany by the mid-1950s.)

15. B. Chertok, "Leader" (in Russian), *Aviatsiia i kosmonavtika*, No. 1, January 1988, p. 31.

16. William P. Barry, "How the Space Race Began: The Origins of the Soviet Space Program," paper delivered to the Society for History in the Federal Government conference, April 3, 1997.

17. N. Dombkovskii, "October-April-Universe" (in Russian), *Sovetskaia Rossiia*, April 12, 1989, p. 3.

18. Korolev was one of the many weapons designers who had become corresponding (junior) members of the Academy of Sciences in the post-Stalin thaw.

19. See the paper by Asif Siddiqi, "Korolev, Sputnik, and the International Geophysical Year," Chapter 2 in this volume.

20. Sergei N. Khrushchev, *Nikita Khrushchev: krizisy i rakety*, Vol. 1 (Moscow, Russia: Novosti, 1994), p. 111.

21. This quote taken from the remarkably low-key TASS press release announcing *Sputnik 1*'s launch, October 4, 1957.

22. James Harford, *Korolev: How One Man Masterminded the Soviet Drive to Beat America to the Moon* (New York: John Wiley, 1997), p. 121.

23. This organization later became known as NPO Applied Mechanics. See Appendices 1–3.

24. This organization became known as TsSKB-Progress. See Appendices 1–3.

25. See: US Central Intelligence Agency, *National Intelligence Estimate 11-8/1-61 Strength and Deployment of Soviet Long Range Ballistic Missile Forces*, (Top Secret) Washington, DC, September 21, 1961, as declassified by the CIA Historical-Review Program.

26. During the 1950s the names and some of the functions of the defense industrial ministries changed several times. In 1960 the Ministry of Aviation Industry was known as the State Committee on Aviation Technology (GKAT). The Ministry of Armaments had become the State Committee on Defense Technology (GKOT).

27. Nikita S. Khrushchev, *Khrushchev Remembers: The Last Testament*, ed. and trans. by Strobe Talbott (Boston: Little, Brown, 1974), p. 35.

28. S. N. Khrushchev, *Nikita Khrushchev: Krizisy i Rakety*, Vol. 1, pp. 371–72; N. S. Khrushchev, *Khrushchev Remembers: The Last Testament*, pp. 44–45.

29. OKB-23 was the organization responsible for the "Bomber Gap" scare in 1953 when its M-4 *Bison* bombers were miscounted by Western attaches. The Miasishchev production facility was subsequently named the Khrunichev plant in honor of the post-war Minister of Aviation Industry.

30. Semen Lavochkin had been a highly successful designer of fighter aircraft. After World War II he turned his attention to building a long-range supersonic cruise missile. Successes with ballistic missiles led to the cancellation of this program.

31. Roald Z. Sagdeev, *The Making of a Soviet Scientist* (New York; John Wiley, 1994), p. 202.

32. OKB-10 later became a major communication satellite developer.

33. Kosmos satellites were not all designed by Yangel's bureau. The name was used as a cover for a number of unsuccessful missions developed by Korolev.

34. For example, Chelomei's space-plane project (analogous to the U.S. Dyna-Soar) was turned over to the MiG OKB.

35. One of the best discussions of the military Saliuts can be found in: Phillip Clark, *The Soviet Manned Space Programme: An Illustrated History of the Men, the Missions and the Spacecraft* (London, England: Salamander Books, 1988), pp. 66–75. See also, Dennis Newkirk, *Almanac of Soviet Manned Space Flight* (Houston, TX: Gulf Publishing Company, 1990), pp. 112–53.

36. One of the more egregious examples of this is the Gas Dynamics Laboratory Museum in St. Petersburg. Another example that caused considerable semi-public acrimony was the

publication of the *Encyclopedia of Cosmonautics* in 1985. See, V. P. Glushko, ed., *Kosmonavtika Entsiklopediia* (Moscow, USSR: Sovetskaia Entsiklopediia, 1985).

37. NPO Iuzhnoe, located in the Ukraine, remains an important producer of launch vehicles, notably the Zenit booster. Although still known as "Iuzhnoe," this organization has been re-named the Dneprovskii Rocket-Space Center.

38. This was after the military space station program had been terminated, so Filial #1 was primarily building space station hardware for Energiia and continuing work on the Proton launch vehicle.

39. Ustinov's promotion to Minister of Defense and Glushko's promotion to the Central Committee coincide with Chelomei's loss of control over Filial #1. Note also that Chelomei and Ustinov's deaths in 1984 and Glushko's death in 1989 coincide with other changes.

40. "Khrunichev State Research and Production Space Center," International Launch Services homepage, http://www.lmco.com/ILS/text_html/facts_KhSC.html, downloaded October 16, 1996.

41. In March 1997, the Russian Ministry of Defense Industry was abolished and its responsibilities were assumed by the Ministry of Economics.

42. Interfax report (in English), Moscow, April 12, 1997.

43. The hiring and rapid promotion of relatives of Politburo members was a particularly notable habit. Although Chelomei was the first to do so, other space chief designers (including Mishin and Glushko) followed suit.

44. There were some space projects that were developed by aircraft design bureaus. However, none of these efforts have been put into operational use.

45. The April 12, 1997, announcement of RKA control over the space industry stated that, after restructuring, the core space industry organizations would be Energiia, Energomash, the Khrunichev space center, and the Samara space center (TsSKB-Progress). (Interfax report (in English), Moscow, April 12, 1997.)

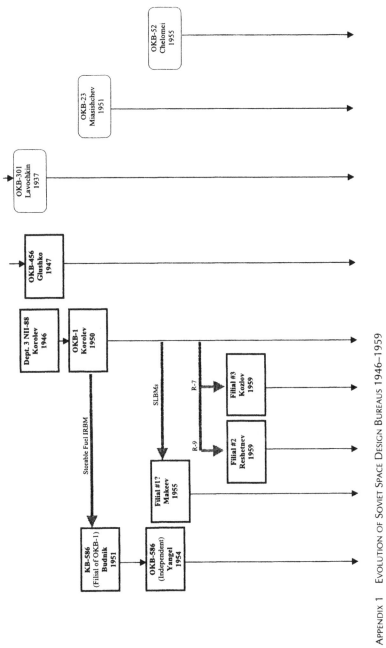

APPENDIX 1 EVOLUTION OF SOVIET SPACE DESIGN BUREAUS 1946–1959

APPENDIX 2 EVOLUTION OF SOVIET SPACE DESIGN BUREAUS 1959–1967

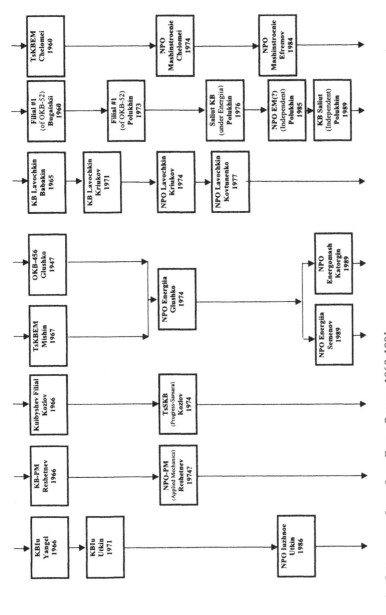

APPENDIX 3   EVOLUTION OF SOVIET SPACE DESIGN BUREAUS 1968–1991

Part 2
A Setting for the International Geophysical Year

A Setting for the International Geophysical Year

Robert W. Smith

The launch of *Sputnik* evoked a range of responses in the United States. Scientist James Van Allen—soon to win fame through his participation in the first successful U.S. satellite, *Explorer I*—could not suppress his excitement at the technological feat, marking it down in his notebook as a "Brilliant Achievement."[1] He and many other people decided straight-away that a new age had dawned, the Space Age.

The 184-pound Soviet satellite orbited the Earth every ninety minutes. Marked by its persistent "beep, beep" radio signals, mocking the United States as Secretary of State John Foster Dulles put it,[2] *Sputnik's* journey had enormous repercussions. Perhaps its most far-reaching effect was that it provoked a wave of near-hysteria in the United States, for as Van Allen and many others had realized immediately, here was a remarkable propaganda coup for the USSR.

But how could the Soviets have outpaced the United States? Some staff at the newly founded *National Review*, for example, even suspected a hoax.[3] Once it became clear to everyone that *Sputnik* was not a hoax, and the launch of *Sputnik 2* underlined it was no fluke, this indeed became the leading question asked about *Sputnik* in the U.S. It is this question, or close versions of it, that still dominates the historical writings on U.S. space related activities and plans in the period immediately before *Sputnik*. How, indeed, could the United States, with all its resources and talents in the sciences and engineering, and prowess in the development of new technologies, have been beaten into orbit by the Soviet Union?

One part of the context for the answer to this question comes, as Rip Bulkeley reminds us, from the facts that *Sputnik* was launched as part of the International Geophysical Year (IGY) of 1957 and 1958 and that the U.S. *Vanguard* program was also designed to be part of the IGY. The IGY entailed a very large-scale international enterprise centered on geophysical Investigations, probably the largest scientific endeavor conducted to that date. First proposed in 1950, it was pursued by scientific teams from over sixty nations. The great majority of its programs caused little public stir, but the satellite programs proposed in 1955 by the United States and the Soviet Union were judged in ways very different from the rest of the IGY. The most important was that *Sputnik* was public in ways that other parts of the IGY were not. As far as the U.S. public was concerned, the Soviet satellite's "beep, beep" radio signals inaugurated the "Space Race" and underscored that space was a highly visible and extremely important theater of the Cold War, one in which the two rivals battled aggressively, and at great cost in national treasure, for prestige.

The IGY was nevertheless designed to be a cooperative scientific endeavor. Reaching agreements on cooperation between the two satellite programs, however, was a painful process, full of mutual recriminations. As Bulkeley notes, the "satellites portion of the IGY was at best a mating of scorpions, conducted in a fog of mutual prejudice and mistrust which the rational ideals of international scientific cooperation could only partly dispel."[4] In the end, the Soviet Union *did* share some scientific data from *Sputnik* with the United States, and there *were* scientific interactions. It is these points, not the limited nature of the exchanges, that Bulkeley finds really striking, and that his painstaking research makes evident.

Direct answers to why the *Sputnik* beat a U.S. satellite into orbit come from the three other chapters in this section. In "Cover Stories and Hidden Agendas," Dwayne A. Day in part expands on the researches and interpretations of Walter A. McDougall on the centrality of establishing the principle of overflight of satellites above other countries, especially the Soviet Union, as key in the formulation of U.S. space policy in the mid-1950s.[5] Day in particular stresses the importance of the Technological Capabilities Panel, its 1955 report on "Meeting the Threat of Surprise Attack" (one of the most influential documents of the Cold War) and the links between that report and the National Security Council's top-level policy document known as NSC 5520, "U.S. Scientific Satellite Program." NSC 5520 emphasized that a small scientific satellite would provide a test of the principle of the freedom of space, and directed that the scientific satellite be associated with the IGY. The IGY, then, would provide the cover story, while the real intent was to establish the principle of satellite overflight. One of the most valuable aspects of Day's chapter is his stress on the pervasiveness of the Eisenhower administration's use of science as cover for intelligence purposes.

Day also argues that being first with a satellite in orbit was far from the be all and end all of the U.S. satellite program. In fact, Day cites the director of the National Science Foundation who argued that it would be possible to launch the first satellite, but lose the scientific competition if the results achieved were not scientifically noteworthy. Hence the Navy's *Vanguard* satellite, chosen as the U.S.'s IGY satellite program, was, in Day's reading, intended to establish a legal precedent *and* produce worthwhile scientific results that would enhance U.S. prestige.

Notions of, and calculations about how to secure, prestige are central to Kenneth Osgood's "Before Sputnik." In 1953, for example, Aristid V. Grosse had predicted that the Soviet Union had trailed the U.S. in the development of the atomic and hydrogen bombs and so it might now be prompted to try to take the lead in the development of a satellite. An "enormous advantage" of such a venture was that it would influence the "minds of millions of people the world over." If the Soviets were to be first, "it would be a serious blow to the technical and engineering prestige of America the world over. It would be used by Soviet propaganda for all its worth.[6]

Osgood, in fact, reads his evidence differently from Day, and argues strongly that prestige, not overflight, is key to understanding the United State's IGY satellite program. This program was "from the outset a defensive rear-guard action designed to protect the United States from a Soviet propaganda broadside. The United States had to get a satellite into space before—or at least not much later than—the Soviets, and it had to design its program in such a way that would deflect potentially damaging Soviet propaganda."[7] For policy makers, programs like the scientific satellite were valuable because they "reinforced confidence in the capacity of the

United States to resist Communist threats through the visible display of technological prowess.[8]

In Osgood's interpretation, President Eisenhower's well-known comments at a May 1957 National Security Council meeting make eminent sense. Eisenhower complained that the *Vanguard* satellite had grown too expensive and that the scientists were gold-plating their instrumentation. He wanted, for reasons of prestige, the emphasis squarely on getting the satellite into orbit, not on the quality of the instrumentation and the scientific results to be secured.

Osgood also places U.S. space efforts in the context of a broad strategy to demonstrate not only the nation's technological and military superiority, but also its peaceful intentions. Hence, for him, *Sputnik* was very much a product of the changing nature of the Soviet-American rivalry. As he concludes, "Sputnik simply provided the defining image of a struggle already underway and a race already being run."[9]

Michael Neufeld, in "*Orbiter*, Overflight, and the First Satellite," concentrates on why the Navy's *Vanguard* program was chosen over the Army's *Orbiter* as the first U.S. satellite project, a decision generally seen as crucial in the "race" to orbit the first satellite. If only Von Braun's army group had been given the go-ahead, so the conventional wisdom goes, the U.S. would very likely have gotten a satellite into orbit first.

Neufeld agrees with McDougall (and more recently R. Cargill Hall and Dwayne Day) that, at least from the level of Donald A. Quarles Assistant Secretary of Defense for Research and Development up to the President, there was a hidden agenda behind the Eisenhower administration's decision to support the flight of a scientific satellite during the IGY. Such a satellite would indeed establish the principle of overflight with a "peaceful" scientific satellite.

But McDougall went further. He speculated that the decision to choose which satellite the U.S. would attempt to fly in the IGY was at least heavily influenced, and perhaps determined, by the overflight arguments. Neufeld's careful research demolishes the latter possibility. He also undermines the currently widely held belief that *Vanguard* won out, and had a very strong advantage in the competition, because it appeared to outsiders to be more of "civilian" program than the Army's

Orbiter. In fact, Neufeld demonstrates that the decision was a close-run thing, that the committee was not loaded so that it would rubber stamp the idea of *Vanguard,* and that the decision was nearly overturned shortly after it was made.

Neufeld pays very careful attention to the technologies involved for, as he argues persuasively, *Orbiter* was not *Explorer. Orbiter* entailed an extremely simple, inert 5 pound sphere as its payload, and so promised relatively little scientific payoff. Also, the basic *Orbiter* proposal envisaged 37 Loki solid rockets in three stages, spun-up before launch for stability. Neufeld argues that this part of the proposal was not seen as fully convincing, given that the Loki I was not especially reliable.[10] The Stewart committee judged that the *Vanguard* proposal offered a better satellite on a similar timetable to *Orbiter.* For the Stewart committee, solid technical and scientific reasons shaped the *Vanguard* decision, not the need to establish overflight with a scientific satellite.

Neufeld therefore adopts a middle position between those of Day and Osgood on overflight. For Quarles and his superiors, overflight, along with the desire to generate international prestige while at the same time not slowing down the critical ballistic missile programs, *were* all important issues. For those lower down in the system, such as the members of the Stewart committee, overflight was not a major factor in their decision making.

With the declassification of ever more relevant documents and the growth of archive-based studies of the *Sputnik* era, our understanding of what might be called the "top-down" aspects of U.S. planning and activities in space before October 4, 1957, has increased greatly in the last decade, particularly in terms of the place of overflight in decision making, as the four chapters in this section witness. These chapters also show that researchers are still able to disagree on this fundamental issue.

But there are too very important "bottoms-up" issues that can fruitfully be taken into account, as the chapters by Bulkeley and Neufeld suggest. It is perhaps here that the biggest opportunities for researchers are to be found. As yet, for example, we do not know in detail what was going on in the laboratories at the Naval Research Laboratory and the Army Ballistic Missile Agency in the mid-1950s. As historians and policy-analysts reach deeper into, as well as examine further the links between, the institutions involved in the early history of U.S. space endeavors, and thereby provide finer-grained studies, we can expect a more fully-rounded account of these undertakings to emerge. There are, no doubt, more surprises to come.

1. James Van Allen, *Origin of Magnetospheric Physics* (Washington, DC: Smithsonian Institution Press, 1983), 46.
2. John Foster Dulles, "The Role of Negotiation," *Department of State Bulletin*, 38(1958), 159.
3. Sam Tanenhaus, *Whittaker Chambers. A Biography* (New York, The Modern Library, 1998), p. 506.
4. Bulkeley, p. 74.
5. Walter A. McDougall, ... *The Heavens and the Earth: Political History of the Space Age* (New York: Basic Books, 1985), chapter 5.
6. Quoted in Kenneth A. Osgood, "Before Sputnik: National Security and the Formation of U.S. Outer Space Policy."
7. Ibid.
8. Osgood, Ibid.
9. Osgood, Ibid.
10. Michael J. Neufeld, "*Orbiter*, Overflight, and the First Satellite: New Light on the *Vanguard* Decision."

The Sputniks and the IGY

Rip Bulkeley

The programs which launched the first artificial earth satellites were conducted by the Soviet Union and the United States during the second half of the 1950s, within the overall framework of the International Geophysical Year (IGY). The IGY was an ambitious international undertaking, comprising a network of planet-wide geophysical studies, which was proposed in 1950 and carried out in 1957–1958. Sixty-seven national scientific teams, from countries with widely dissimilar political systems and levels of economic development, participated in different ways and to differing degrees. In view of the generally hostile relations between the countries of the North Atlantic Treaty Organization (NATO) and those of the Warsaw Treaty (signed only two years before the opening of the IGY) it is particularly remarkable that nearly all the members of the two rival alliances took part in the Year, and that the two nuclear superpowers led the way in this cooperation, in spite (or perhaps because?) of the fact that many of the scientific topics to be studied were of considerable military as well as scientific significance.

The proposal to develop and use the daring technological innovation of artificial satellites for IGY experiments was a late addition to the program and like one or two other areas with similar dramatic appeal, such as Antarctic exploration, it became a focus for informal but intense rivalry between the Soviet Union and the United States, as the only two nations with the technological capacity to enter such a "race." At first glance, the rival IGY satellite programs recall the old dictum that nations which challenge one another over prestige will usually need to "cooperate in order to compete." However, the precise extent and nature of the IGY cooperation in this area has been little studied to date.

THE U.S. PROPOSAL

The use of artificial satellites for IGY experiments was proposed unilaterally by the U.S. IGY Committee in October 1954. At that point the Soviet Union had not yet officially joined the Year, which had been established as an international scientific committee, by the International Council of Scientific Unions (ICSU), only two years previously. An informal Soviet delegation was however present at the successive Assemblies, in Rome, of the International Union of Geodesy and Geophysics (IUGG) and the Comité Scientifique de l'AGI (CSAGI), the annual coordinating meeting for the IGY, at which the satellites idea was proposed and then, in principle, accepted. And the USSR Academy of Sciences announced in Rome that it had decided to join the Year, and was about to form the requisite national IGY committee through which to do so. For its part, the U.S. IGY Committee was now obliged to "take back" the CSAGI satellite resolution, which it had written itself, for further domestic consideration. It would be necessary, first, to determine whether the idea was technically feasible and scientifically desirable, within the short timeframe of the IGY, and second, in the event of a positive determination on this point, to persuade the economically conservative and militarily preoccupied Eisenhower administration to divert the necessary resources from its urgent missile program. This process took several months, but finally, on June 29, 1955, U.S. IGY scientists took part in the official White House announcement of the world's first satellite program, appropriately but unfortunately designated "*Vanguard*" by its planners.

For the next fourteen months there was only one, very public, national program for a scientific IGY satellite, that of the United States.

THE SOVIET RESPONSE BEFORE *SPUTNIK 1*[1]

The USSR Academy of Sciences was obliged to begin its belated participation in the IGY by forming a national committee, reviewing the entire IGY program as formulated to that date, determining what additional activities it might like to suggest, and starting to prepare its scientific teams, instruments and experimental stations.[2] In March 1955, the Soviet Academy informed the Secretary of CSAGI, Marcel Nicolet, that its Committee President would be Academician I. P. Bardin,[3] and in April, that its representatives on the CSAGI committee would be Academicians V. V. Beloussov and N. V. Pushkov. On April 12, *Pravda* told the Soviet public, in general terms, that there would be a Soviet IGY program. Further details were anxiously awaited, and wherever possible requested, by Western scientists, including the overall president of the IGY, Sydney Chapman.[4] But the record of the informal Rome conversations indicates that the Soviet delegation could reasonably have formed the impression that national reports were to be prepared for presentation at the next annual CSAGI Assembly, scheduled for Brussels in September 1955, and not before. This was, in any case, the procedure which they followed, to the extent of bringing multiple copies of their report to Brussels, rather than forwarding it to Nicolet for circulation in advance.[5]

In a separate development, the official Soviet media announced in April 1955 the formation of an "Interdepartmental Commission on Interplanetary Travel." The U.S. committee prepared a major presentation of its recently announced satellite program for Brussels, and hoped for news of Soviet plans, despite the fact that no announcement linking the new Commission with the Soviet IGY program had yet been made from Moscow. The Americans were disappointed when no leading Soviet experts in the rocket sciences even came to Brussels, and the single Soviet delegate to attend the relevant working group adopted an entirely passive role.[6]

The CSAGI working groups were responsible for coordinating the global scientific effort within the IGY disciplines, both at the annual international meeting and through subordinate meetings and correspondence, insofar as was possible for busy working scientists with many other current commitments, during the rest of the time. An overall "reporter" for each discipline was tasked with the preparation of a manual or group of manuals, for distribution to the participating stations, which should embody the detailed experimental plans and guidelines drawn up in the preparatory consultations. In the satellites part of the Rockets and Satellites discipline, the preparation of instruments, launch vehicles and tracking stations was of course fraught with the additional burden of novelty. The late arrival of the Soviet Union in the IGY, and its continued silence on the question of satellite experiments, left the U.S. scientists frustratingly unable to conduct the necessary dialogue. In February 1956 further details of the U.S. program were circulated to all national committees, and in the same month the first informal indications were received that the Soviet satellite program would include scientific experiments of the sort appropriate for the IGY.[7] Lloyd Berkner, Vice-President of the IGY and reporter for *Rockets and Satellites*, proposed that a full two days of the next CSAGI Assembly, scheduled for September in Barcelona, should be given over to a symposium on the satellite program.[8] Meanwhile, he continued to urge that the Soviet Union be pressed to join its satellite program with the IGY.[9] The Soviet IGY committee, however, remained uncommunicative in general and completely silent on the matter of satellites. No response having been received to the proposal for a two-day symposium, this was curtailed to a single day. The only information given about Soviet plans was a terse announcement from Bardin, to the effect that there would indeed be a Soviet IGY satellite program and that information about proposed launch schedules and experiments would be supplied in due course. One resolution passed by the rockets and satellites working group at Barcelona, and not contested by the Soviet delegates, recommended "that for all IGY satellites the radio systems employed for tracking and telemetering be compatible with those which have been announced at the current CSAGI meeting in order that the same ground-based receiving equipment can be used throughout."[10] The "announced" systems were of course those planned for the U.S. *Vanguard* program, using a frequency of 108 megahertz, which had been described in U.S. documentation distributed before the meeting.[11] Other resolutions at Barcelona called for "countries having satellite programs" [i.e. the Soviet Union] to supply information about their tracking equipment and launch schedules. However, it was to be a further nine months before any such information was officially supplied by the Soviet IGY committee.

The problem for IGY planners was exacerbated, rather than eased, by the amount of unverified information about the likely shape of the Soviet satellite program that had appeared by late 1956, including the disturbing claim that Soviet scientists might attempt to place a satellite in orbit during 1956 (which quite apart from any prestige considerations would have implied a general disregard for the IGY framework).[12] This informal material was compiled and assessed by various bodies to the best of their ability.[13] But although some of the estimates derived from it were later proved to have been remarkably accurate, it could not be used or even referred to within the CSAGI structures, because none of it had been supplied or endorsed by the Soviet IGY committee.

On March 6, 1957, Radio Moscow confirmed that the size of the first Soviet satellite would be around 50 kilograms, as had already been suggested by unofficial sources. But it was only in June, less than a month before the start of the IGY, that Soviet scientists began to provide any detailed information about their rocket and satellite programs. A seven-page document was sent to Berkner, giving brief indications of the types of experiments to be carried on sounding rockets and satellites, but with no technical information about the measuring instruments or telemetry. Three "zones" for rocket launches were listed as "Franz Joseph Land," "the Antarctic, mainly in the area of Mirny," and "middle latitudes of the SSR." The site for satellite launchings was not specified.[14] In the same month the Soviet journal *Radio* published two articles giving detailed explanations of how the satellite's radio telemetry could usefully be observed by amateurs on frequencies of about 20 and 40 megahertz; a further two articles appeared in the same journal in July, another in August, and another in September.[15] The Central Amateur Shortwave Radio Station of the Soviet Union went on to broadcast specimens of the planned satellite signals "several times a week" during August and September, without any notice being taken by the outside world.[16]

Berkner's first draft proposal on the interchange of rocket and satellite data, intended for a future manual, had been written, with no Soviet input, in December 1956.[17] In July 1957 it was rewritten, without any significant changes, by Admiral Day, the IGY's Coordinator, to form the relevant chapter for the first edition of *The CSAGI Guide to IGY World Data Centres*.[18] At that stage, the only material taken from the Bardin document was the designation of rocket launching zones.

Expressing his dissatisfaction with the lack of detail in the Bardin document, on July 23, 1957, the British space scientist Harrie Massey drew particular attention to the absence of any specification of a telemetering frequency.[19] Three days later, however, Massey attended a meeting with a party of Soviet rocket scientists who were on a two weeks' visit to their British counterparts. On this occasion A. M. Kasatkin gave a detailed description of one of the principal Soviet meteorological sounding rockets, including its telemetering frequency of 22 megahertz.[20] Then on August 16, Bardin wrote again to Berkner, this time giving the exact frequencies to be used in Soviet satellite telemetry. A note on the copy of this letter in Chapman's files states that the U.S. IGY office forwarded the original to Berkner at Boulder, where he would have been attending the Assembly of the International Radio Science Union (URSI).[21] However, the letter appears to have miscarried, and its importance was missed by the U.S. IGY staff, with the result that U.S. scientists preparing their country's satellite-tracking stations were left ignorant of the Soviet decision on frequencies until shortly before the launch of *Sputnik 1* on October 4, 1957.

The Bardin letter had not been copied to Nicolet at the international IGY office in Brussels, where its significance would not have been overlooked. But, either in response to a recent pressing inquiry about the telemetry frequency from the British, or else by accident, a copy was sent to the Royal Society.[22] Although the British did not set the probability of a Soviet satellite high enough to make any advance preparations for tracking it,[23] one British scientist put his mind to devising a method for studying the distribution of ionization in the upper atmosphere by comparative observation of the signals on the two frequencies, and presented this at the Washington meeting, three days before the launch of *Sputnik 1*.[24]

The final source of information of the Soviet satellite program, prior to *Sputnik 1*, was the September issue of the Soviet Academy's journal *Achievements of Physical Science*, a special two-volume issue devoted exclusively to articles about past or planned experiments to be carried on rockets and satellites.

It is worth mentioning here that the contrast which is usually drawn between American openness and Soviet secrecy about their IGY satellite plans, prior to October 1957, has probably been overdrawn in the literature. As late as mid-June of 1957 W. T. Blackband, a senior physicist with the Royal Aircraft Establishment, was still urging that both the United States and the Soviet Union should be pressed to provide full details of their telemetry systems. He continued: "Of course both have agreed to standardize, but it has been so hard for us to learn details of American plans that it is likely that the Russians do not know much, and, knowing little they will go their own way."[25]

PROBLEMS OF THE SOVIET IGY COMMITTEE

By comparison with the wealth of advance information that was distributed about the U.S. *Vanguard* satellite program, that provided by the Soviet IGY committee about what was nominally "their" satellite program can fairly be described as "too little too late." But several considerations are worth mentioning at this point. First, nations joining in the IGY were not thereby committing themselves to adopt U.S. values or policies in the conduct of their scientific programs. In the Antarctic, for example, nations maintaining territorial claims, such as Argentina, Britain, or Chile, frequently pointed out that their acceptance of international scientific activities within "their" sectors during the Year should not be regarded as establishing any sort of political precedent.[26] Next, as has been made clear, no procedures for the conduct of satellite experiments, including the nature and extent of information to be released in advance, had been mutually discussed, elaborated and agreed, under IGY auspices, prior to the Washington conference which convened only a few days before *Sputnik 1*. Doubtless the lack of such a meeting was largely due to the nonattendance of Soviet rocket scientists at Brussels and Barcelona, but it seems certain that the participation of such personnel in IGY meetings was not something controlled by the Soviet IGY committee. Even in the United States, a satellite program which had been initiated, and in its early phases planned, entirely by the national IGY committee showed signs of reverting to more direct government control after the national furor occasioned by the first two Sputniks. In the Soviet case, it seems doubtful that the Academy of Sciences ever had much say in any aspect of the first Sputniks, other than their scientific instrumentation.

The USSR Academy of Sciences may have been a *de facto* department of the Soviet government, but the IGY committee formed under its auspices probably carried little political weight. Only one member of its inner circle, Pushkov, was actually a member of the Communist party.[27] Bardin's two communications to Berkner, in June and August 1957, contain clear and probably deliberate indications of his helplessness in matters pertaining to the Soviet satellite. The first points out that he is merely *forwarding* the document, adds that he hopes it is what Berkner expected, and goes on to apologize for sending it so late. In the second he once again emphasizes that the frequency information comes from "the specialists" and he is merely passing it on.[28]

It is also reasonable to assume that the Soviet decision to use the prototype R-7 military launcher for their satellite program placed them under severe constraints in respect of security, of a kind also experienced, though to a lesser degree, in the United States. Influential Soviet officials may well have had some inkling of Lloyd Berkner's excellent connections with the political leadership of the United States, as well as its intelligence community.[29] If so, they would have mistrusted the repeated attempts at obtaining information closely related to their ICBM program that were channeled through the IGY, such as Berkner's proposal for a two-day conference at Barcelona.[30] Such fears, if they existed, may or may not have been justified. There were always excellent scientific reasons for coordinating the two countries' satellite programs as early as possible. But it is certainly noticeable that nothing like the same pressure for advance information of Soviet plans was generated by the U.S. IGY leadership in any of the thirteen other subject areas of the IGY.

In June 1957 the political leadership of the Soviet Union was rocked by a crisis in which Khrushchev, who had strongly backed the development of a satellite-launching potential within the Soviet rocket program, was hard put to retain political power. While the crisis and its resolution in Khrushchev's favor came after Bardin had sent the seven-page description of the Soviet satellite program to Berkner, and after the first *Radio* articles had gone to press, it may be presumed that if matters had gone the other way, and Khrushchev's rivals, who were largely opposed to the diversion of military resources for a satellite program, had succeeded in overthrowing him, then even the small amount of information which began to flow that summer would have dried up.[31] If the Soviet satellite program had been closed down completely, of course, we would not be sitting here today. Conversely, by rewarding the Soviet leader's gamble with a political triumph, the first two successful Sputniks probably made it easier for some Soviet scientists to develop interactions with their Western colleagues.

In any event, the Soviet reluctance to share advance information about their satellite plans was not immediately harmful to their scientific or political interests. The tracking stations of other IGY nations were able to be adapted to the required frequencies and orientations within days, sometimes hours, of the launching of *Sputnik 1*. It can be argued, however, that the lack of a cooperative international network of tracking stations, such as might have been built up by following a more open policy, cost the Soviet scientists the prestige of one of the most impressive discoveries in the early years of space science, the Van Allen radiation belts, which could not be clearly detected from the partial orbital and experimental data which were all that they could obtain with their own resources.[32]

TO COOPERATE, OR NOT TO COOPERATE?—WRITING THE BOOK

The CSAGI Conference on Rockets and Satellites, held in Washington from September 30 to October 5, 1957, was in effect the first real joint discussion between Russian and American scientists about the arrangements for satellite experiments within the IGY. While minutes of some of its working groups survive, no full record of the conference proceedings appears to be extant. There seems to have been little disagreement, at this stage, about the arrangements for the exchange of experimental results. What proved controversial was the nature and extent of the "operational data" to be made available by those launching a satellite for observers in other countries. The Americans at first wanted flash announcements of all launchings to be broadcast within one hour, and precise details of launch sites and orbital inclination angles to be given. This was modified at Washington to allow a two-hour notice period, and for successful launchings only. Reference to the launch site was deleted, and the phrase "Complete orbital elements at an instant near the time of launching" was substituted, presumably because such data could be packaged in such a way as to protect the location of the launch site or at least, appear to do so, for reasons of "face."

The issue of telemetry frequencies was a subject of intense discussion. Kasatkin defended his country's choice on the grounds of its scientific value for the study of ionospheric refraction and suggested that the tracking precision, for which the Americans had opted with their higher frequency, might be attainable through statistical treatment of a large number of observations of a lower frequency beacon. It might also be practicable to include a tracking signal at the U.S. frequency in the "next series" of Soviet satellites. In reply to a cross-examination from Porter and others about the design of the Soviet transmitter, the polarization of its signal, and the design of tracking station antennas, he stated that the Soviet delegation "had not been prepared to give detailed information of this nature at this conference."[33] Western participants repeatedly underlined the undeniable inference that, if the Soviet scientists wanted observational support for their satellite program from scientific institutions and amateurs around the world, as they appeared sincerely to do, then they would have to provide considerably more information on some of its technical aspects than had been given so far. On the other hand, Kasatkin's reference to the articles in *Radio* may have come as a surprise to most Western delegates. The telemetry issue was resolved by endorsing both the American and the Russian choice, and the potential value of the Soviet 40 megahertz frequency for ionosphere measurements was also accepted.[34] The draft *Guide to IGY World Data Centres* had proposed that orbital predictions (ephemerides) be distributed to observing stations via the IGY's global telecommunications network, known as Agiwarn.[35] This was accepted, but with a rider that "additions" to the text might be needed after further study. In general, the conference recognized that much remained to be done in building effective global cooperation for the satellites portion of the IGY.

The urgency of these tasks was underlined by the news, on the evening of October 4, 1957, of the launching of *Sputnik 1*. At the last session of the conference on the following day, Blagonravov volunteered a few more details of the spacecraft, namely its size, weight and the expected battery life for its transmitter. Chapman closed the meeting with a speech congratulating the Soviet scientists, but drawing attention to the indirect way in which CSAGI had been informed of the launch (through the news media) and contrasting the openness of the American program with the "silence" of the Russian effort. He ended with a plea "that our resolutions concerning timely announcements and adequate information will be fully observed."[36]

Blagonravov had also remarked that *Sputnik 1* was not a properly instrumented satellite of the kind which Soviet scientists were planning for the IGY, but merely a preliminary test vehicle.[37] This distinction, which had also been floated within the *Vanguard* program, raised concerns that *Sputnik 1* might never be registered and reported as an IGY experiment, despite the worldwide publicity which it had generated for the Year. After a ten-day interval, during which the Soviet IGY committee showed little sign of conforming with Chapman's recommendations, CSAGI coordinator Day cabled Moscow to request information about the satellite's orbit for distribution on Agiwarn, "while understanding from press reports that satellite now orbiting is not part of IGY program." Yu. D. Boulanger, one of the vice presidents of the Soviet committee, replied that, on the contrary, *Sputnik 1* was "launched ... in accordance with the Soviet IGY program."[38] The U.S. committee responded by distributing its own ephemerides for the two objects then in orbit, through the Agiwarn system.[39]

There followed a period of acute misunderstanding and dis-agreement. Replying to a further telegram from Day, Boulanger listed 26 foreign observing stations to which his committee was already supplying ephemerides for *Sputnik 1*. This was not yet being done through Agiwarn, but Boulanger stated that the IGY network would be used from the beginning of November.[40] At about the same time, the deputy president of the USSR Academy's Astronomical Council, A. G. Massevitch, wrote to Fred Whipple, director of the Smithsonian Astrophysical Observatory (SAO), who had proposed the first international nomenclature for satellites and whose observatory was coordinating the Moonwatch program for optical satellite observations by voluntary groups. The purpose of her letter was to inform Whipple, "according to our agreement with Mr. L. Campbell in Barcelona," of the telegraphic codes which the Soviet committee proposed to use when transmitting ephemerides, and which it desired observing stations to use when sending data to Moscow. She added that stations in the United States, Japan, and South America (areas not yet represented in the Soviet distribution list) would also be supplied with predictions, about 48 hours in advance of the relevant transmit, as soon as their precise geographical coordinates and IGY station numbers were sent to Moscow.[41]

Seemingly unaware of these favorable, if tardy, developments, Berkner wrote to Chapman on November 7 with his considered opinion as to the degree of compliance and non-compliance of the Soviet committee with the existing IGY agreements about coopera-tion in respect of satellites. The Soviet committee, he thought, had probably met the requirements of the IGY in respect of advance notice and early publication of orbital data, especially in the case of *Sputnik* 2, which had been launched a few days earlier on November 3. But little of that information, such as it was, had been sent through IGY channels. He accepted, also, that it would be unreasonable to expect the results of onboard experiments to be communicated internationally for some time. However, "we should reasonably expect ... under the CSAGI agreements to have early and frequent information from the Soviet Union on its orbital observations and on its telemetry code in order that scientific information from the satellite could be observed and interpreted by observers of other nations. In my opinion, failure to provide this information promptly seems inexcusable on their part, and with this I am sure you must agree." He concluded by expressing strong support for an attempt which was currently being mounted by Day to secure the attendance of Soviet scientists at a meeting on satellites to be held at the Royal Society at the end of November, which could be extended for a second day for the express purpose of hammering out a clear arrangement on the exchange of satellite information, by which the Soviet committee might thereafter feel itself to be bound.[42] Chapman responded on November 20. While he agreed with much that Berkner had said, he disagreed strongly with his interpretation of the IGY rules on the matter of telemetry codes. As Chapman pointed out, the ability of foreign observers to record but not interpret the telemetry from the Sputniks was no different in principle to that which "was expected to arise from the U.S. satellites"—perhaps a tactless remark, in view of the fact that none had yet been launched.[43] Nicolet had already sent Chapman his own opinion to the same effect, namely that Berkner was reading far too much into the IGY rules. In Nicolet's typically brusque opinion, the rules had been written by the Americans and were now being followed by the Russians.[44]

Apparently as unaware as Day of what Massevitch was doing, V. A. Troitskaya, the secretary of the Soviet IGY committee, cabled Day on November 14 to the effect that precise orbital data could not be supplied "as satellites still move about in space." Such information could only be supplied at some indefinite future date, "after final reduction of observation results."[45] A week later, Beloussov informed Day that the Soviet committee would not now be sending a representative to the second part of the Royal Society meeting, which was therefore canceled.[46]

A few days later the cooperation pendulum swung yet again, when Bardin sent Day a detailed proposal for the transmission of ephemerides from the Moscow Research Institute of Terrestrial Magnetism, Ionosphere, and Radio-Wave Propagation (NIZMIR) to observers around the world, and the transmission of data on Soviet satellites from foreign observers to Moscow. Though Day was skeptical at first, he soon realized that the new Soviet proposal would have to be taken seriously.[47] In December he replied to Bardin with a letter agreeing that the interchange of visual observations, both between observers and between the two IGY World Data Centers (WDCs), one in the Soviet Union and the other in the United States, and between those Centers themselves, to which a third was about to be added in Britain, was now a priority. The second type of data exchange had not yet begun, but should do so immediately.[48]

Day first tried, unsatisfactorily, to incorporate the new Soviet communication codes into the existing text of the *Guide*.[49] Then in January he went to Moscow for a full discussion with the Soviet committee on all aspects of data exchange. The meeting was conducted amicably, despite the continuing dispute between the U.S. and Soviet academies over the latter's claim that parts of the upper stage (Alpha-1) of the *Sputnik 1* carrier rocket had crashed on U.S. territory after reentry from orbit and should therefore be returned to the Soviet authorities.[50] At this meeting the chairman of the Soviet committee's working group on rockets and satellites, E. K. Fedorov, presented Day with a complete redraft of the *Guide* chapter which proposed several important changes. The timeframe for a launch announcement and initial details of orbit was relaxed still further to "the day of launching." Orbital predictions should be supplied only "to the institutions participating in observations," in other words, in return for data supplied by them. Preliminary accounts of satellite observations and a description of the onboard experiments should be published several weeks after a launch. Full scientific reports should be published within a year. No explicit provision was made for observers to send their observations or recordings of one country's satellite to a WDC located in another country, although this was already being done. In a subsequent letter repeating the Soviet proposals, Fedorov categorically rejected the possibility of sharing the telemetry codes, and hence the unreduced experimental data, with other scientists: "All investigation work and the whole analysis of observational data should be done by the launching country." The letter underlined the impression already received by Day that the Soviet proposals were not intended for further discussion.[51]

Undismayed, Day presented a further revision of the *Guide*, now incorporating what he saw as "the two programme principle" insisted on by the Soviet committee, to a meeting of the CSAGI Bureau in Brussels on February 27. This document was rejected by Beloussov in March, with a statement that he could not accept anything beyond the Fedorov draft. A series of amendments proposed by the U.S. committee, however, were incorporated into the February draft. Berkner, as the IGY discipline reporter, then authorized this version both for distribution by Day and for inclusion in the *Manual on Rockets and Satellites*, which was to be published in July in time for the Moscow CSAGI Assembly. At Berkner's suggestion, a statement was added to this section of the *Guide*, explaining that, because of the novelty of satellite experiments, certain aspects of the data exchange could not yet be finalized, but that meanwhile Berkner's own proposals on the exchange of precision observations, and on depositing the reduced satellite data at the WDCs, were being appended as a basis for discussion at the forthcoming CSAGI conference in Moscow.[52]

At the CSAGI Bureau meeting in February, Berkner reported that the three IGY satellites launched to date, two Soviet and one American, had yielded "very substantial scientific information." There had been "some delay and problems" with the exchange of orbital observations of *Sputnik 1*, caused by "understandable failure to foresee the exact problems of encoding and communicating the necessary information," but a better system had now been devised "through close collaboration with the nations concerned." "Certain necessary revisions in the World Data Guide" were being worked out with Day and "Certainly future problems remain to be solved as might be expected in dealing with such a new form of scientific activity."[53]

On May 20, Fedorov wrote again to Day, flatly rejecting the revised version of the *Guide* that had been circulated in April.

Perhaps because of the obscurity of the circumlocution "primary prediction authority," which Day had started using to refer to the national committee that a satellite belonged to, Fedorov mistakenly supposed that there were now no provisions requiring observers to send their data "to the country that launched the satellite." He also rejected out of hand any obligation to pass "to certain scientific institutions ... the unprocessed data of registration of different measurements carried out on rockets and satellites."[54] Since Berkner's suggestion, appended to the April version of the *Guide*, had carefully explained that the data in question should first "be reduced and corrected as may be necessary to put them in the form of physically significant parameters useful for scientific analysis," it seems likely that Fedorov had recast this as an unreasonable request for raw data, simply in order to reject it. Drawing attention to the likelihood of linguistic confusions, Day wrote to Berkner that the completion of the *Guide* would now have to be left to the CSAGI Assembly, due to open in Moscow on July 29.[55] More than two-thirds of the way through its course, the IGY was still without an agreed structure for scientific cooperation on satellites.

THE MOSCOW ASSEMBLY

At the CSAGI Assembly in Moscow, discussions about scientific cooperation on satellites were distracted both by positive and by negative developments. On the positive side, scientists were naturally more interested in sharing and discussing some of the earliest results of satellite experiments, than in discussing a set of bureaucratic rules about future exchanges of data. On the negative side, the conference was racked with severe political disagreements between Western delegations and the Soviet hosts, which placed considerable tensions on all participants, and which sometimes meant that experienced delegates were unable to attend a formal session, because they were caught up in something going on elsewhere behind closed doors.[56]

The working group on rockets and satellites was chaired by Homer Newell, convenor of the U.S. committee's panel on IGY rocket experiments, in his capacity as Berkner's alternate. At the end of the meeting, Newell was obliged to report that it had been impossible for American and Russian scientists to reach agreement on the outstanding issues of data exchange in respect of satellites. In his view, the resolutions passed by the working group "are so general that they do not guarantee an automatic and adequate flow of the kinds of data needed to make it possible for researchers in countries other than launching countries to conduct research on artificial satellites and the radio signals from them."[57] In a subsequent confidential report he summarized these disagreements as arising from "a firm party line: ... don't give away any basic rocket and satellite data ... . They refused to: (a) provide orbital elements for Soviet satellites during the satellites' lifetimes; (b) provide precision radio tracking data for satellites; (c) agree to an automatic dispatch of basic data to the World Data Centers."[58] Berkner's suggestions for the *Guide*, in particular, seemed "to upset them."[59]

Another type of data which Soviet delegates at the Moscow Assembly refused to supply were the physical characteristics of rocket stages placed in orbit, on the grounds that studying such objects had "no scientific value." When the Americans proposed to cease sending orbital predictions for such objects, however, Massevitch reportedly contradicted herself by stating that after all, they were "of some scientific value." The line on withholding information about operational aspects of the Soviet satellites was firmly held, and Newell formed the impression that "the day's work was being reviewed each evening, and specific orders issued with regard to the position to take up on the following day."[60]

Other U.S. delegates also noted "a rather sharply drawn line between the research effort and the 'hardware' by means of which the research was accomplished." Logical contortions, or simple silence, were resorted to whenever the discussions in Moscow showed signs of crossing this line.[61] Discussing this distinction, Newell emphasized that "basic data" were usually essential to establish the methodological credentials of a piece of research. His impression was that such material was provided in some of the papers on upper atmosphere rocket experiments, but was rigorously withheld in respect of satellite experiments. American attempts to acquire operational data about the Russian satellite program beyond the confines of the conference, such as from the Sputnik Exhibition which was staged in Moscow to coincide with the CSAGI and IAU Assemblies, were also frustrated.[62]

A careful reading of the actual amendments to the *Guide* which were agreed by the Moscow working group, however, suggests that Newell's reproaches, though not without grounds, may have been overdrawn. On the third point, section 17 of the *Guide*, the Soviet delegates did indeed refuse to send such data automatically. But on the first and second points, sections 8, 9 and 13 in the Moscow version of the *Guide* committed every launching authority to give the approximate orbital characteristics of its satellite within 24 hours of the launch, to relay orbital predictions to observing stations, and to lodge reduced results of precision observations of the orbit with World Data Centers within six months. The main effect of the Moscow amendments was to delete all references to "complete" data, and to substitute such phrases as "results ... necessary for processing scientific experiments," or "the observational scientific data concerned" with a particular experiment.[63] The Soviet scientists were not, then, refusing to hand over *any* basic data from their satellite observations and experiments. But they were certainly refusing to hand over *all* their data and reserving the right to choose what they would hand over and to whom. It is relevant to note that the U.S. State Department may not have given its final approval to Berkner's proposal for exchanging the complete precise orbital data either, prior to the Moscow Assembly.[64]

The working group meetings in Moscow appear to have ignored the slightly variant version of the *Guide* chapter given in the official *Manual on Rockets and Satellites*. The Moscow revision of Day's April draft thus became the final text of the *Guide*. It was circulated in October 1958, shortly before the end of the Year, and published in the *Annals of the IGY* in 1959.[65] The U.S. and Soviet IGY committees both agreed to an *ad hoc* extension of the IGY for the calendar year of 1959, a period known as the International Geophysical Cooperation (IGC). But there was no meeting of the full CSAGI Assembly or an equivalent successor body during this period, nor of its working group on rockets and satellites. The ICSU Assembly of October 1958 in Washington had created a new Committee on Space Research (COSPAR), which was intended to take over the functions of the IGY working group. However, Soviet objections to its constitution meant that it was unable to contribute to the development of the principles of international cooperation on space science until the early 1960s.

Chapter XI of the *CSAGI Guide*, including the important disagreements and reservations entered by the Soviet Union and the United States, therefore remained the last official word on this subject for some time.

EXCHANGING DATA

Whatever may have been the contents and status, at any particular date, of the arrangements for cooperation on scientific satellites proposed by the *CSAGI Guide*, the overt purpose of including a rockets and satellites discipline within the IGY, as with all its other disciplines, was to promote the exchange of relevant geophysical data and research findings between the participating committees. The effectiveness of this part of the IGY can therefore be assessed, at the cost of a little repetition, from the extent to which such exchanges actually took place. Unfortunately most of the primary evidence on this aspect of the IGY derives from internal documents or public exchanges in which either the United States or the Soviet Union, but most often the United States, tended to assume that what it wanted to happen was what had been positively agreed to by the other party, instead of listening carefully to what the other party had actually said it would or would not do.

The accusations of bad faith leveled against the Soviet IGY committee by Odishaw and by the director of the U.S. National Science Foundation, Alan Waterman, within less than three weeks of the opening of the space age on October 4, 1957, were not a good omen for how such matters would be handled over the next two years.[66] One typical U.S. document, written around November 18, 1957, contrasted the detailed descriptions of *Vanguard* instrumentation, telemetry and experimental plans which had been circulated during 1956 and 1957, with the absence of such information from Bardin's document of June 1957. It went on to characterize Soviet information on the first two Sputniks as "belated and somewhat incomplete" and "ineffective," by comparison with the "detailed orbital data" on *Sputnik 1* which had been sent by the U.S. committee to all the others, and specifically to the Soviet Academy on October 27, 1957. Although it noted that the proposals on data exchange in the *CSAGI Guide* had not been formally adopted by the Washington conference, the document tended throughout to treat them as a set of mutually agreed "rules" which the Soviet Union was infringing.[67]

The early, unilateral versions of the *Guide* had included a recommendation, later dropped, that a launching authority should announce the "approximate period of launchings" in advance. With the exception of *Sputnik 1*, both the United States and the Soviet Union were generally held to have complied with this, usually through press statements or news broadcasts.[68] In respect of prompt announcements of a successful launch, both sides also complied, except that the Soviet authorities used the official IGY Agiwarn channels only for one satellite, *Sputnik 3*, on the day after its launching, and for two lunar probes launched in 1959.

The hasty condemnations of their Russian colleagues by a few American scientists are all the more puzzling, in retrospect, in the light of early moves by the Soviet committee to establish a modest level of cooperation. As noted above, *Sputnik 1* was confirmed as an IGY experiment on October 18. The delay seems to have been a matter of confusion rather than deliberate obstruction. Then on October 31, Boulanger sent Day the list of 26 stations outside the Soviet Union to which predictions were already being sent, adding that predictions would be circulated on the Agiwarn network from the beginning of November.[69] Meanwhile Massevitch had already written directly to Whipple on October 26, arranging to send predictions to the SAO, and promising to send them to observatories in the United States, South America, and Japan, as soon as the relevant coordinates were provided.[70] Predictions for *Sputnik 2* began to be received at the SAO on November 5.[71] Thereafter something seems to have gone wrong, perhaps as a result of Odishaw's bilateral intervention into a process that was being fostered with some success by Day from Brussels, in his role as IGY Coordinator, or perhaps because the Soviet committee was rethinking the proposed arrangements. For whatever reason, the Soviet committee never did distribute its predictions through Agiwarn, and did not even copy to the CSAGI office those which were already being distributed to a limited number of stations. There also seems to have been an interruption in the supply of predictions to the SAO.[72]

The Soviet committee continued to show an interest in *receiving* observational data on their own satellites from abroad. At the end of

November, Bardin sent Day the proposed codes for such messages, and Massevitch sent Shapley a list of 68 satellite-observing stations in the Soviet Union, complete with coordinates.[73] But that this was not expected to be a one-sided arrangement is shown, first, by the fact that Bardin included outward codes for the transmission of ephemerides to foreign observers of the Sputniks, and second, by the explicit Soviet proposal, in January, to amend the *Guide* so as to underline the reciprocity of such exchanges.

In December the U.S. committee responded by sending a list of visual (Moonwatch) stations to Massevitch. This was followed in March with a combined list of over 100 stations, which included not only U.S. stations but also those established by other national IGY committees in cooperation with the U.S. committee and the SAO.[74] But the Soviet committee apparently made few if any changes to its existing distribution of orbital predictions.[75] By July 1959 the flow of predictions to the SAO had dried up, but visual and photographic observations of Soviet satellites from Soviet stations were still arriving at a rate of about two messages a week.[76] In May 1959 an *Izvestiya* article stated that over 70,000 "tables of observations," from all IGY fields, had been supplied to foreign investigators from the World Data Center B in Moscow.[77]

While the Soviet committee generally declined to publish the codes for interpreting telemetry from the early Sputniks, the code for the cosmic ray detector on *Sputnik 3* was given to an American scientist at the Moscow Assembly in August 1958, and was regularly used by foreign scientists thereafter.[78]

In the sphere of scientific results, the Soviet committee's *Preliminary Report* on *Sputniks 1* and 2 was published at the end of January 1958, much as required by the provisional IGY arrangements. The first report on work with meteorological rockets in the Arctic and Antarctic followed in March.[79] A preliminary scientific report on the first two Sputniks appeared in August 1958 and a final one in March 1959. Although this was after the one-year deadline specified for the IGY, the stipulation could not be said to have been seriously breached, in view of the numerous scientific papers based on observations and measurements with the first two Sputniks which had already been presented at the Moscow Assembly, some of which were also published in Russian journals during 1958.[80] The *Sputnik*

3 experiments were also covered in some of this literature, and studies of the orbit of its upper stage (1958 1) were published from December 1958. A final scientific report on *Sputnik 3* appeared in May 1959, carefully meeting the one-year deadline.[81]

Just as the U.S. committee had given a far more detailed account of its satellite-launching plans in advance, including technical details of the launch vehicle which were never matched by any information from the Soviet Union, so too the data and reports on satellite observations and experiments which were circulated by the U.S. committee, on both Russian and American satellites, were both prompt and abundant. The set of documents drawn up by the U.S. committee in 1959 omits some of the earliest Soviet information, and includes on the American side of the expression several items which hardly amount to satellite data or results, such as bibliographies. But it should be added, in fairness, that if the lists were to be completed with the advance descriptions of satellite experiments and equipment supplied by either side, the balance would certainly tip still further in the Americans' favor. Nor do the summary entries convey the real extent of the American cooperation; for example, on every working day between May 16, 1958, and July 16, 1959, the SAO sent about eight to ten observations of all satellites to the Agiwarn center at Fort Belvoir for onward distribution to the global IGY network, including of course the Soviet Union.[82] (The Fort Belvoir facility may not have forwarded them daily, but it certainly did so several times a week.) One noticeable feature is the regular use of the Agiwarn network, for launch announcements, that was made by the U.S. committee, in accordance with the *Guide*. Another point worth making is that some of the fully processed scientific observations took many months to complete and publish, just as they did in the Soviet Union.

The U.S. satellites made more use of interrogated and less of continuous telemetry than the early Sputniks, with the result that Soviet tracking stations were less able to record U.S. telemetry and therefore less interested in acquiring the codes for processing it, even though they were made available.[83] From Berkner's original draft onwards, all versions of the *CSAGI Guide* included a clause which went unchallenged, and which read in its final version as follows: "18. Raw data whether film records of optical observations, primary records of radio observations and telemetered signals are not suitable for exchange but it is expected that they will be available for consultation through WDCs when so requested."[84] The continuous radio telemetry of *Sputnik 1* was broadcast for 23 days, that of *Sputnik 2* for seven, and that of *Sputnik 3* for 691, thanks to its use of solar-powered batteries. Very large amounts of the raw data were thus able to be recorded by tracking stations outside the Soviet Union. One of the bitterest disputes between the U.S. and Soviet committees, in respect of this part of the IGY, began with a Soviet claim that such recordings "are not forwarded to the Soviet Union immediately," "in spite of the existing thesis of the CSAGI Guide for WDC."[85] References in the document to the U.S. Atlas communications experiment in December 1958 and to unpublished information from *Explorer 4*, namely its Project Argus data, suggest that it was probably intended to make anti-American propaganda at a time when the disputes in COSPAR and the U.N. Space Committee (COPUOS) were extremely severe. There was however a genuine issue at stake, and one which had not been resolved by the IGY discussions, because they had tended to assume that the raw data from a satellite would be in the hands of its launching authority, and because they had overlooked the fact that the WDCs would be expected to harmonize their holdings by copying material between themselves.[86] The Soviet committee could not in fact point to any provision within the IGY arrangements which required the transfer, let alone the immediate transfer, of such material, but their moral case was superficially strong. The propaganda problem was, however, that such recordings had already been handed over by U.S. scientists to their Soviet colleagues, as the Soviet note itself remarked.

By May 1958 Donald Menzel, director of the Harvard Observatory, had already been in correspondence with Fedorov about the handling of recordings of signals from the Sputniks. He had also already sent "some tapes" directly to the Soviet Union. Under Odishaw's guidance, the U.S. IGY office told Menzel that they were "interested in getting these tapes for transmittal to the USSR as a part of the IGY program," and asked how much material might be involved. Menzel estimated that at that point there were between 50 and 100 tapes in the hands of various groups.[87] Odishaw wrote to Beloussov offering to send tapes of Sputnik signals, and asking for tapes of *Explorers 1* and *3*. Beloussov replied thanking him for the offer and regretting that the Soviet committee had no tapes of the Explorers. At the end of July, six tapes of

Sputnik 1 telemetry were delivered to the Soviet committee at the CSAGI Assembly in Moscow, and a further six, of *Sputniks 1* and *2*, were delivered to the Soviet Embassy in Washington in August. In September and October the Soviet committee sent requests for tapes of *Sputnik 3*. Odishaw replied asking for tapes of *Explorer 4*, which "you must have" in view of its higher inclination. The Soviet committee simply repeated its request, and the Americans handed over 34 tapes from *Sputnik 3* on December 31, 1958. A Soviet request for more *Sputnik 3* tapes, in February 1959, was countered with another request for *Explorer 4* tapes in April.[88]

Although the record of these exchanges dries up with Odishaw's understandably curt reply to Beloussov at the end of July 1959, it seems safe to conclude that the Americans never received any tapes of their satellites' telemetry from the Soviet Union, and that the last tapes sent to the Soviet Union by the U.S. committee were those dispatched at the end of 1958.[89] But in late July or early August 1959 one further tape (at least) was probably sent directly from the University of Alaska's Geophysical Institute to the Soviet physicist M. G. Kroshkin, who had sent the request in February 1959 (repeated in March) referred to above. Once again Odishaw expressed a preference for placing such exchanges on an official basis, preferably through the World Data Centers, but did not finally intervene to prevent a direct transfer, which presumably ensued.[90] The sending of this tape is of particular interest, in view of Beloussov's accusation in May, that "The most valuable in scientific respect part of these records made in Alaska has not been received at all."[91]

The Soviet satellite tracking stations were all sited within the USSR, and hence unable to record the continuous telemetry from their satellites while over the southern hemisphere. Berkner, for one, had long recognized the potential scientific value of such data.[92] A Russian request to the Australians for records of the southern telemetry of *Sputnik 3* is known to have been refused.[93] It is not known whether any of the tapes of Sputnik telemetry given by American scientists to their Soviet colleagues contained such data.

The official picture of the extent of scientific cooperation in respect of the IGY satellites is given by interim and final catalogues of the holdings of the WDCs in January 1959 and December 1962. The first of these was compiled from the WDC A report, with some input from

WDC C in Britain, but none from WDC B in Moscow. It includes only two Soviet documents, and omits the "Preliminary Report" on *Sputniks 1* and *2* which had reached Washington in February 1958, and seems to be unreliable for that reason alone.[94] The holdings of Soviet material in WDC A must surely have increased quite rapidly during 1959. From one source or another, at least, Homer Newell was able to produce a comprehensive comparative survey of the space science achievements of the United States and the Soviet Union, by the end of the year.[95] The final catalogue of WDC holdings, drawn up in consultation with all three WDCs, was published in the *Annals of the IGY*. It lists extensive series of Soviet orbital reports on Soviet satellites, as well as hundreds of scientific papers from Soviet scientists. None of the handful of raw data records held in WDCs A and C was of Soviet origin, and no such holdings are recorded for WDC B. Using a crude measure of page-length, American contributions of all types out-numbered those from the Soviet Union in an approximate ratio of 3 to 2. No Russian scientific papers discuss the orbits or experimental data from American satellites. Dozens of papers by American and other Western scientists discuss the orbital data, and some the experimental data, from Soviet satellites.[96]

DISCUSSION

The events related in this paper show that the IGY-IGC period was not some golden age of cooperation between the early satellite programs of the Soviet Union and the United States. It is probably more useful to see the IGY interactions over satellites as continuous with the frictions in COSPAR and COPUOS which immediately followed them. Such agreements over satellite cooperation as it was possible to reach within the IGY structures were arrived at only tardily, in the middle of 1958. They were imperfectly devised, incomplete, and marred with get-out clauses. Even such as they were, the Soviet committee failed to observe the clauses which had apparently been agreed, for example in the matter of distributing orbital predictions through the Agiwarn network. The last part of this paper will summarize what the author suggests were the obvious and less obvious causes for things having been so.

The first consideration must be the unusual properties of satellites in respect of international scientific cooperation, especially as it was understood at that time. The remainder of the IGY was conducted, for the most part, within a traditional structure of national compartmentalization. The plans of national committees were strongly coordinated for each discipline or sub-discipline, above all in respect of making similar, often synchronous, observations with similar instruments. But the actual work of taking such measurements was not done publicly, in the direct presence of foreign colleagues. They were taken at the various stations equipped and manned for that aspect of the discipline by a team sponsored by a national committee. Data were subsequently "worked up," reduced and tabulated, and only deposited with a World Data Center in due course, when they were thought to be ready. By general consensus, the initial scientific analysis of the data was held to be the privilege of the primary observers, or their close institutional or national colleagues. Multinational or transnational scientific enterprises were still exceptional during the IGY. Some scientific teams from relatively developed countries, such as Poland, the Netherlands and the Soviet Union, mounted major or minor expeditions to less developed countries or colonial territories, such as Vietnam, Curaçao and Egypt, respectively. Oceanographic expeditions naturally visited many countries and sometimes internationalized their personnel to a small degree. In Antarctica there was an internationally-staffed Weather Central under U.S. overall control, and some other exchanges of personnel. The Trans-Antarctic Expedition was staffed by several countries, but only on an "old Commonwealth" basis. The strongest internationalization in Antarctica was that addressed to meteorology, both for immediate practical reasons and also because of the long transnational tradition in that science, itself the product of obvious scientific requirements.[97]

The conduct of the early Soviet space program shows many signs of insisting that the separation of national efforts, which was the dominant pattern elsewhere in the IGY, should be applied to satellites also. They would launch the scientific packages, collect the data, tabulate the data, and write the papers. They would then publish the papers and deposit the reduced data at WDCs in their own time. They were not interested in receiving raw satellite data from U.S. satellites or providing their own to other scientists. However, Soviet scientists were powerless to alter the fact that their satellites were public objects. The physical characteristics of their orbits and continuous radio signals were exposed to anyone, from school children to intelligence agencies. But in Western countries, that did not mean that data downloaded by a military service, for example, would automatically have been placed at the disposal of the local IGY committee.

The attempt to develop cooperation in respect of the IGY satellites also suffered from the close involvement of the new technology with the top secret missile programs of the Soviet Union and the United States. In the prevailing circumstances, the U.S. demand for a full and immediate disclosure of all the "operational data" from the Soviet satellites was certain to be perceived, and rejected, as an attempt to acquire military intelligence. U.S. IGY officials certainly did consult regularly with the State Department and occasionally with the CIA, and U.S. scientists attending foreign conferences were routinely required to report back on any matters that might be of interest to their government, as, doubtless, were Soviet scientists also. There were of course perfectly legitimate scientific reasons for American scientists to wish to discover, if they could, how their Soviet colleagues were adapting their instrumentation to cope with the stresses of the onboard and near-space environments. But the fact was that many of those problems also had to be addressed in missile and warhead development, and they could not be freely discussed without careful scrutiny by the relevant security agencies on each side. Soviet scientists were probably entitled to be suspicious in this area. There were no similar demands from the U.S. committee about the specifications and performance of Soviet lorries taking scientific instruments into the Libyan desert, even though the lorries, too, may have affected the eventual scientific results. One other aspect of this part of the problem, already mentioned above, is that the Soviet IGY committee probably had no control over, and not much liaison with, the group of scientists and engineers under Korolev who were actually charged with launching the first Sputniks. (A somewhat similar situation developed in the United States when the U.S. Army's Explorer project was added to the IGY program after *Sputnik 1*.)

There must also have been widespread considerations of "face." The Soviet Union came late into an IGY program that was dominated by the United States and its allies. The U.S. scientific agenda had become the IGY agenda; their norms were assumed to be "the" norms of international science. They already held the key positions and were writing the rules. Prestige plums, such as the South Pole station, were already spoken for. Intentionally or not, the United States also became the Year's chief paymaster. By contrast, it is evident to anyone reading between the lines of the few contemporary accounts of the internal Soviet preparations for the IGY that are available, that there were enormous organizational, logistical, and financial problems to be overcome. Much of the infrastructure for the Soviet IGY program was simply not in place until the middle of 1958 or later.[98] (This single fact, more than any other, explains the Soviet determination to extend the Year through 1959.) Faced with difficulties arising from the relative backwardness of their country,[99] Soviet scientists may sometimes have needed to conceal the fact that they were *unable* to do something behind arguments which stated that they could not *agree* to doing it. With the benefit of hindsight, it is possible to see that if the Americans had had a better appreciation of how long the sort of things they were looking for were bound to take in the Soviet Union, and had exercised a little more patience even while letters slowly limped from one country to another (by modern standards), there might have been far less friction. On the other hand, exactly the same point can sometimes be made about the Soviet committee. Kroshkin wrote asking for further data from *Sputnik 3* in March 1959; by April the Americans were moving to meet his request, but in May there came Beloussov's intemperate intervention into the process. A little more patience in Moscow and Kroshkin would have received the Alaskan data far sooner than, presumably, he did, and much heat and fury would have been avoided.

Also in the matter of "face," the willingness of the leadership of the U.S. delegation to rain on the Soviet parade, at the 5th CSAGI Assembly in Moscow, can hardly have smoothed the way for negotiations in the rockets and satellites working group or any other part of the meeting.

Considerations of prestige were also present on the American side, of course, especially between October 4, 1957 and January 31, 1958. The hasty American accusations of bad faith, over *Sputnik 1*, were not only churlish but also short-sighted, since they probably did much to set back negotiations over space cooperation which were still at a very early stage at the end of 1957.

Finally, however, it is important to remember that the satellites portion of the IGY was at best a mating of scorpions, conducted in a fog of mutual prejudice and mistrust which the rational ideals of international scientific cooperation could only partly dispel. From time to time in their internal discussions, American IGY scientists consciously reminded each other that they were dealing with foreigners, and should make due allowance for the strange expectations and ways of doing things of such people. In the case of the feared and detested Communist enemy, however, it must have been enormously difficult for scientists passionately committed to Western ideals, as nearly all of them were, to go on making such allowances indefinitely. And the same could doubtless be said, *mutatis mutandis*, about the obverse relationship. The feelings of triumph, pride, and sheer relief must have been overwhelming for Russian scientists who worked on the early Sputniks, and the temptation to snub the Americans *for a change* must sometimes have proved just as irresistible to others, as it did to Fedorov in January 1958.

Taking all these difficulties into account, however, what is really striking about the Soviet IGY satellite program is not that the level of cooperation achieved was so imperfect, but that there was after all some degree of useful scientific interaction and exchange, even on such a militarily sensitive topic in the depth of the Cold War. As one optimistic scientist told the U.S. IGY committee, and anyone else who may have been reading over its shoulder, in his confidential report on the Moscow CSAGI Assembly:

> The most important overall conclusion to be reached from the Moscow conference, however, is not that cooperation was sometimes difficult and incomplete, but on the contrary that there was indeed more cooperation than ever before and that with patience and understanding it may yet be possible to achieve a working relationship among scientists as far apart as the United States and Soviet Union.[100]

Abbreviations (a) General—B: = Box; D: = Drawer; F: = File; CF = Central Files; IGY = IGY Archive.
(b) Archives—CNA = Canadian National Archives, Ottawa; CP = Sydney Chapman Papers, University of Alaska; DDE = Dwight D. Eisenhower Library, Abilene; SP = John Simpson Papers, University of Chicago; NAS = U. S. National Academy of Sciences; NHO = NASA History Office; RS = Royal Society.

1. At the time of the launch of the first Soviet satellite and in the years since then, Soviet and Russian sources have generally referred to the satellite as simply *Sputnik*. In this and subsequent chapters, the more common *Sputnik 1* has been used since that is more familiar in Western literature.

2. "Draft Report of the Meeting of the CSAGI Bureau with the Soviet Delegation at the IUGG Assembly," Rome, September 25, 1954—CP, B: 62, F: 254.

3. This was also announced in *Soviet News* on March 11, 1955.

4. Chapman-Bardin, April 4, 1955—CP, B: 62, F: 254; Chapman-Beloussov, May 8, 1955—*ibid*.

5. Nicolet-Bardin, August 5 and September 1, 1955—CP, B: 62, F: 254, and B: 52, F: 51.

6. Wyckoff-Odishaw, September 26, 1955, "Report of Activity at CSAGI Meeting in Brussels"—NAS IGY, D: 18, F: TPESP Sat. Corr. Jan.-Sep. 1955.

7. A full and clear account of the Porter-Sedov encounter at a conference in Freudenstadt, which was the principal source of this information at the time, is contained in "The USSR Satellite Program for the International Geophysical Year," DSI Report No. 8/56, Canadian Department of National Defence, July 1956.

8. Berkner-Chapman & Nicolet, February 29, 1956—CP, B: 53, F: 65.

9. Berkner-Nicolet, April 9, 1956, *ibid*.

10. *Annals of the IGY*, 6:457. Only one Soviet rocket scientist, S. M. Poloskov, went to Barcelona, and nothing is known about the extent of his participation in the relevant discussions.

11. John Hagen, the director of the *Vanguard* program, sent details of the American satellite, including its telemetry frequency, to the British IGY committee on August 1, 1956 (RS, NGY/74 (56)), and the same document was sent to all national committees by Nicolet a few days later—CN-CIR-15-568/6, August 6, 1956. There were also indications in J. Van Allen, ed., *Scientific Uses of Earth Satellites* (1st ed.), published at about this time, that the Americans were considering the use of frequencies of the order of 100–200 megahertz, but no precise figure was given.

12. "Soviets Plan to Launch Space Satellite This Year," *New York Daily Worker*, January 3, 1956.

13. The U.S. IGY Committee produced one such compilation for internal use: "A Summary of Some Recent Public Information on USSR Rocket and Satellite Developments," January 1956—NAS IGY, D:18, F: TPESP Sat. Corr. Jan. 1956; see also the excellent Canadian DND/DSI report cited above (n. 7).

14. "USSR Rocket and Earth-Satellite Program for the IGY," June 10, 1957—NAS IGY, D: 19, F: TPESP Sat. Corr. June 1957.

15. Translations of five articles were published in 1958 in the *IGY Manual on Rockets and Satellites, Annals of the IGY*, 6:222-54. One article contains the intriguing statement that: "As a last resort, the resonant circuits may be tuned to a frequency of 40 Mc/s by using the harmonics of the generator GSS-6 (see *Radio*, No. 5, 1956)"!—O. Rzhiga and A. Shakhovskoy "Use of UHF receiver to monitor the satellite," *Radio*, No. 7, July 1957, *Annals of the IGY*, 6:253. Until this reference has been checked, the author assumes that the May 1956 article gave details only of the general tuning method, not of the actual frequencies or the intention to use them for satellites.

16. A. M. Shakhovskoy, "USSR Amateur Radio Observations of Signals from the Soviet Artificial Satellites," *Annals of the IGY*, vol. XII, pt. II, p. 916.

17. Berkner, "Draft of Preliminary Proposal" etc., December 19, 1956—NAS IGY, D: 18, F: TPESP Sat. Corr. Dec. 1956.

18. *The CSAGI Guide to IGY World Data Centres*, 1st ed., June 7, 1957, Ch. XI (added 10 July 1957). A draft version of the "IGY Manual on Rockets and Satellites" was also circulated by Berkner at about this date.

19.   Fraser-Berkner, July 23, 1957—NAS CF, F: International Relations 1957: IGY General; Berkner-Fraser, July 29, 1957—*ibid*. The comments of Richard Porter, convenor of the satellite technical panel of the U. S. IGY committee, on what was missing from the Bardin document, put telemetry in fourth place, after launch site, launching schedule, and program agency. MS notes on Reid circular, June 24, 1957—NAS IGY, D: 19, F: TPESP Sat. Corr. June 1957.

20.   "Report of a Meeting at University College London—Friday July 26th, 1957," Guided Weapons Department, Royal Aircraft Establishment, August 13, 1957—Australian National Archives, Adelaide, Ref. D174/T1, Item A750/1/1 Pt. 1. The Soviet scientists had also visited the Cranfield College of Aeronautics, where, on July 19, 1957, another member of their delegation, B. N. Petrov, gave a detailed account of the scientific program planned for Soviet satellites: U.S. Secretary of the Navy to U.S. Secretary of Defense, "The U. S. Satellite Program," n.d. but early November 1957—NHO, F: 006596. According to Harrie Massey the account of a Soviet meteorological rocket, given to the IGY Conference on Rockets and Satellites in Washington two months later, was not substantially different from that given to the British. Massey, Report on the Washington Conference, RS, NGY/117 (57). Interestingly, the Soviet scientists in Britain attributed the choice of the 22 megahertz frequency [sic] to "equipment inherited from a wartime application in other work." This suggests that it may have been a frequency in use for sounding rocket experiments for some years, but if so it does not appear to have been detected, or noticed, by the network of U.S. monitoring stations around the external borders of the Soviet bloc.

21.   Bardin-Berkner, August 16, 1957. CP, B: 62, F: 257.

22.   The British sub-committee on satellites first formally discussed the Soviet frequencies on September 9, 1957, three weeks before the Washington conference. RS, NGY/93 (57). The sub-committee was "profoundly disappointed that the resolution adopted at Barcelona whereby both types of satellites would radiate the same frequency has been disregarded." Nicolet seems to have first learned of the existence of Bardin's letter giving the frequencies from a press report at the end of October or early November, and only then to have obtained a copy of it from David Martin, the Assistant Secretary of the Royal Society. Nicolet-Berkner, November 5, 1957. CP, B: 62, F: 257. In replying to Nicolet on this point, Berkner stated that the information on frequencies had already appeared in *The New York Times* at the end of July; the author has not yet been able to verify this. Berkner-Nicolet, November 7, 1957. CP, B: 62, F: 257.

23.   H. Massey and M. O. Robbins, *History of British Space Science* (Cambridge, England: Cambridge University Press, 1986), p. 39.

24.   W. T. Blackband. See "Draft Proceedings of the First Session of the Working Group on IGY Satellite Internal Experiments and Instrumentation Program," October 1957. NAS IGY, F: unknown. It transpired that features of the upper atmosphere, unknown at the time the proposal was drafted, rendered the method unworkable.

25.   Blackband-Martin, June 17, 1957. RS, NGY/76 (57).

26.   That it did in fact do so is beside the present point.

27.   V. A. Troitskaya, secretary of the committee, described this inner circle, to which she had considerable access, especially during foreign conferences, because of her competence with French and English, as comprising Bardin and three of the committee's vice presidents, Beloussov, Pushkov, and Yu. D. Boulanger. The role and status of the fourth vice president, F. F. Davitaya, was not touched on. Interviews with the author.

28.   Notes 13 and 20 above.

29.   Allan A. Needell, *Science, Cold War, and the American State: Lloyd V. Berkner and the Balance of Professional Ideals*, forthcoming.

30.   The proposal for a satellite conference at Barcelona actually originated from Richard Porter, chairman of the U.S. IGY committee's panel on satellites, after discussions with members of the International Astronautical Federation (a body which had enjoyed more

success in "drawing out" Soviet space scientists, to that date, that had the CSAGI). But Nicolet presented it to the Soviet committee as Berkner's idea.

31. Brian Harvey, *Race Into Space: the Soviet Space Programme* (London: Ellis Horwood, 1988), pp. 28–29.

32. Rip Bulkeley, *The Sputniks Crisis and Early United States Space Policy* (Bloomington: Indiana University Press, 1991), pp. 117–18.

33. Draft Minutes, Working Group on Satellite Launching, Tracking and Computation, CSAGI Conference on Rockets and Satellites, October 1, 1957. NAS IGY, D: 72, F: Porter 16 C (continued).

34. *Annals of the IGY*, vol. VI, pp. 458–9.

35. A full description of the mature Agiwarn system, including its satellites component, is given in *Annals of the IGY*, vol. VII, pt. 1.

36. Chapman, Closing Remarks, quoted in W. Sullivan, *Assault on the Unknown* (London: Hodder & Stoughton, 1962), pp. 69–70; see also pp. 1–2.

37. *Ibid.*

38. Day-Chapman, et al., October 19, 1957. CP, B: 62, F: 257.

39. Chapman's MS notes of telephone conversation, Shapley-Chapman, November 4, 1957. CP, B: 62, F: 257.

40. Day-Chapman et al., October 31, 1957, *ibid.*

41. Massevitch-Whipple, October 26, 1957—NAS IGY, D: 19, F: TPESP Sat. Corr. Oct. 1957 (2). The reference was to the conference of the International Astronautical Federation, Barcelona, October 6–12, 1957, not to the CSAGI Assembly of 1956. By November 5 the SAO had begun to receive the Soviet ephemerides. *New York Times*, November 8, 1957.

42. Berkner-Chapman, November 7, 1957. CP, B: 62, F: 257. The accusation, that the Soviet committee was refusing to meet an alleged obligation to provide the telemetry codes for *Sputnik 1*, had already been sent directly to Bardin by Hugh Odishaw, the secretary of U. S. IGY committee, on November 1, and this may well have been the source of Berkner's confident indignation here. The author has not yet traced a copy of Odishaw's letter, but however it was worded it seems unlikely to have eased the path to better cooperation. Quite independently, Day sided with Chapman and Nicolet in rejecting the American interpretation of the relevant passage in the *CSAGI Guide*. Day-Odishaw, November 8, 1957. NAS IGY, D: 34, F: CSAGI General Part 5. But even a week later Day remained unaware that Soviet ephemerides were now being sent to Harvard. Day-Chapman et al., November 14, 1957, *ibid.* In his opinion, the aim of the Royal Society meeting would be "an attempt to get things back on to the IGY rails whence they seem to have strayed." Day-Berkner, November 14, 1957. CP, B: 62, F: 257.

43. Chapman-Berkner, November 20, 1957, *ibid.*

44. Nicolet-Chapman, November 13, 1957, *ibid.*

45. In Day-Chapman et al., November 14, 1957, *ibid.*

46. Day-CSAGI Bureau, November 22, 1957, *ibid.*

47. Bardin-Day, November 25, 1957; Day memo, November 28, 1957; Day-Bardin, December 6, 1957, all at *ibid.* Day later told Odishaw that his conversation with Troitskaya in November, at which she handed over the new Soviet proposal, had helped him to appreciate that "their IGY admin difficulties ... are apparently little different from those pertaining in most countries. Almost everywhere a handful of workers is taking on extra IGY tasks without sufficient assistance and is striving to get organised." Day-Odishaw, December 7, 1957. NAS IGY, D: 34, F: CSAGI—Earth Satellites, Part 1.

48. Day-Bardin, December 6, 1957. NAS IGY, D:19, F: TPESP Sat. Corr. Dec. 1957. Day's wording tactfully ignored the fact that visual observations and predictions were already being sent to the Soviet committee by American and British scientists. See for example, R. W. Porter, "Summary of Report on the Earth Satellite Program," by October 30, 1957,

p. 2. NAS IGY, D:3, Attachment to Minutes, 24th Mtg, USNC-IGY Executive Committee. Day was careful to apologize for this deliberate omission in a letter to Alan Shapley, the CSAGI reporter for World Days who was responsible for the IGY World Warning Agency, based at the Fort Belvoir installation of the National Bureau of Standards' Central Radio Propagation Laboratory outside Washington, and therefore closely involved with the whole IGY communications system for data exchange. Day hoped that Shapley had "been able to understand the objective in my letter ... to Bardin," which had of course been to draw the Soviet committee into cooperation rather than to confront them with offensive accusations: Day-Shapley, December 16, 1957. NAS IGY, D: 19, F: TPESP Sat. Corr. Dec. 1957.

49. Day-Berkner, January 28, 1958. CP, B: 62, F: 252.

50. *Science*, vol. 126, December 27, 1957; Nesmeyanov-Bronk, January 7, 1958, and Bronk-Nesmeyanov, January 29, 1958, in Odishaw, memo, February 15, 1958. NAS IGY, D: 1, F: USNC Corr. (January–July 1958).

51. Day-Berkner, January 20, 1958. CP, B:62, F:252; "Amendments to CSAGI Guide to IGY WDCs suggested by Soviet IGY Committee, 6/7 Jan 1958"–attached to the former; Fedorov-Day, January 20, 1958, *ibid*. See also Day-Berkner, January 28, 1958, *ibid*.

52. The February 27 text, referred to in Berkner's letters of March 21, 1957, to Day and Beloussov (NAS IGY, D: 19, F: TPESP Sat. Corr. Mar. 1958), has not been found by the author. It can safely be assumed that it was equivalent to that in the fourth issue of *Amendments to the CSAGI Guide to IGY World Data Centres*, April 2, 1958, minus the amendments which were incorporated from Odishaw-Day, March 19, 1958. NAS IGY, D:72, F: Porter 16A. There were in fact several differences between Day's version and that which appeared in the *IGY Manual on Rockets and Satellites*, *Annals of the IGY*, vol. VI, three months later. The main one was an additional "note" from Berkner which largely contradicted the "two programme principle" conceded by Day, by stating that observers might send their data to any computing center they liked, and that the U.S. committee definitely wished to receive observations of Soviet satellites in order to do its own computations of their orbits. *Annals*, vol. VI, p. 468, para. (A).

53. Berkner, "Report on Rockets and Satellites," February 1958. CP, B: 62, F: 252.

54. Fedorov-Day, May 20, 1958. NAS IGY, D: 34, F: CSAGI—Manuals and Annals, Part 1.

55. Day-Berkner, May 23, 1958, *ibid*.

56. Berkner was obliged to miss the conference for personal reasons. For a graphic account of the political problems as they affected one distinguished American scientist, see J. A. Simpson, "An Account of Experiences in the Soviet Union, 1958," n.d. SP, B: 95, F: 1.

57. Newell, "To CSAGI Bureau," ca. August 9, 1958. CP, B: 58, F: 162. This was, of course, a flat rejection of the Soviet view that telemetered data, in particular, was not intended for use by researchers from other countries, and that they themselves had no wish to receive such data from American satellites.

58. Newell, "Impressions Received During the Fifth General Assembly of the Comité Speciale pour l'Année Géophysique Internationale," August 18, 1958. NAS IGY, D: 73, F: IGY: CSAGI Moscow Mtg. 1958.

59. Newell, "Items of non-agreement," MS notes, n.d., *ibid*.

60. Newell, "Impressions." n. 58 above.

61. "Comments of Delegates on the Scientific Meetings," n.d., p. 9. NAS IGY, D: 2, F: USNC Agenda, 15th Meeting; see also P. Hart, "Report on Fifth Assembly of CSAGI, Moscow, August 1958," January 29, 1959. NAS IGY, D: 35, F: Delegates' Reports, Moscow.

62. Newell, "Impressions," loc. cit. n. 58; by contrast, two American scientists were easily able to buy Geiger counters from a scientific equipment shop in Leningrad, of a type that would not have been publicly available in the United States at that date. Simpson, "An Account," n. 56, pp. 45–46.

63. "Amendments to Chapter XI," n.d., attachment 5 to Document 145, 5th CSAGI Assembly—NAS IGY, D: 35, F: CSAGI. 5th Assembly, Moscow, 1958, Documents.

64. P. Hart, untitled minutes of a meeting between Hugh Odishaw and State Department officials, July 14, 1958—NAS IGY, D: 35, F: CSAGI 5th Assembly, Correspondence.

65. *Annals of the IGY*, vol. VII, pt. II, ch. XI.

66. For the Odishaw-Bardin letter of November 1, 1957, see nn. 42–44 above. An anonymous memorandum, noting Waterman's proposal that "we can point out to good advantage the Soviet's lack of frankness in making available information on their earth satellite," was drafted with his approval on October 23, 1957. This suggests that Odishaw may have been responding to political guidance from Waterman, even if he ignored the latter's recommendation that the issue should be taken up through CSAGI, rather than bilaterally. "Soviet Non-cooperation with IGY," October 23, 1957. DDE, White House Office Papers, Staff Research Group Series, B: 15, F: N. S. F.

67. MS text, 10 pp., n.d., no title, ca. November 18, 1957. NAS IGY, D: 20, F: TPESP Sat. Corr. Undated.

68. Berkner's first draft ("Preliminary Proposal," December 19, 1956. n. 17 above) made explicit provision for the use of the news media in this way, but Day dropped the reference from his revised version in July 1957.

69. Text of cable is in Day-Chapman et al., October 31, 1957. CP, B: 62, F: 257.

70. Massevitch-Whipple, October 26, 1957. NAS IGY, D: 19, F: TPESP Sat. Corr. Oct. 1957 (2).

71. *New York Times*, November 8, 1957. Soviet ephemerides and observations were first sent directly to the SAO on November 5, 1957. After an interruption of unknown date and duration (see n. 72 below), they continued from January 1958 to at least the middle of 1959. They are another striking omission from the listing of satellite data received from the Soviet Union, that was prepared for Odishaw in 1959 (see Appendix A).

72. Shapley-Odishaw, March 24, 1958, states that the SAO *began* receiving regular station predictions from Moscow only on January 17, 1958. Presumably those received in November 1957 did not at first become a regular service. NAS IGY, D: 19, F: TPESP Sat. Corr. Mar. 1958.

73. Bardin-Day, November 25, 1957. CP, B:,62, F:,257; Massevitch-Shapley, November 26 1957, ibid. In a subsequent interview with *Izvestiya*, April 2, 1958, Bardin amended this figure to 66.

74. Odishaw-Beloussov, March 19, 1958. CNA, RG77, vol. 26, F: 4-G7-S7-17 pt. 2.

75. The minutes of the British subcommittee on artificial satellites show that orbital predictions for Soviet satellites continued to be received from the Soviet committee through 1958 and on into 1959, and that visual observations and, occasionally, recordings of Sputnik telemetry, were routinely dispatched to Moscow from British observatories.

76. Boggs-Weldon, July 21, 1959. NAS IGY, D: 20, F: TPESP Sat. Corr. Oct. 1959.

77. "In World Centre 'B'," *Izvestiya*, May 20, 1959.

78. Richter-Reid, September 10, 1958—NAS IGY, D: 20, F: TPESP Sat. Corr. Sep. 1958; Space Science Board Committee on International Relations, Minutes of 2nd Meeting, October 15, 1958, p. 3—SP, B: 96, F: 6; British IGY Committee, Artificial Satellites Sub-Committee, Radio Methods Working Group, Minutes, December 4, 1958. RS IGY.

79. "Preliminary Report on Launching in the USSR of the First and Second Artificial Earth Satellites (1957 and 1957)," received in Brussels, February 3, 1958. *Annals of the IGY*, vol. VI, p. 488; "Information Nr.1 on the USSR meteorological rocket firings under the IGY programme," March 13, 1958, attachment to Beloussov-Odishaw, March 13, 1958. CP, B: 62, F: 252. The preliminary report on the Sputniks was received by the U.S. IGY committee on February 11, 1958, but surprisingly, and tactlessly, neither the date nor even its reception was acknowledged by the latter in the compilation "Material on IGY

Rockets and Satellites sent to USA from USSR," attachment A, Odishaw-Beloussov, July 31, 1959. NAS IGY, D: 20, F: TPESP Sat. Corr. July 1959. The early orbital predictions sent to the SAO were also ignored, or rather, their publication and re-transmission to the Soviet Union were listed as data flowing in the *opposite* direction. See item A:25 in Appendix A.

80. For details, see *Annals*, vol. XII, pt. 1, p. ix.

81. *Artificial Earth Satellites*, #2, Moscow, 1959.

82. Mechau-Hart, July 16, 1959. NAS IGY, D: 23, F: Rockets & Satellites. Shapley-Odishaw, March 24, 1958, n. 72 above, shows that the first observation of a Soviet satellite (1957) was sent to Agiwarn from the SAO on March 19, 1958. Many of the early U.S. IGY satellites orbited at low angles of inclination, which probably meant that their ephemerides were of little interest to Soviet observers. Massevitch-SAO, n.d. but ca. June 1959, *ibid.*

83. For evidence that Soviet satellites may have had some interrogated telemetry from as early as *Sputnik 2*, see Richter-Reid, September 10, 1958, n. 78 above.

84. *CSAGI Guide to World Data Centres*, Ch. XI, first circulated October 22, 1958.

85. Beloussov-Newell, "Fulfillment of the V CSAGI Resolutions," May 15, 1959. NAS IGY, D:73, F: IGY: CSAGI Corr.

86. In June 1959 a document prepared for the U.S. committee's satellite panel recognized that "there has not as yet been any definite guidance [from CSAGI] on whether this [data for exchange] should be raw or processed data." "The US-IGY Earth Satellite Program," June 1959. NAS IGY, D: 73, F: TPESP Meeting, July 21, 1959.

87. Reid, memo for record, May 27, 1958. NAS IGY, D: 19, F: TPESP Sat. Corr. May 1958. The Harvard Observatory had a long history of good relations with Soviet astronomers under its previous director, Harlow Shapley, both directly and through the International Astronomical Union.

88. Anon., "Notes on Beloussov Communication to Newell," May 1959. NAS IGY, D: 49, F: 646 etc. The date of March 1959 for Kroshkin's request, given in the attachment to this document, has been corrected to February in the light of: JCT-AWF, February 11, 1959. NAS IGY, D: 49, F: 630.

89. Odishaw-Beloussov, July 31, 1959 and attachments. NAS IGY, D: 20, F: TPESP Sat. Corr. July 1959.

90. Odishaw-Elvey, April 30, 1959, and Elvey-Odishaw, July 7, 1959. NAS IGY, D: 49, F: 646 etc.

91. Beloussov, "Fulfillment," n. 85 above.

92. Berkner-Fraser, July 29, 1957. n. 19 above.

93. J. W. Dungey, "The Radiation Belts," in D. R. Bates, ed., *The Planet Earth*, (2nd ed.), (New York: Pergamon Press, 1964), p. 311.

94. Office of Coordinator, *Catalogue of IGY Data received at WDCs by January 1959*, Ch. XI, March 1959. It seems scarcely credible, but might WDC A have refused to "count" any Soviet material not addressed directly to it? At all events, the omission of the Soviet "Preliminary Report" from WDC A's listing might account for its omission from the U.S. committee's compilation later in the year (Appendix A).

95. H. E. Newell, "U.S. and USSR Space Science Results," December 15, 1959. NASA Release No. 59–277.

96. *Annals of the IGY*, vol. XXXVI, 1964, ch. XI, "Rockets and Satellites—Catalogue of Data in the World Data Centers," January 1963. "All papers received through December 1962 ... are included." p. 578.

97. If the sciences of the upper atmosphere and space had been able to develop harmoniously out of meteorology, instead of breaking away from it, they might perhaps have inherited a more securely internationalist cultural viewpoint.

98. U.S. Department of Commerce, *Soviet Bloc International Geophysical Year Information*, November 16, 1957, pp. 1–12; *ibid.*, April 25, 1958, pp. 1–7; idem, *Information on*

Soviet Bloc International Geophysical Cooperation, June 12, 1959, pp. 1–2. Also Hart, "Report," n. 60 above.

99. One possibly apocryphal rumor has it that stocks of duplicating paper for the 5th CSAGI Assembly in Moscow had to be flown in from Brussels, either because of a physical shortage or else because existing supplies could not be "released" to the Soviet Academy even for such a prestigious event.

100. "Comments of Delegates," n. 61 above, p. 8.

CHAPTER 6

Cover Stories and Hidden Agendas: Early American Space and National Security Policy

Dwayne A. Day

It was language that was perfect in keeping with the popular conception of space at the time, but which was not well-received in the White House. On February 19, 1957, before a collection of spaceflight enthusiasts gathered in San Diego, Major General Bernard Schriever delivered a speech entitled "ICBM—A Step Toward Space Conquest." Schriever was the head of the Western Development Division (WDD), the Air Force development office which was assigned the task of building an operational ICBM as quickly as possible. Schriever's role in developing the ICBM was well known. What was unknown, however, was the fact that WDD was also responsible for developing intelligence satellites. The Air Force planned to launch 92 intelligence satellites over the next five years, culminating in a large signals intelligence satellite in 1962. It was an ambitious plan, but one still lacking money and unknown to the outside world.

Schriever could not speak in San Diego of the classified mission he oversaw. Instead he talked about how WDD's experience in developing the Atlas ICBM could be applied to space. He also spoke of the value of space to the military, and he did not mince words:

> In the long haul our safety as a nation may depend upon our achieving "space superiority." Several decades from now the important battles may not be sea battles or air battles, but space battles, and we should be spending a certain fraction of our national resources to insure that we do not lag in obtaining space supremacy.[1]

The next day, arriving at his headquarters in Los Angeles, Schriever had a telegram from the Office of the Secretary of Defense waiting for him. It was explicit: do not use the word "space" in any more speeches. Remembering back on this event, Schriever's frustration still showed. "I was angry," he recalled over 40 years later while sitting in his office in Washington. But he was an Air Force officer and the Secretary of Defense was his boss. There was never any question about what he would do: "I didn't use the word space anymore in my speeches."[2]

The language in Schriever's speech was not unusual for the times. Wernher von Braun, who had far more visibility as a spokesperson for space than Schriever, had used the term "conquest" often when discussing space, including a series of high-profile articles for *Collliers'* magazine. But where von Braun saw space as new territory to be conquered like the Old West, Schriever saw it as new territory from which to wage battle. He had not been reticent to talk about it in these terms, and despite the muzzling from the Pentagon, the message had already gotten out. On March 4, *Newsweek* ran an article about the American *Vanguard* space program. A small accompanying article included a picture of a stern-looking Schriever and referred to him as "one of the major prophets of space warfare." It quoted the General as stating that "the compelling motive for the development of space technology is ... national defense."[3]

Schriever was only mentioned in the sidebar article, but the main story was unlikely to be any more soothing to the Eisenhower White House, for it was titled "Race Into Space: Can We Win?" The story likely rattled the White House because President Eisenhower and his advisors had explicitly chosen a space policy and a public posture that attempted to downplay the idea the United States was attempting to "race" the Soviets into space. They had chosen a policy which portrayed the fledgling American space program as a small, scientific effort—the development of a rocket and satellite combination referred to as *Vanguard*. Official Pentagon policy was to make no mention of a race and no mention of a military space program.

The only public space program was in actuality a cover, a deception. It shielded what the White House and Pentagon leadership actually cared about, the political vulnerability of its defense programs. This was not unique to the space program. It was a common strategy for the way that Eisenhower approached many sensitive national security missions—hiding them behind open programs, usually scientific and civilian in appearance, so that they would not be challenged openly by the Soviets. It was to become a common strategy for the later military space program, but it all started with *Vanguard*.

THE KILLIAN REPORT

In September 1954, the Science Advisory Committee of the Office of Defense Mobilization, under orders from President Eisenhower, began a scientific study of the problem of surprise attack.[4] Eisenhower was concerned about recent Soviet developments in various strategic weapons systems and asked the Committee to evaluate American capabilities to respond to them using the latest advances in science and technology. This study was headed by James Killian, then president of the Massachusetts Institute of Technology. The group was known as the Technological Capabilities Panel (TCP); it issued its report, "Meeting the Threat of Surprise Attack," on February 14, 1955. This was often referred to as the "Killian Report" and it greatly impressed Eisenhower. It convinced him that science and technology could be used to defeat an enemy.

During the course of its deliberations, the study's intelligence panel, headed by Polaroid's Edwin "Din" Land, became aware of two advanced proposals for intelligence collection. One was a nuclear powered reconnaissance satellite using a television camera. This was outlined in a report by the Air Force's think-tank, the Rand Corporation, known as *Project Feed Back*. The other idea that Land's panel investigated was a U.S. Air Force development program called BALD EAGLE which was intended to develop a high-altitude reconnaissance aircraft.

While investigating the aircraft program, the panel became aware of another proposal by the Lockheed Skunk Works for its own high-flying strategic reconnaissance aircraft, known as the CL-282. Land brought this proposal to the attention of President Eisenhower in November 1954. Unlike the Air Force program to develop a reconnaissance aircraft, the CL-282 would be configured to carry out strategic reconnaissance prior to hostilities—so-called pre-D-Day reconnaissance.

This was a mission that the Strategic Air Command had previously rejected, but one that those on the TCP thought vital.

Eisenhower approved the CL-282 project and placed it under the charge of the Central Intelligence Agency. The plane eventually became known as the U-2. The program was managed by a newcomer to the CIA, Richard Bissell. The aircraft was never mentioned in the TCP report itself, but was described in a highly classified annex. This annex was for the "Eyes Only" of President Eisenhower, and was probably destroyed by him along with another classified annex on the submarine-launched ballistic missile program.[5]

According to Eisenhower's staff secretary, then-Colonel Andrew Goodpaster, Eisenhower assigned the U-2 mission to the CIA for three reasons. First, he thought it would be less provocative if a civilian pilot, rather than a military one, flew the aircraft into foreign territory. Second, he wanted the product—the reconnaissance photographs—to be evaluated at the national leadership level, as opposed to being evaluated within the military services. Based upon his long years of experience in the military, Eisenhower knew that the military had an incentive to interpret intelligence to its advantage. Finally, he was concerned about not antagonizing the Soviets by pursuing a provocative program in the open. He was concerned that the military would pursue the program in a way that would only escalate tensions between the superpowers.[6] According to Goodpaster, Eisenhower was always distressed to see magazines carrying advertisements of bombers and fighter aircraft. He felt that the military encouraged defense contractors to place such ads and that they projected an image of a society preparing for war with the Soviet Union. This could prompt the Soviets to respond the same way, leading to a spiraling arms race. For Eisenhower, stealth and concealment were the preferable approach, hence the U-2 was given to an organization that could operate in secret.

Those involved in the TCP report realized that overflight of another nation's territory by such an aircraft would constitute a clear violation of international law and could be viewed as a hostile act by the Soviets. In fact, such issues were not abstract, since American aircraft flying missions on the periphery of the Soviet Union were being fired upon on a regular basis and even shot down. The consequences of violating national airspace was clearly a major concern for those planning aircraft reconnaissance missions. But a satellite, flying much higher, would not necessarily violate international law since no clear definition existed of where "airspace" ended and "space" began. Realizing this, Land and the others on the panel decided to attempt to strongly influence the *establishment* of international law.

The intelligence committee recommended that the United States develop a small artificial earth satellite to establish the right of "freedom of space" for future larger intelligence satellites.[7] Doing so would allow the United States to distinguish between "national airspace" and "international space." By establishing the definition itself, the United States could later use international space to its advantage by flying intelligence satellites over the Soviet Union.[8]

Land, Killian, and others in 1955 considered the reconnaissance satellite to be technologically unrealistic in the near future. They believed that the CL-282 and an Air Force reconnaissance balloon program (later known as *Genetrix*) were more realistic near-term possibilities worthy of the most effort. But they felt that the United States should begin work toward establishing a precedent to enable future satellite reconnaissance missions. The TCP advocated using an ostensibly civilian program to clear a path for a future military program.

THE SCIENTIFIC SATELLITE PROGRAM

In parallel to the deliberations of the TCP, serious proposals for an initial U.S. scientific satellite were emerging. For example, Wernher von Braun and his colleagues at the Army Ballistic Missile Agency in Huntsville, Alabama, teamed up with the Office of Naval Research to propose a satellite called *Orbiter*. Later in the year, the American Rocket Society prepared a detailed survey of possible scientific and other uses of a satellite and proposed it to the National Academy of Sciences' U.S. National Committee for the International Geophysical Year. The IGY was to run from June 1957 to December 1958 and involve international cooperation on the study of the earth.

As it was, 1954 proved to be a very significant year for the generation of ideas concerning scientific and intelligence collection systems. In addition to both the *Feed Back* and CL-282 proposals, the National Academy was now considering a scientific satellite as well. These projects became inextricably politically linked.

The *Feed Back* and the Killian reports were both highly secret, although the *Orbiter* proposal was not. The CL-282 proposal, in particular, was known to only a handful of people.[9] One person who did know of all three projects, as well as of the TCP report, was the Assistant Secretary of Defense for Research and Development, Donald Quarles.

Quarles was also no stranger to proposals for Earth-orbiting satellites. In 1952, Aristid Grosse, a physicist at Temple University, had been asked by President Truman to prepare a report on "the Present Status of the Satellite Problem." The report was not ready before Truman left office and instead the White House forwarded it to Quarles, who as Assistant Secretary of Defense for Research and Development was in charge of virtually all defense-related research projects.

On March 14, 1955 (the same day as the presentation of the TCP report), the U.S. National Committee on the International Geophysical Year of the National Academy of Sciences (NAS) presented a recommendation to Alan Waterman, the Director of the National Science Foundation (NSF). The Committee recommended that a scientific satellite be developed as part of the IGY.[10] After the TCP report came out, Quarles asked Waterman to formally suggest the National Academy's idea to the National Security Council. Four days later Waterman sent a letter to Robert Murphy, Deputy Under Secretary of State, proposing that the United States conduct just such a scientific mission.[11] Those at the National Academy would not have been privy to information about *Feed Back*, CL-282, or the TCP report, so they were unlikely to recognize the strategic objectives behind Quarles' support of their satellite proposal.

Four days later Murphy met with Waterman, Detlev Bronk, President of the NAS, and Lloyd Berkner, one of the Academy's influential members, to discuss the issue. In a letter one month later, Murphy stated that such a proposal would "as a matter of fact, undoubtedly add to the scientific prestige of the United States, and it would have a considerable propaganda value in the cold war."[12] Having gained the support of the Department of State, Waterman then discussed the issue once again with Quarles, who suggested that he consult the Director of Central Intelligence, Allen Dulles, on how best to proceed. Waterman did so and gained Dulles' support for the program. He also spoke with Percival Brundage, the Director of the Bureau of the Budget, to gain his cooperation when needed. Thus, the scientific satellite proposal now had the support of the Departments of State, Defense, and the Central Intelligence Agency, as well as the Bureau of the Budget. Waterman also agreed to formally propose the full program to an executive session of the National Science Board on May 20.[13] Waterman was lining up bureaucratic support for the project at Quarles' urging.

Whether any of these other people were aware that Quarles' interest in the proposal had been prompted by a classified national security report is unknown. But clearly Quarles was on a mission to enact one of the TCP's recommendations and he worked quickly.

NSC 5520

The speed at which these events took place is startling. Only nine weeks after the TCP report recommended the approval of a scientific satellite program, on May 20, 1955, the National Security Council approved a top-level policy document known as NSC 5520, "U.S. Scientific Satellite Program." The document stated that the United States should develop a small scientific satellite weighing 5 to 10 pounds. It was very clear about why the United States should do this:

> The report of the Technological Capabilities Panel of the President's Science Advisory Committee recommended *that intelligence applications warrant* an immediate program leading to a very small satellite in orbit around the earth, and that re-examination should be made of the principles or practices of international law with regard to "Freedom of Space" from the standpoint of recent advances in weapon technology.[14]

The document continued:

> From a military standpoint, the Joint Chiefs of Staff have stated their belief that intelligence applications strongly warrant the construction of a large surveillance satellite. While a small scientific satellite cannot carry surveillance equipment and therefore will have no direct intelligence potential, it does represent a technological step toward the achievement of the large surveillance satellite, and will be helpful to this end so long as the small scientific satellite program does not impede development of the large surveillance satellite.[15]

NSC 5520 also stated:

> Furthermore, a small scientific satellite will provide a test of the principle of "Freedom of Space." The implications of this principle are being studied within the Executive Branch. However, preliminary studies indicate that there is no obstacle under international law to the launching of such a satellite.
>
> It should be emphasized that a satellite would constitute no active military offensive threat to any country over which it might pass. Although a large satellite might conceivably serve to launch a guided missile at a ground target, it will always be a poor choice for the purpose. A bomb could not be dropped from a satellite on a target below, because anything dropped from a satellite would simply continue alongside in the orbit.

Although the document correctly noted the limited utility of satellites as active military offensive threats, this was not the purpose of the surveillance satellite program. In fact, deploying weapons systems in orbit has never been a significant aspect of American military space policy.

A year later, another White House document stated that the scientific satellite program would have the target date of 1958 "with the understanding that the program developed thereunder will not be allowed to interfere with the ICBM and IRBM programs but will be given sufficient priority by the Department of Defense in relation to other weapons systems to achieve the objectives of NSC 5520."[16] In other words, the scientific satellite would not be allowed to interfere with ballistic missile programs like Atlas and *Jupiter*, but it also would not be allowed to languish without attention while the Department of Defense focused on other programs with more direct military utility. Such inattention would have effectively undercut the entire purpose of NSC 5520. At the same time, the launching of such a satellite was *not* to interfere with the development of the intelligence satellite.

NSC 5520 specifically directed that the scientific satellite program be associated with the International Geophysical Year. "Freedom of space" could be challenged by the Soviets more easily if the U.S. simply launched a satellite at any time. The IGY provided an international context for conducting the program—an excuse. It was a legitimate cover story even better than the civilian auspices of the existing program. Even if the Soviets had not announced their intention to build a satellite for the IGY, the United States would be able to use the IGY to justify its program if the Soviets complained. Quarles himself specifically noted this reality only a week after the approval of NSC 5520.[17]

The TCP report was perhaps one of the most influential documents of the Cold War. It served as the starting point of a number of major American defense programs in the 1950s. Not only did it recommend the development of what became the U-2, the scientific satellite program, and reconnaissance satellites, but it also recommended the development of what became the Polaris submarine-launched ballistic missile, the *Sosus* underwater sonar array, and the extension of the Distant Early Warning (DEW) Line. The freedom of space recommendation was therefore one among many, and it was in no way particularly special. It was merely one that was unlikely to make progress without specific high-level involvement. Quarles gave it the start that it required.

DEFINING THE HEAVENS

The TCP's extensive list of recommendations required a concerted implementation effort by the various agencies and military services that were affected. It required an extensive oversight effort as well and this task fell to the Operations Coordinating Board of the National Security Council which produced a progress report in June 1955 on the status of the TCP recommendations. This progress report, known as NSC 5522, also included reports from various government departments on their views on the Panel's recommendations. In response to the TCP recommendation concerning freedom of space, NSC 5522 noted that a program had already been started to meet the recommendation. Further it stated:

> State, Treasury, Defense, and Justice Comment: Any unilateral statement by the U.S. concerning the freedom of outer space is unnecessary. It is clear that the jurisdiction of a state over the air space above its territory is limited and that the operation of an artificial satellite in outer space would not be in violation of international law. State and Justice point out that by the convention on international civil aviation of 1944 (to which the U.S. is a party, but the USSR is not) and by customary law every State has exclusive sovereignty "over the air space above its territory." However, air space ends

with the atmosphere. There has been no recognition that sovereignty extends into airless space beyond the atmosphere.[18]

This statement may have appeared self-evident to those involved, but in effect it begged the question: if space was not sovereign territory (a question for which there was no precedent), how was a nation to determine where the atmosphere—which *was* sovereign territory—ended and space began? The edge of the atmosphere was not a clearly defined boundary and, indeed, could not be defined until satellites began orbiting the Earth to measure the extent of the atmosphere. Still, it made sense for the Departments to advise that the United States not issue a unilateral statement concerning freedom of space, for such a statement would only draw more attention to the subject at a time when the whole point of the scientific satellite program was to establish freedom of space in as inconspicuous a manner as possible. For the government lawyers as well as the policy makers, it was better to simply do it and talk about it later.

The CIA also expressed its opinion on the subject in a commentary that proves highly illuminating and remarkably prescient, especially in light of the world reaction to *Sputnik 1* two years later:

The psychological warfare value of launching the first earth satellite makes its prompt development of great interest to the intelligence community and may make it a crucial event in sustaining the international prestige of the United States.

There is an increasing amount of evidence that the Soviet Union is placing more and more emphasis on the successful launching of the satellite. Press and radio statements since September 1954 have indicated a growing scientific effort directed toward the successful launching of the first satellite. Evidently the Soviet Union has concluded that their satellite program can contribute enough prestige of Cold War value or knowledge of military value to justify the diversion of the necessary skills, scarce material and labor from immediate military production. If the Soviet effort should prove successful before a similar United States effort, there is no doubt but that their propaganda would capitalize on the theme of the scientific and industrial superiority of the communist system.

The successful launching of the first satellite will undoubtedly be an event comparable to the first successful release of nuclear energy in the world's scientific community, and will undoubtedly receive comparable publicity throughout the world. Public opinion in both neutral and allied states will be centered on the satellite's development. For centuries scientists and laymen have dreamed of exploring outer space. The first successful penetration of space will probably be the small satellite vehicle recommended by the TCP. The nation that first accomplishes this feat will gain incalculable prestige and recognition throughout the world.

The United States' reputation as the scientific and industrial leader of the world has been of immeasurable value in competing against Soviet aims in both neutral and allied states. Since the war, the reputation of the United States' scientific community has been sharply challenged by Soviet progress and claims. There is little doubt but that the Soviet Union would like to surpass our scientific and industrial reputation in order to further her influence over neutralist states and to shake the confidence of states allied with the United States. If the Soviet Union's scientists, technicians, and industrialists were apparently to surpass the United States and first explore outer space, her propaganda machine would have sensational and convincing evidence of Soviet superiority.

If the United States successfully launches the first satellite, it is most important that this be done with unquestionable peaceful intent. The Soviet Union will undoubtedly attempt to attach hostile motivation to this development in order to cover her own inability to win this race. To minimize the effectiveness of Soviet accusations, the satellite should be launched in an atmosphere of international good will and common scientific interest. For this reason, the CIA strongly concurs in the Department of Defense's suggestion that a civilian agency such as the U.S. National Committee of the IGY supervise its development and that an effort be made to release some of the knowledge to the international scientific community.

The small scientific vehicle is also a necessary step in the development of a larger satellite that could possibly provide early warning information through continuous electronic and photographic surveillance of the USSR. A future satellite that could directly collect intelligence data would be of great interest to the intelligence community.

The Department of Defense has consulted with the [Central Intelligence] Agency, and we are aware of their recommendations, which have our full concurrence and strong support.[19]

The Department of Defense in mid-1955 was not pursuing an active intelligence satellite program. All it had at the time was a series of small Air Force technology development efforts. But the scientific satellite program was underway and the lawyers were now debating the legalities of the program in secret.

If those involved with the legal issues needed their thinking sharpened about issues of international airspace, they could ask for no better lesson than that provided less than a year later by the *Genetrix* reconnaissance balloon program, which had also been endorsed by Killian and Land. Hundreds of balloons were launched beginning in January 1956 so that they would drift over the Soviet Union, their cameras photographing the countryside. Officially, the balloons were part of a scientific program intended to photograph clouds. In reality, they carried sophisticated reconnaissance equipment.

The majority of the balloons were never seen again and American officials knew that they had come down (or been shot down) inside the Soviet Union. This was in reality a rather ironic development; because of the planned future flights of the U-2 at 70,000 feet, the balloons had been ballasted to fly lower than their operational altitude. This was done so that the Soviets would not be provoked into quickly developing a capability to track and shoot down objects at that extreme altitude. The result was that as the balloons cooled at night, they sank below 40,000 feet and well within range of Soviet aircraft, which shot them down in the early morning hours before they rose back to their operational altitude.

On February 7, anticipating the Soviet response, Eisenhower suggested to Secretary of State John Foster Dulles that the operation be suspended and "we should handle it so it would not look as though we had been caught with jam on our fingers." On February 9, the Soviets held a press conference outside Spridonovka Palace. About fifty balloons and instrument containers were placed on display. The balloons, the Soviets said, were part of an espionage project and had been a clear violation of their airspace. It was a major embarrassment for the United States, but the U-2 was still under development and promised far greater capabilities.

Meanwhile, legal issues concerning exactly what "freedom of space" meant occupied U.S. government lawyers. The State Department's Policy Planning Staff was assigned the task of reporting on-going progress on achieving several of the TCP report's recommendations. On October 2, 1956, the Policy Planning Staff reported some preliminary thinking on the issue of freedom of space:

> So far as law is concerned, space beyond the earth is an uncharted region concerning which no firm rules have been established. The law on the

subject will necessarily differ with the passage of time and with practical efforts at space navigation. Various theories have been advanced concerning the upper limits of a state's jurisdiction, but no firm conclusions are now possible.

A few tentative observations may be made: (1) A state could scarcely claim territorial sovereignty at altitudes where orbital velocity of an object is practicable (perhaps in the neighborhood of 200 miles); (2) a state would, however, be on strong ground in claiming territorial sovereignty up through the "air space" (perhaps ultimately to be fixed somewhere in the neighborhood of 40 miles); (3) regions of space which are eventually established to be free for navigation without regard to territorial jurisdiction will be open not only to one country or a few, but to all; (4) if, contrary to planning and expectation, a satellite launched from the earth should not be consumed upon reentering the atmosphere, and should fall to the earth and do damage, the question of liability on the part of the launching authority would arise.[20]

Genetrix had provided a good warning of what could happen if an intelligence mission went wrong and if a program's cover was blown. A different lesson was provided in July, when the U-2 made a half dozen sorties over the Soviet Union. The U-2, like *Genetrix*, also had a cover story. It was portrayed as a civilian weather reconnaissance aircraft. But the aircraft went virtually unnoticed outside of the aviation trade press, where some reporters accurately surmised its true purpose. Although the Soviets tracked the plane immediately, they did not protest. Lacking the ability to knock the U-2 out of the sky, they chose not to publicly admit their vulnerability to American reconnaissance technology.

The U-2 demonstrated a number of things to Eisenhower and his advisors. First and foremost, it demonstrated the immense value of overhead reconnaissance systems. The photographs returned by the U-2 were stunning in their quality. The U-2 also demonstrated that both the United States and the Soviet Union had a stake in keeping certain things private. Finally, it demonstrated that a cover story could work. Concealment and obfuscation of intelligence missions had a benefit for superpower relations even when the other side suspected, or actually knew, what was happening.[21] This was the guiding principle of the satellite program adopted under NSC 5520. It was to be a guiding principle for other space programs as well.

SELECTING A SATELLITE PLAN

After the scientific satellite concept had received approval with NSC 5520 in May 1955, it had to be turned into programmatic reality. Not surprisingly, this task fell to Donald Quarles. On June 8, 1955, Secretary of Defense Charles Wilson wrote a memo placing Quarles in charge of starting the scientific satellite program. Quarles then created an Advisory Group on Special Capabilities, chaired by Homer J. Stewart, to select a scientific satellite and a method for launching it. They held hearings and listened to presentations by the Army, Navy and Air Force. The Army's proposal was basically a variation of von Braun's earlier Orbiter concept. The Air Force submitted a proposal from Schriever's Western Development Division, but did not make a concerted effort to win the program. The Navy Research Laboratory proposed using a modified upper atmosphere research rocket. The Advisory Group selected the Navy proposal, which as named *Vanguard*.[22]

Many involved, particularly the Army group, which consisted primarily of Wernher von Braun's "rocket team," were surprised that the Army proposal had lost out. The Army team had extensive experience with ballistic missiles and had begun developing their proposal a year before. When explaining the decision years later, members of the Advisory Group gave a number of reasons, including Navy developments in satellite instrumentation, the supposed lower costs and growth potential of the *Vanguard*, and the desire to avoid using a ballistic missile for the program.[23] The latter point was the most relevant one: NSC 5520 was quite explicit in its statement that the satellite program should not interfere at all with ICBM and IRBM programs. There was no way that the Army proposal could be selected without interfering with Army development of the *Redstone* booster as the basis of an IRBM to be called *Jupiter*—at least without significant additional cost.

The Advisory Group's decision in favor of the Navy program did not settle the issue. The Huntsville group, justifiably, felt that they deserved to win. "There wasn't much confidence, frankly, in the *Vanguard*," said Fred Durant, one of the original members of the *Orbiter* group. The Navy proposal was too new and the team unproven. But according to Durant it was not necessarily arrogance that caused the Army team to be so stunned. The Huntsville group was more confident because of their longer experience launching rockets and making mistakes.[24] The *Vanguard* team had none of that experience and lacking that, would have to make their own mistakes before achieving success. From the perspective of the Army, that meant that the Soviets were more likely to reach the goal first.

Throughout 1955 and 1956, the Army team proceeded with development of a *Jupiter/Redstone* rocket designed to conduct re-entry tests of ballistic nosecones while the Navy developed the *Vanguard* and the NSF and National Academy of Sciences developed the satellite for flying atop it. But the Army continued to lobby behind the scenes to obtain permission to proceed with their own satellite launching program.

On April 23, 1956, the newly formed Army Ballistic Missile Agency, under the command of Major General John B. Medaris, informed the Office of the Secretary of Defense that a *Jupiter* missile could be fired in an effort to orbit a small satellite as early as January 1957. It proposed that this be considered an alternate to the Navy *Vanguard* program. On May 15, the Secretary of Defense's Special Assistant for Guided Missiles disapproved ABMA's proposal. On May 22, the Assistant Secretary of Defense informed ABMA that no plans or preparations should be initiated for using either *Jupiter* or *Redstone* missiles as satellite launch vehicles.[25]

But all good military officers and bureaucrats know that there is no such thing as a "no" in Washington. ABMA took its plea to Deputy Secretary of State Hoover, who was in charge of the Operations Coordinating Board which was responsible for over-seeing the implementation of the NSC 5520 decision. In June 1956, then-Colonel Andrew Goodpaster, who was serving as Eisenhower's staff secretary, reported on an incident that demonstrated both the Huntsville group's lobbying and their perception of the prejudice against them. Goodpaster stated:

On May 28th Secretary Hoover called me over to mention a report he had received from a former associate in the engineering and development field regarding the earth satellite project. The best estimate is that the present project would not be ready until the end of '57 at the earliest, and probably well into '58. *Redstone* had a project well advanced when the new one was set up. At minimal expense ($2–$5 million) they could have a satellite ready for firing by the end of 1956 or January 1957. The *Redstone* project is one essentially of German scientists and it is American envy of them that has led to a duplicative project.

I spoke to the President about this to see what would be the best way to act on the matter. He asked me to talk to Secretary Wilson. In the latter's absence, I talked to Secretary [Deputy Secretary of Defense] Robertson today and he said he would go into the matter fully and carefully to try to ascertain the facts. In order to establish the substance of this report, I told him it came through Mr. Hoover (Mr. Hoover had said I might do so if I felt it necessary).[26]

This incident started a new evaluation of the satellite selection process which ultimately rejected the Huntsville group once again. On June 22, 1956, Homer J. Stewart, who was Chairman of the Advisory Group on Special Capabilities, reported the results of two meetings held by the Group in late April to Charles C. Furnas, Assistant Secretary of Defense for Research and Development who had replaced Quarles (who was now Secretary of the Air Force). Stewart reported that although Project *Vanguard* was suffering some minor setbacks and was short of highly capable people, in general the project was on a satisfactory schedule and "one or more scientific satellites can be successfully placed in orbit during the IGY."[27] Stewart also stated:

> *Redstone* re-entry vehicle No. 29, now scheduled for firing in January 1957, apparently will be technically capable of placing a 17 pound payload consisting principally of radio beacons and doppler-equipment in a 200-mile orbit, even with the degradation in performance below the present design figures which might reasonably be expected, but without any appreciable further margin. This capability will depend upon successful accomplishment of several developments, such as the use of a new fuel in the *Redstone* booster, and the spinning cluster of fifteen solid propellant motors. The probability of success of this single flight cannot be reliably predicted now, but it would doubtless be less than 50 per cent.

Stewart explained why the Army proposal should be rejected once again:

> In any case, such a single flight would not fulfill the Nation's commitment for the International Geophysical Year because it would have to be made before the beginning of that period. Adequate tracking and observation equipment for the scientific utilization of results would not be available at this time. Moreover, any announcement of such a flight (or worse, any

leakage of information if no prior announcement were made) would seriously compromise the strong moral position internationally which the United States presently holds in the IGY due to its past frank and open acts and announcements as respects *VANGUARD*.

Stewart mentioned that a *Redstone* could be used as a backup later in 1957 if *Vanguard* fell behind schedule and that No. 29 was the only vehicle which could be used without interference in the *Redstone* program. Finally, he concluded, "At the present time, therefore, the Group does not recommend activating a satellite program based on the *Redstone* missile, but will reconsider this question and the possibilities of the ICBM program at its subsequent meetings when the critical items of the *VANGUARD* program are further advanced."[28]

On July 5, 1956, E.V. Murphree, DoD's Special Assistant for Guided Missiles, wrote to Reuben B. Robertson, Jr., the Deputy Secretary of Defense. Murphree stated that he had looked into the matter of using a *Jupiter* re-entry test vehicle for possibly launching a satellite into orbit. He also stated that the January 1957 test could be adapted to this purpose with little effort and no impact on the program and that an attempt could be made as early as September 1956, although this *would* affect the *Jupiter* program. Both dates were before the June 1957 start of the International Geophysical Year.

Murphree further noted that proposals for using the *Jupiter* were not new and that the original *Redstone* satellite and re-entry test vehicle proposals resulted from a common study (the early Orbiter proposal) which argued that the same vehicle could be used for both. He also stated that the first two tests of the *Jupiter* were essentially propulsion system tests and could accomplish much of their goals for that program even if used for satellite launch attempts. He continued: "There is, however, room for serious doubt that two isolated flight attempts would result in achieving a successful satellite, and the dates of such flights would be *prior* to the Geophysical Year for which a satellite capability is specifically required, and prior to the time when tracking instrumentation will be available." (emphasis added)[29]

Murphree said these facts had been taken into consideration at the time that the Office of the Secretary of Defense reviewed the satellite program and decided to assign the mission to the Navy group. He stated: "That decision was based largely on a conviction that the

VANGUARD proposal offered the greater promise for success. The history of increasing demands for funds for this program confirms the conviction that this is not a simple matter. I know of no new evidence available to warrant a change in that decision at this time."

The rest of Murphree's memorandum is extremely interesting and is reprinted below in full:

> While it is true that the *Vanguard* group does not expect to make its first satellite attempt before August 1957, whereas a satellite attempt could be made by the Army Ballistic Missile Agency as early as January 1957, little would be gained by making such an early satellite attempt as an isolated action with no follow-up program. In the case of *Vanguard*, the first flight will be followed up by five additional satellite attempts in the ensuing year. It would be impossible for the ABMA group to make any satellite attempt that has a reasonable chance of success without diversion of the efforts of their top-flight scientific personnel from the main course of the *Jupiter* program, and to some extent, diversion of missiles from the early phase of the re-entry test program. There would also be a problem of additional funding not now provided.
>
> For these reasons, I believe that to attempt a satellite flight with the *Jupiter* re-entry test vehicle without a preliminary program assuring a very strong probability of its success would most surely flirt with failure. Such probability could only be achieved through the application of a considerable scientific effort at ABMA. The obvious interference with the progress of the *Jupiter* program would certainly present a strong argument against such diversion of scientific effort.
>
> On discussing the possible use of the *Jupiter* re-entry test vehicle to launch a satellite with Dr. Furnas, he pointed out certain objections to such a procedure. He felt there would be a serious morale effect on the *Vanguard* group to whom the satellite test has been assigned. Dr. Furnas also pointed out that a satellite effort using the *JUPITER* re-entry test vehicle may have the effect of disrupting our relations with the nonmilitary scientific community and international elements of the IGY group.
>
> I don't know if I have a clear picture of the reasons for your interest in the possibility of using the *Jupiter* re-entry test vehicle for launching the satellite. I think it may be helpful if Dr. Furnas and I discuss this matter with you, and I'm trying to arrange for a date to do this on Monday.[30]

Robertson's special assistant, Charles G. Ellington, forwarded the memos to White House Staff Secretary Andrew Goodpaster. Goodpaster wrote on the bottom of Ellington's cover memo, "Secy.

Robertson feels no change should be made—per Mr. Ellington. Reported to President."[31]

On September 20, 1956, *Jupiter*-C rocket number RS-27 was launched from Cape Canaveral. It flew 3,355 miles, reached an altitude of 682 miles, and achieved a velocity of Mach 18. It carried a deactivated fourth stage that was filled with sand under specific orders from Major General Medaris. Medaris had received specific orders himself. He had been instructed by someone in the Pentagon to ensure that the *Jupiter*'s upper stage was not fueled and did not "accidentally" place a satellite in orbit before the International Geophysical Year. Wernher von Braun had been told to keep quiet, and did so.[32]

Thus, despite the obvious potential to launch a satellite sooner to beat the Soviets into space or simply to maximize the possibility that the United States would place something in orbit, no such effort was made. Despite much high-level discussion of the psychological impact of being first into space, neither the White House or Pentagon officials pursued this feat with great commitment. Indeed, there was no clear desire to simply be first.

HOW TO "BEAT" THE SOVIETS

As one would expect, the scientists involved in creating the program stressed that a scientifically useful program would enhance U.S. prestige. National Science Foundation Director Alan Waterman specifically noted that the schedule was less important than the prestige to be gained from a program that produced a major scientific breakthrough. In other words, being first but not scientifically significant *wasn't* as important for national prestige as accomplishing something scientifically noteworthy.[33] *One could win the race, but lose the scientific competition.*

Rather surprisingly, this was also the conclusion of the National Security Council's Planning Board. NSC 5520 had mentioned the possibility of the Soviets developing a satellite, but took no position on whether the U.S. should attempt to launch before the Soviets. The Defense Department's response in NSC 5522 stated with more urgency that the United States should be first, but must do so carefully. The NSC Planning Board, in November 1956, made an amazing statement:

> The USSR can be expected to attempt to launch its satellite before ours and to attempt to surpass our effort in every way. It is vitally important in terms of the stated prestige and psychological purposes that the United States make every effort to (1) make possible a successful launching as soon as practicable and (2) put on as effective an IGY scientific program as possible. The prestige and psychological set-backs inherent in a possible earlier success and larger satellite by the USSR could at least be partially offset by a more effective and complete scientific program by the United States. Even if the United States achieved the first successful launching and orbit, but the USSR put on a stronger scientific program, the United States could lose its initial advantage.[34]

Thus, one year before Sputnik, the NSC's official position was that the United States could lose prestige *even* if it launched a satellite first but the Soviets developed a better science program. Science, therefore, was a higher priority than schedule.

The issue of the propaganda value of launching a satellite had been mentioned in numerous documents, including Rand reports, intelligence assessments, and NSC 5520 and NSC 5522. In NSC 5522, the CIA equated it with the psychological impact of the atomic bomb. Clearly, many top U.S. policy makers felt that it was a major issue.

But Eisenhower always dismissed these concerns. He just did not believe that it would be that important.[35] His concern was with the political vulnerability of reconnaissance systems, not a race for prestige. He proved to be dreadfully wrong from a domestic political standpoint, but not necessarily from the standpoint of international law.

SPUTNIK CRISIS?

By the fall of 1957, the *Vanguard* satellite program was proceeding roughly on schedule, with first launch anticipated for late 1957 or early 1958, during the International Geophysical Year. No one, not the Huntsville team or the *Vanguard* engineers themselves, expected the first launch to be successful. All rockets blew up the first time they were launched. But *Vanguard* had a year and a half to place a satellite in orbit. A year and a half to fail. The primary criteria was that *Vanguard* achieve orbit before the American intelligence satellite was ready to fly.

The satellite reconnaissance program, however, was underfunded and not making significant progress. Rather surprisingly, this was the fault of Donald Quarles, who had served as Secretary of the Air Force and then risen to become Deputy Secretary of Defense. Although Quarles had immediately appreciated the potential for using a civilian scientific satellite to establish freedom of space for future military satellites, he was skeptical of the reconnaissance satellite then under development by Schriever's Western Development Division. Quarles felt that the program was unlikely to produce anything in the short term and he therefore refused to provide significant funding for development. As many of those involved in the program noted, Quarles created a self-fulfilling prophecy: Quarles thought that the program was not advanced to deserve funding, but the program was not advancing *because* it was under-funded.

Quarles had ordered those in charge of it to not build any actual hardware. Schriever's underlings essentially ignored the order, pushing ahead with a satellite design. But they lacked money. They had asked for $117 million in 1956 and got almost nothing. General Schriever was bewildered. "Now what can I do?" he asked later. "I can't understand it. One of the great problems is surprise attack, and here we're saying we can do a satellite reconnaissance program, and we get $3 million ... ."[36]

Eisenhower's scientific advisors Din Land and James Killian took renewed interest in the reconnaissance satellite by the fall of 1957 and the Science Advisory Committee even sponsored a special briefing for the White House on the subject on September 20.[37] A proposed interim reconnaissance satellite was gaining increased attention from Washington at the same time that the small *Vanguard* satellite was taking shape. The strategy was proceeding, but by October it was overtaken by events.

On October 4, 1957, the Soviet Union launched a 184 pound metallic ball called *Sputnik 1*. It was instantly major news around the world. On October 7, acting on the publicity generated in the wake of *Sputnik 1*, President Eisenhower asked Donald Quarles, by then Deputy Secretary of Defense, to explain why the United States was in the position it was. Goodpaster recorded Quarles' explanation in the minutes of the meeting: "The Science Advisory Committee had felt, however, that it was better to have the Earth satellite proceed separately from military development. One reason was to stress the peaceful character of the effort, and a second was to avoid the inclusion of material, to which foreign scientists might be given access, which is used in our own military rockets." Furthermore, "He [Quarles] went on to add that the Russians have in fact done us a good turn, unintentionally, in establishing the concept of freedom of international space—this seems to be generally accepted as orbital space, in which the missile is making an inoffensive passage."[38]

Two days later, Eisenhower mentioned the subject of getting "beaten" in another staff meeting. "When military people begin to talk about this matter, and to assert that other missiles could have been used to launch a satellite sooner, they tend to make the matter look like a 'race,' which is exactly the wrong impression."[39] As far as Eisenhower and Quarles were concerned, the United States had never been in a race and had certainly not been "beaten." The complex strategy had not gone entirely according to plan, but the end result was the same for them. But Eisenhower also had a blind spot. While he was concerned that taking certain actions like publicly racing the Soviets into space would prompt them into escalating the Cold War, he failed to understand that failing to take these actions could allow others to establish the agenda. He and his advisors were caught in a web of political manipulation that may have been too subtle and complex for their own good.

The true purpose of the early American space program, however, remained shrouded in secrecy for decades. It was only one of many secrets surrounding the program. There were others.

THE CIA AND THE SCIENTIFIC SATELLITE

When Alan Waterman met with Allen Dulles in early May 1955 to discuss NSF sponsorship of a scientific satellite, he also met with Richard Bissell, whom he described as "the one in Central Intelligence who is following this closely."[40] Bissell had reason to follow it closely, since he was then in the middle of managing the newly created U-2 reconnaissance aircraft program. Use of the U-2 hinged on issues of international law and the CIA therefore had good reason to pay attention to any subject involving international airspace.

But even more surprising is the discovery that, as the scientific satellite program continued and ran into significant cost overruns, the CIA actually provided money for it to continue. In April 1957, the Director of the Bureau of the Budget, Percival Brundage, sent a lengthy memo to Eisenhower on the *Vanguard* cost overruns. *Vanguard* was initially supposed to cost 15–20 million dollars. By spring 1957, it was projected to cost ten times that amount. Brundage recounted the funding difficulties of the program and stated, "Apparently, both the Department of Defense and the National Science Foundation are very reluctant to continue to finance this project to completion. But each is quite prepared to have the other do so." He also noted that the National Science Foundation had contributed an extra $5.8 million in funds to the Department of Defense to fund the program and that the CIA had contributed $2.5 million of its own money to the program as well.[41]

Why was the CIA, which had no official stake in the scientific satellite program and no space programs underway or even under study, willing to spend its own money on a civilian scientific space program? The answer is unknown. But it was likely due to Bissell's intervention, since he was the person delegated to follow the program and he was also in charge of the U-2 reconnaissance aircraft.[42] By April 1957, when the CIA provided the money to the scientific satellite program, the U-2 had already made nearly a dozen flights over the Soviet Union, each protested vigorously, but quietly, by the Soviets. Although the CIA did not have a reconnaissance satellite program at this time, it would be given one by Eisenhower less than a year later, a program code-named *Corona*. Not surprisingly, Richard Bissell was placed in charge of that endeavor as well. Bissell was in control of a substantial amount of funding for covert and technical operations at CIA and it is likely that the money to support the scientific satellite program came from these funds or from Director of Central Intelligence Allen Dulles' substantial discretionary account (which may have accounted for up to a sixth of the CIA's budget at the time).

Brundage concluded his memo to Eisenhower by noting that the Air Force had already started its own, much larger, reconnaissance satellite. "Therefore, whether or not the International Geophysical Year satellite project is completed, research in this area will not be dropped." But this missed the point, since the scientific satellite program had less to do with research than with establishing legal precedent.

FURTHER HIDDEN AGENDAS

The Soviets launched *Sputnik 2* on November 5 and it had an even more profound public relations impact than its predecessor. *Sputnik 2* was not only far larger than the *Sputnik 1*, but it also carried a dog, demonstrating a sophistication that belied early administration attempts to downplay the Soviet achievement. The size of the payload was sufficient for the carrying of an atomic bomb, which heightened Americans' fear that the Soviets could now effectively attack the United States. A little over a month later, the U.S. attempt to launch the *Vanguard* satellite ended in embarrassing failure. Clearly, in the realm of space exploration, the Soviets had taken a substantial lead over the United States. From an international legal standpoint, this should have rendered the subject of freedom of space moot, but it did not. Furthermore, the Eisenhower administration continued its practice of using deception to cover its real goals.

In November, newly-appointed Secretary of Defense Neil McElroy proposed centralizing control of the various American space projects then underway, such as *Vanguard* and WS-117L, along with advanced ballistic missile development. McElroy proposed that they be put in a Defense Special Projects Agency (DSPA), which would be responsible for whatever projects the Secretary would assign to it. The idea for the DSPA apparently came from the Science Advisory Council in mid-October, just days after both Sputnik and McElroy's nomination.[43] Eisenhower himself expressed the opinion that a fourth service should be established to handle the "missiles activity."[44] McElroy said that he was weighing the idea of a "Manhattan Project" for antiballistic missiles. The president thought that a separate organization might be a good idea for this problem.[45] In testimony before Congress, Quarles, whom some regarded as an Air Force partisan, stated that long-range, surface-to-surface missiles had been assigned to the Air Force because it possessed the targeting and reconnaissance capabilities to use them, *not* because it was uniquely an Air Force mission.[46] Space could conceivably be treated in the same way.

Although the plan for incorporating ballistic missile development in the new agency was eliminated, the new space agency idea proceeded. The Special Projects Agency would act as a central authority for all U.S. space programs and would essentially contract out missions to the separate services, civilian government agencies, and even universities and private industry. "Above the level of the three military services, having its own budget, it would be able to concentrate on the new and the unknown without involvement in immediate requirements and inter service rivalries." McElroy stated in front of Congress that "the vast weapons systems of the future in our judgment need to be the responsibility of a separate part of the Defense Department."[47] This proposal was placed in a DoD reorganization bill. At this point, it was still assumed that the entire American space program would remain under military control, although at the level of the Secretary of Defense in an office specially created to manage it.

Discussion of the Defense Special Projects Agency continued within the Administration. Its name was changed to the Advanced Research Projects Agency (ARPA) and Eisenhower sent a message to the Congress on January 7, requesting supplemental appropriations for the agency.[48] In early January the newly created President's Science Advisory Committee addressed the issue of ARPA.

On February 7, 1958, James Killian and Din Land, who was also a member of the President's Board of Consultants on Foreign Intelligence Activities, met with Eisenhower and his staff secretary, General Andrew Goodpaster. There they briefed him on the potential of both a recoverable space capsule and supersonic reconnaissance aircraft program, suggesting that in order to speed up the development of a reconnaissance satellite, the U.S. should pursue the recoverable capsule idea as an "interim" solution. Eisenhower accepted this recommendation at that time and the satellite program was soon named "*Corona.*"

An equally important result of this first meeting was the decision to finalize Secretary of Defense McElroy's proposal and create the Advanced Research Projects Agency (ARPA) to house highly technical defense research programs. General Electric executive Roy Johnson was named as its director. Eisenhower decided to give

ARPA control of all military space programs. The military man-in-space program, meteorological programs, and WS-117L would all be turned over to ARPA.

Sputnik also led to the creation of NASA to manage a civilian space program. Although Eisenhower had initially felt that the military could handle the task of space science, he was eventually persuaded that a civilian space agency was needed, in part by Vice President Nixon, who argued that a civilian agency was important for international prestige purposes. A military space agency could not be used as an effective means to "show the flag," particularly if the President was interested in keeping the Cold War from heating up. But a civilian space agency could be used to great effect for psychological purposes. The creation of these new bureaucracies proceeded apace, drawing much attention from the Congress and the public. ARPA was officially in charge of the nation's WS-117L reconnaissance satellite program, whose existence had leaked to the press in the aftermath of Sputnik.

While all this was being done, Eisenhower had also directed that the recoverable satellite program—the program which eventually became known as *Corona*—be handled by a covert team involving the Air Force and the CIA. For many of those who had been involved in it, including those who had proposed it, the recoverable satellite program simply ceased to exist. As far as they knew, WS-117L was the only satellite program that the United States had.

Those in charge of WS-117L eventually split it off into several programs, including the reconnaissance program known as *Sentry* (later *Samos*), an early warning satellite known as Midas, and Discoverer, an engineering and development program which was to include the launching of biomedical payloads such as mice and monkeys. In reality, Discoverer was nothing more than a cover for the *Corona* program, and *Sentry* appears to have been continued at least in part as a sleight of hand—to distract attention away from Discoverer. Canceling it would have been too suspicious and would have raised the ire of the Air Force. Continuing it kept attention elsewhere. While the American public paid attention to NASA and ARPA and *Sentry*, another program materialized into existence unnoticed. It was just like the strategy used earlier with the scientific satellite program.

FREEDOM OF SPACE MARCHES ON

By the first half of 1958 the issue of the militarization of space flared up again due to the expanding space programs. The National Security Council addressed it initially in a classified document in August 1958. Known as NSC 5814/1, the document stated that the United States must "seek urgently a political framework which will place the uses of U.S. reconnaissance satellites in a political and psychological context more favorable to the U.S. intelligence effort." Responding to this, the State Department declared that one of the priorities for the United States was establishing "an acceptable policy framework for the WS-117L program ... ."[49] Even within this highly-classified document, *Corona* was nowhere mentioned.

Protection of WS-117L then became the U.S. goal, and the State Department debated it internally and eventually brought it to the United Nations. At the suggestion of the United States, the United Nations created the U.N. Ad Hoc Committee on the Peaceful Uses of Outer Space (COPUOS). The COPUOS became the source of much heated debate over the next several years, as the Soviets refused to participate and instead complained about American nuclear weapons and overseas bases, stating that concessions on these issues were a prerequisite to their participation in COPUOS. Calls by COPUOS for cooperation between the superpowers on space projects were met with derision from the Soviets, who feared that such cooperative efforts would reveal the limitations of their ICBM technology. Much controversy was generated, but no actual policy. All the time that this was being discussed in the United Nations, the newspapers, and in classified State Department meetings, *Corona* continued its development totally unknown to even those at high levels of the government, who only discussed the legal protection of the overt WS-117L program.

The advent of military reconnaissance satellites themselves created their own legal issues. In January 1959, O.G. Villard, Jr., of Stanford University and a member of the National Academy of Sciences and the Space Science Board (SSB) wrote Lloyd V. Berkner, Chairman of the SSB. Villard expressed concern that the U.S. military might attempt to portray its satellite launchings as scientific in nature. This deception could have negative effects on American space science, particularly with regards to international cooperation. Villard stated that it was in the best interests of U.S. scientists that such deception not occur.

On January 28, 1959, Berkner brought the issue to the attention of Killian, who was by now Eisenhower's science advisor. On February 13, Killian in turn mentioned it to NASA Administrator T. Keith Glennan, and Gordon Gray, Eisenhower's Special Assistant for national security affairs. Gray felt that it might require further action by the National Security Council.[50] The results of their discussion of the issue are not known, but this discussion took place over a year after Eisenhower had directed that a small reconnaissance satellite program be peeled away from the bigger reconnaissance program (the one that Villner referred to in his letter) and be conducted covertly. Indeed, the first launch was scheduled to take place later in the month—under the cover of an engineering and scientific program.

The first Discoverer was launched on February 28, 1959. It was in actuality a test of equipment for the *Corona* program. Although the satellite did not reach orbit, the Soviets still protested the flight, decrying its military nature. This prompted Richard Leghorn, the architect of Eisenhower's 1955 "Open Skies" proposal and the president of Itek, which was then managing the manufacture of *Corona* cameras, to draft a proposal for the president titled "Political Action and Satellite Reconnaissance." In it Leghorn stated:

> The problem is not one of technology. It is not a problem of vulnerability to Soviet military measures. The problem is one of the political vulnerability of current reconnaissance satellite programs.
>
> For many years the U.S. has had overflight capabilities (aircraft and balloons) which have been substantially invulnerable to Soviet military countermeasures, but very vulnerable politically. Already the Communists (East Germany) have attacked Discoverer I as an espionage activity, and we can anticipate powerful Soviet political countermeasures to the Discoverer/*Sentry* series.

Leghorn continued:

> What is needed is a program to put reconnaissance satellites "in the white" through early and vigorous political action designed to:
> 1.   blunt in advance Soviet political countermeasures;

2. gain world acceptance for the notion that the surveillance satellites are powerful servants of world peace and security, and are not illegitimate instruments of espionage;
3. regain the political initiative of the "open skies" proposal.[51]

Leghorn was fully aware of all American reconnaissance satellite efforts and it is impossible to believe that his proposal was anything other than a continuation of his earlier thinking on overflight. But at this time the United States still had not orbited a covert reconnaissance satellite (and would not for over a year, suffering a string of failures with *Corona*) and was not *overtly* anywhere near flying a reconnaissance satellite. Taking any public position at all on the subject seemed premature at best and foolish at worst. From the White House's point of view it was best to let the political deliberations at the United Nations run their course. Leghorn's proposal went nowhere.

Gary Powers' U-2 reconnaissance aircraft was shot down on May 1, 1960, creating yet another public embarrassment for Eisenhower that in fact revealed a carefully planned and executed strategy to gather intelligence on the Soviet Union—a strategy that like *Genetrix*, *Corona* and *Vanguard* also relied upon a scientific cover story to mask the true mission of the program. Although the downing ruined the upcoming summit, the U-2 had proved an intelligence bonanza for the United States, something which Eisenhower did not wish to emphasize even in his television address to the nation following the incident. [52] He had no desire to reveal America's intelligence coup and no desire to inflame relations with the Soviet Union.

Soviet Premier Khrushchev used the event to maximum propaganda effect, rejecting Eisenhower's renewed proposal for "Open Skies." Khrushchev declared, "As long as arms exist our skies will remain closed and we will shoot down everything that is there without our consent." France's President de Gaulle then asked whether this would include satellites, noting that Soviet satellites had already carried cameras into space. Khrushchev then replied, "As for sputniks, the U.S. has put up one that is photographing our country. We did not protest; let them take as many pictures as they want."[53]

The State Department continued to discuss the subject in secret. By mid-1960, the Bureau of European Affairs at the State Department had even drafted a policy paper concerning the planned upcoming launch of the *Samos* reconnaissance satellite. One option proposed was an "open approach" which advocated the sharing of all reconnaissance photographs taken by *Samos*. This approach was considered "more likely to facilitate wide acceptance of photographic satellites than the 'closed' approach."[54] While the State Department was debating the merits of sharing reconnaissance satellite photographs with the world community (something that the United States had not done even with U-2 photographs), *Corona's* engineers were preparing for another launch. The memo was written August 12. Six days later *Corona* returned its first images of the Soviet Union. There was no talk in the White House of sharing the photos with anyone.

The State Department continued to debate the merits of the American space program, but gradually even *Samos* itself was enveloped within the dark cloak of secrecy that surrounded *Corona*. State Department discussions in general, and public statements in particular, ceased. "Freedom of space" had essentially been achieved. Further discussion was largely irrelevant. And *Corona* began to chalk up success after success, as the press focused on *Samos* and the public paid attention to NASA and its daring Mercury astronauts.

CONCLUSION

Any close study of the Eisenhower presidency reveals a president who frequently rejected overt and provocative policies in favor of covert actions intended to achieve the same result but without heightening Cold War tensions. It also reveals a President who saw little separation between national security and science and who was fully willing to use allegedly "civilian" science programs to cover military operations. Reconnaissance is perhaps the premier example of this. First with the U-2, then with *Corona*, Eisenhower chose to change direction and place the programs under different management primarily to avoid public scrutiny. He allowed the existing programs to continue before they slowly disappeared. He also used scientific cover stories extensively to cover the reconnaissance programs. Few expected these cover stories to last forever. They were, in the language of the intelligence community, "melting assets." But they worked for a time. Rather surprisingly, they did not appear to have any negative effects for the civilian scientific community.

There is no evidence among Eisenhower's scientist advisors of any real debate over this practice of using science as cover for intelligence programs. They seem to have not recognized any line between the scientific establishment and the national security establishment. The idea of the scientific establishment deliberately separating itself from the military establishment did not come until later. And perhaps most ironically of all, this deliberate blurring of the lines between scientific and military establishments that Eisenhower did so much to create was one of the things he warned of in his farewell address.

1. Major General Bernard A. Schriever, Commander, Western Development Division, "ICBM—A Step Toward Space Conquest," February 19, 1957.
2. General Bernard A. Schriever, interview by Dwayne A. Day, July 17, 1997.
3. "For Defense," *Newsweek*, March 4, 1957, p. 66.
4. J.R. Killian, Jr., to General Curtis E. LeMay, September 2, 1954, Papers of Curtis LeMay, Box 205, Folder B-39356, Manuscript Division, Library of Congress.
5. Information on the classified annexes comes from an interview by Donald E. Welzenbach with James Killian and is referenced in: Donald E. Welzenbach, "Science and Technology: Origins of a Directorate," *Studies in Intelligence*, Vol. 30, Summer 1986, RG 263, National Archives and Records Administration (hereafter referred to as NARA). Although the intelligence section of the TCP report remains classified and awaits review, the index has been declassified. It includes the word "satellites," but apparently in the context of satellite countries of the USSR "The Report to the President by the Technological Capabilities Panel of the Science Advisory Committee, February 14, 1955, Office of the Staff Secretary: Records of Paul T. Carroll, Andrew J. Goodpaster, L. Arthur Minnich and Christopher H. Russell, 1952–61, Subject Series, Alphabetical Subseries, Box 16, "Killian Report—Technological Capabilities Panel (2)", Dwight D. Eisenhower Library (hereafter referred to as DDE).
6. General Andrew Goodpaster, interview by Dwayne A. Day, March 19, 1996. Goodpaster went to the White House in October 1954 as a Colonel and was promoted to Brigadier General while there. He eventually rose to the rank of General and assumed command of Supreme Headquarters Allied Powers Europe (SHAPE) in 1969.
7. Although the majority of the contents of the intelligence section of the TCP report remain classified, crucial portions were printed in other documents which have now been declassified. For instance, a cover letter accompanying the report upon its delivery to the Policy Planning Staff at the Department of State declares that one of the intelligence panel's conclusions was the need for the "re-examination of the principles of freedom of space, particularly in connection with the possibility of launching an artificial satellite into an orbit about the earth, in anticipation of use of larger satellites for intelligence purposes." See "Report to the President on the Threat of Surprise Attack," Policy Planning Staff, Department of State, March 14, 1955, General Records of the Department of State: Records Relating to State Department Participation in the Operations Coordinating Board and the National Security Council, 1947–1963, Box 87, "NSC 5522 Memoranda", RG 59, NARA.
8. Another document, dated only two weeks later, called for the Department of State to discuss a TCP recommendation on freedom of space. It stated: "The present possibility of launching a small artificial satellite into an orbit about the earth presents an early opportunity to establish

a precedent for distinguishing between 'national air' and 'international space,' a distinction which could be to our advantage at some future date when we might employ larger satellites for intelligence purposes." Robert R. Bowie, "Memorandum for Mr. Phleger," Policy Planning Staff, Department of State, March 28, 1955, Department of State Central Files, 711.5/3–2855.

9. In fact, Schriever, who was responsible for one of the most advanced weapons development projects in the United States military, did not learn of it until later, when he overheard someone discussing it on the phone and attempting to conceal its identity.

10. Joseph Kaplan, Chairman, United States National Committee, International Geophysical Year 1957–58, National Academy of Sciences, to Dr. A.T. Waterman, Director, National Science Foundation, March 14, 1955.

11. Alan T. Waterman, Director, Memorandum for Mr. Robert Murphy, Deputy Under Secretary of State, March 18, 1955.

12. Robert Murphy, "Memorandum for Dr. Alan T. Waterman, Director, National Science Foundation," April 27, 1955.

13. Alan T. Waterman, Director, to Donald A. Quarles, Assistant Secretary of Defense (Research and Development), May 13, 1955.

14. The newly released part of the document is in italics. NSC 5520, May 20, 1955, General Records of the Department of State: Records Relating to State Department Participation in the Operations Coordinating Board and the National Security Council, 1947–1963, Box 112, "NSC 5520", RG 59, NARA.

15. Ibid. This portion of the document remained classified until 1995.

16. "James S. Lay, Jr., Executive Secretary, National Security Council, Memorandum for the National Security Council, "U.S. Scientific Satellite Program," November 9, 1956, General Records of the Department of State: Records Relating to State Department Participation in the Operations Coordinating Board and the National Security Council, 1947–1963, Box 86, "NSC 5520—US Scientific Satellite Program (Memoranda)", RG 59, NARA.

17. Memorandum, "Discussion at the 250th Meeting of the National Security Council, Thursday, May 26, 1955," May 27, 1955, Files of Dwight D. Eisenhower as President, Ann Whitman File, NSC Series, Box 6, "250th Meeting of NSC, May 26, 1955," DDE.

18. National Security Council NSC 5522, June 8, 1955, Comments on the Report to the President by the Technological Capabilities Panel, p. S-5, White House Office, Office of the Special Assistant for National Security Affairs: Records, 1952–61, NSC Policy Papers, Box 16, Folder NSC 5522 Technological Capabilities Panel, DDE.

19. National Security Council NSC 5522 June 8, 1955 Comments on the Report to the President by the Technological Capabilities Panel, p. A-55-6.

20. Robert R. Bowie, Policy Planning Staff, Department of State, "Recommendations in the Report to the President by the Technological Capabilities Panel of the Science Advisory Committee, ODM (Killian Committee): Item 2—NSC Agenda 10/4/56," General Records of the Department of State: Records Relating to State Department Participation in the Operations Coordinating Board and the National Security Council, 1947–1963, Box 87, "NSC 5522 Memoranda," RG 59, NARA.

21. These "gentlemen's agreements" were a tenet of the Cold War. They were certainly common in the field of espionage, where the superpowers frequently chose to keep quiet when they caught the other side spying. They were also part of a larger unwritten code of cooperation between the two nations, such as the tacit, unspoken agreement not to seek the assassination of the other side's leaders.

22. Constance McLaughlin Green and Milton Lomask, Vanguard: A History (Washington, DC: Smithsonian Institution Press, 1971), pp. 34–56.

23. Ibid., p. vi. Green and Lomask added: "To these observations, I can add from my own experience that inter-service rivalry exerted strong influence; also, that any conclusion

drawn would be incomplete without taking into account the antagonism still existing toward von Braun and his co-workers because of their service on the German side of World War Two."

24. Fred Durant, interview by Dwayne A. Day, July 27, 1997.

25. *Redstone Arsenal Complex Chronology, The ABMA/AOMC Era, 1956–62*, Redstone Arsenal webpage.

26. Colonel A.J. Goodpaster, "Memorandum for Record," June 7, 1956, White House Office, Office of the Staff Secretary: Records, 1952–1961, Box 6 "Missiles and Satellites," DDE.

27. Homer J. Stewart, Chairman, Advisory Group on Special Capabilities, Memorandum for the Assistant Secretary of Defense (R&D), "VANGUARD and REDSTONE," June 22, 1956, White House Office, Office of the Staff Secretary: Records, 1952–1961, Box 6, "Missiles and Satellites," DDE.

28. Stewart's memorandum was stamped "SECRET," but there is some doubt as to whether it was actually written in *May* 1956 instead of June. It is rare for a report of a meeting to be written two months after the meeting. Furthermore, the memo also mentions the Group's *upcoming* meeting on June 19 and 20 concerning the propulsion systems for Vanguard and invites contractor representatives to attend this meeting, which would already have happened by the time the memo was written. The June 22 date may be a typo.

29. E.V. Murphree, Special Assistant for Guided Missiles, Memorandum for Deputy Secretary of Defense, "Use of the JUPITER Re-entry Test Vehicle as a Satellite," July 5, 1956, White House Office, Office of the Staff Secretary: Records, 1952–1961, Box 6, "Missiles and Satellites," DDE.

30. In another brief, one-page memorandum from C.C. Furnas to the Deputy Secretary of Defense, dated July 10, 1956 and stamped "Secret," Furnas mentioned the meeting that he and Murphree had with Robertson on July 9. Furnas used this memorandum as a cover letter to forward the previous report to him by Homer Stewart's Advisory Group on Special Capabilities. He concluded by saying "I trust that this will serve your purpose in reporting your evaluation of the suggestion that a Redstone vehicle will be used." C.C. Furnas, Assistant Secretary of Defense for Research and Development, Memorandum for Deputy Secretary of Defense, July 10, 1956, White House Office, Office of the Staff Secretary: Records, 1952–1961, Box 6, "Missiles and Satellites," DDE.

31. William Ewald, *Eisenhower the President: Crucial Days 1951–1960* (Englewood Cliffs, NJ: Prentice-Hall, 1981), p. 284.

32. Ernst Stuhlinger, letter to Dwayne A. Day, August 23, 1997.

33. Alan T. Waterman, National Science Foundation, Memorandum to Mr. Percival Brundage, Bureau of the Budget, "Funding of Earth Satellite Program, International Geophysical Year," April 7, 1956, General Records of the Department of State: Records Relating to State Department Participation in the Operations Coordinating Board and the National Security Council, 1947–1963, Box 86, "NSC 5520—U.S. Scientific Satellite Program (Memoranda)," RG 59, NARA.

34. James S. Lay, Jr., Executive Secretary, National Security Council, Memorandum for the National Security Council, "U.S. Scientific Satellite Program," November 9, 1956, with attached: "Draft Report on NSC 5520, U.S. Scientific Satellite Program Background," General Records of the Department of State: Records Relating to State Department Participation in the Operations Coordinating Board and the National Security Council, 1947–1963, Box 86, "NSC 5520—US Scientific Satellite Program (Memoranda)," RG 59, NARA.

35. Goodpaster interview.

36. General Bernard Schriever interview by ArcWelder Films.

37. David Z. Beckler, Executive Officer, Science Advisory Committee, Memorandum for General Goodpaster, "Special Briefing," September 19, 1957, Office of the Staff Secretary: Records of Paul T. Carroll, Andrew J. Goodpaster, L. Arthur Minnich and Christopher

H. Russell, 1952–61, Subject Series, Alphabetical Subseries, Box 23, "Science Advisory Committee (2) Sept.-Oct. 1957, " DDE.

38. Brigadier General Andrew Goodpaster, Memorandum of Conference with the President, October 7, 1957, Office of the Staff Secretary: Records of Paul T. Carroll, Andrew J. Goodpaster, L. Arthur Minnich and Christopher H. Russell, 1952–61, Subject Series, Department of Defense Subseries, Box 6, "Missiles and Satellites," DDE.

39. Brigadier General Andrew Goodpaster, Memorandum of Conference with the President, (following McElroy swearing in) October 9, 1957, Office of the Staff Secretary: Records of Paul T. Carroll, Andrew J. Goodpaster, L. Arthur Minnich and Christopher H. Russell, 1952–61, Subject Series, Department of Defense Subseries, Box 6, "Missiles and Satellites," DDE.

40. Waterman to Quarles, May 13, 1955.

41. Percival Brundage, Director, Bureau of the Budget, Memorandum for the President, "Project VANGUARD," April 30, 1957, White House Office, Office of the Staff Secretary: Records, 1952–1961, Box 6 "Missiles and Satellites," DDE.

42. Former CIA Deputy Director of Science & Technology Albert "Bud" Wheelon has speculated that the money probably came from Allen Dulles' substantial discretionary DCI budget.

43. Goodpaster interview.

44. Eisenhower's comments on this subject appear in numerous documents. For instance, in October 1957 Goodpaster reported "The President went on to say he sometimes wondered whether there should not be a fourth service established to handle the whole missiles activity." Brigadier General A.J. Goodpaster, "Memorandum of Conference with the President, October 11, 1957, 8:30 AM," October 11, 1957, Ann Whitman File, DDE Diary Series, Box 67, "Oct. 57 Staff Notes (2)," DDE. In January 1958 Goodpaster reported "In the course of the discussion the President indicated strongly that he thinks future missiles should be brought into a central organization." Brigadier General A.J. Goodpaster, Memorandum of Conference with the President, January 21, 1958," January 22, 1958, Office of the Staff Secretary: Records of Paul T. Carroll, Andrew J. Goodpaster, L. Arthur Minnich and Christopher H. Russell, 1952–61, Subject Series, Department of Defense Subseries, Box 6, "Missiles and Satellites, Vol. II (1) [January-February 1958]" DDE. In February 1958, Goodpaster reported "The President said that he has come to regret deeply that the missile program was not set up in OSD rather than in any of the services." Brigadier General A.J. Goodpaster, "Memorandum of Conference with the President, February 4, 1958 (following Legislative Leaders meeting)," February 6, 1958, Office of the Staff Secretary: Records of Paul T. Carroll, Andrew J. Goodpaster, L. Arthur Minnich and Christopher H. Russell, 1952–61, Subject Series, Department of Defense Subseries, Box 6, "Missiles and Satellites, Vol. II (1) [January-February 1958]," DDE.

45. Brigadier General A.J. Goodpaster, Memorandum of Conference With the President, October 11, 1957, Ann Whitman File, DDE Diary Series, Box 27, "Oct. 57 Staff Notes (2)," DDE. In February, another memo states "The President said that he has come to regret deeply that the missile program was not set up in OSD rather than in any of the services. Personal feelings are now so intense that changes are extremely difficult." Brigadier General A.J. Goodpaster, Memorandum of Conference With the President, February 4, 1958 (Following Legislative Leaders meeting)," February 6, 1958, Office of the Staff Secretary: Records of Paul T. Carroll, Andrew J. Goodpaster, L. Arthur Minnich and Christopher H. Russell, 1952–61, Subject Series, Department of Defense Subseries, Box 8, "Missiles and Satellites, A National Integrated Missile and Space Vehicle Development Program, December 10, 1957," DDE.

46. Robert Frank Futrell, *Ideas, Concepts, Doctrine, Vol. 1*, (Washington, DC: U.S. Government Printing Office, 1989), p. 589. The comments on Quarles' partisanship come from the interview with Goodpaster.

47. *Organization and Management of Missile Programs, Hearings before a Subcommittee of the Committee on Government Operations*, U.S. House of Representatives, 86th Cong., 2d Sess. (Washington, D.C.: U.S. Government Printing Office, 1959), p. 133.

48. Ibid.

49. McDougall, *The Heavens and the Earth*, pp. 182–83.

50. Gordon Gray, Special Assistant to the President, to Brigadier General Andrew J. Goodpaster, February 16, 1959, with attached: James R. Killian, Jr., to Gordon Gray, Special assistant to the President, February 13, 1959; James R. Killian, Jr., to Dr. T. Keith Glennan, Administrator, National Aeronautics and Space Administration, February 13, 1959; Lloyd V. Berkner, Chairman, Space Science Board, National Academy of Sciences, to Dr. James R. Killian, Jr., Special Assistant to the President for Science and Technology, January 28, 1959; O.G. Villard, Jr. Space Science Board, to Dr. L.V. Berkner, President, Associated Universities, Inc., January 22, 1959, Office of the Staff Secretary: Records of Paul T. Carroll, Andrew J. Goodpaster, L. Arthur Minnich and Christopher H. Russell, 1952–61, Subject Series, Department of Defense Subseries, Box 15, "Space [January-June 1959]," DDE.

51. Richard S. Leghorn, Political Action and Satellite Reconnaissance [Draft], April 24, 1959, Office of the Staff Secretary: Records of Paul T. Carroll, Andrew J. Goodpaster, L. Arthur Minnich and Christopher H. Russell, 1952–61, Subject Series, Department of Defense Subseries, Box 15, "Space [January-June 1959]," DDE.

52. The U-2 also raised once again the issue of where airspace ended and space began. At the Eleventh International Astronautical Federation Congress in Stockholm, Sweden, Spencer M. Beresford presented a paper which connected the U-2 and violations of international airspace with the possibility of future flights by military MIDAS and SAMOS flights. A State Department official obtained a copy of Beresford's paper before his presentation and notified the U.S. Information Service in Stockholm that the paper raised a number of "highly sensitive" topics which the United States should not comment. W.E. Gathright, to USIS-Stockholm, "TOUSI II, Joint State USIA Message," August 12, 1960, with attached: Remarks of Spencer M. Beresford, United States of America, at the Eleventh Annual Congress of the International Astronautical Federation, Stockholm, Sweeden, August 16, 1960, General Records of the Department of State, Bureau of European Affairs, Office of Soviet Union Affairs, Subject Files, 1957–1963, Box 6, "12 Satellite and Missile Programs," RG 59, NARA.

53. Khrushchev was referring to the U.S. Tiros weather satellite launched in August.

54. Foy D. Kohler, Bureau of European Affairs, to Phillip J. Farley, "SAMOS," July 18, 1960, General Records of the Department of State, Bureau of European Affairs, Office of Soviet Union Affairs, Subject Files 1957–1963, Box 6, "12 Satellite and Missile Programs," RG 59, NARA.

CHAPTER 7

Before Sputnik: National Security and the Formation of U.S. Outer Space Policy

Kenneth A. Osgood

The successful launch of the first earth satellite by the Soviet Union on October 4, 1957, shocked the world, but nowhere was there greater consternation than in Washington, D.C.[1] The United States had always been first—first to exploit the splitting of the atom, first to detonate a thermonuclear device—in the Cold War's scientific arena. To many Americans, the forces of democracy had been suddenly beaten to the punch by backward Communists in the Soviet Union. In the wake of the announcement from Moscow, American pundits and congressional leaders pressured the Eisenhower administration to redouble its efforts to place an American satellite in orbit. In what has been described as the "Sputnik panic,"[2] Americans demanded that the United States reassert the superiority of American technology by surpassing the Soviet Union in space exploration.

The widespread domestic outcry which followed *Sputnik 1* has diverted attention away from the formative years of the U.S. outer space program. By examining the period from 1953–1957 from a national security planning perspective, this essay contests the view that "political realism" and domestic considerations alone propelled the space race.[3] Further, it challenges the conventional view that the Eisenhower administration failed to recognize the prestige value of space flight prior to the *Sputnik 1* launch. Rather, the American space effort grew logically out of the administration's evolving Cold War strategy. Based to a great extent on psychological considerations, this strategy accorded science and technology a significant and increasingly important role in Cold War planning. The administration believed that the United States needed to demonstrate technological superiority over the Soviet Union or risk forfeiting its position of leadership in the free world. By taking the lead in the scientific exploration of outer space, moreover, the United States could also demonstrate its commitment to scientific progress and the promise of peace. Viewed this way, *Sputnik 1* appears less instrumental than conventional wisdom holds. The American space effort, rather than being viewed as a response to Soviet success, should be seen as part of a broader strategy to demonstrate not only the technological and military superiority of the United States, but also (paradoxically) its peaceful intentions.

Historians have tended to view American space policy largely as a product of the domestic outcry which followed the *Sputnik 1* launch. In most accounts, President Dwight D. Eisenhower, pressured by a panic-stricken press corps and opportunistic politicians, abandoned his policy of fiscal restraint in order to reassure a badly-shaken nation and quiet his political opponents.[4] Most historians, notably Walter McDougall, Stephen Ambrose, and Robert Divine, agree that prior to *Sputnik 1*, Eisenhower gave military and strategic imperatives priority over questions of national prestige.[5] For Eisenhower, according to this interpretation, space-related research served two primary objectives. First, anticipating the day when the high-flying U-2 spy plane would lose its invulnerability to Soviet surface-to-air missiles (SAM), he sought the development and production of spy satellites to monitor Soviet military developments.[6] Second, the President wanted to bolster the American deterrent capability through the development of ballistic missile systems. It was not until after *Sputnik 1*, these authors contend, that the promotion of American prestige became a primary objective of U.S. outer space policy. Russian "firsts" in space and embarrassing American launch failures caused the administration to reorder its priorities. Henceforth, Eisenhower placed a premium on programs which he considered unnecessary from a national security perspective but imperative to promoting American prestige. Ultimately, these authors maintain that Sputnik did more than fire the starter's pistol for the space race. It also ushered in a new emphasis on prestige as a critical component of the Eisenhower administration's foreign policy.[7]

Admittedly, there is much to be gleaned from this interpretation; however, it tells only part of the story. From the decision to pursue the satellite program to the selection of a launch vehicle, psychological considerations permeated U.S. outer space policy. The American effort was predicated from the start on the belief that the nation which first successfully launched a satellite would be in a position to reap considerable prestige and psychological benefits—which could then be used as international currency in the struggle between Moscow and Washington. Because a powerful ballistic missile would hurl the satellite into orbit, a successful launch would signal to world audiences that the United States possessed an effective intercontinental ballistic missile (ICBM) capability. Eisenhower's National Security Council (NSC) believed that the satellite's unambiguous relationship to ICBM technology could affect the will of neutral nations— especially in the developing world—to resist Communist threats. Furthermore, administration officials saw a link between demonstrations of scientific and technical prowess and American credibility. They understood that the inevitable nuclear stalemate that would result from deployment of ICBMs placed a premium on alternative demonstrations of national power.[8]

PARITY AND THE "PEACE OFFENSIVE"

Against a background of steadily increasing Soviet nuclear capabilities, the Eisenhower administration formally committed the United States to the scientific exploration of space in 1955. In 1949 the Soviets shocked the Truman administration with the detonation of an atomic bomb. Four years later the Soviets again demonstrated their technical proficiency by exploding a hydrogen weapon. At the same time, reports of Soviet successes in rocketry provided evidence that soon they would possess ballistic missiles.[9] These developments came at a time when the United States—under the guidance of Eisenhower's "new look" strategy—relied heavily on nuclear weapons to deter Soviet aggression.[10] To the consternation of policymakers, Soviet technological feats suggested that the era of American overwhelming superiority would soon give way to an age of nuclear stalemate.[11] Moscow's conciliatory tactics compounded the situation. In the years following Stalin's death, the new Soviet leadership helped bring the Korean War to a close; negotiated a peace treaty with Austria; pursued improved relations with such countries as Israel, Yugoslavia, and Greece; agreed to a summit with Western leaders; and, in general, emphasized "peaceful coexistence" over confrontation with the West.[12] These developments threatened to render obsolete assumptions on how best to deal with the Soviet threat and forced officials in Washington to reevaluate their strategy for opposing the Soviet Union.

The National Security Council (NSC) warned in 1953 that Soviet nuclear capabilities combined with the new "soft line" coming from Moscow strengthened the Soviet hand by undermining American leadership. National Security Council report 162/2—the policy paper which guided the implementation of the "New Look" as basic national security policy—predicted that American retaliatory power remained strong enough to deter the Soviets from a general attack on the West.[13] Nevertheless, the NSC cautioned that fear of involvement in nuclear war circumscribed the will of American allies to risk war and prematurely led to allied pressure to negotiate. According to this analysis, Soviet peace overtures exacerbated this trend. Because American allies "tend to see the actual danger of Soviet aggression as less imminent than the United States does," the NSC warned that they might fall prey to Soviet tactics designed to undermine the cohesion of the free world coalition. Assuming that Soviet peace gestures were "merely designed to divide the West," the NSC warned that the twin tactics of nuclear terror and peace rhetoric portended serious political and psychological difficulties for United States leadership. "Using both the fear of atomic warfare and the hope of peace," NSC 162/2 concluded, "such political warfare will seek to exploit differences among members of the free world, neutralist attitudes, and anti-colonial and nationalist sentiments in underdeveloped areas."[14]

Articulated in mid-1953, these themes achieved greater significance in the second half of the decade. Although the United States enjoyed a clear preponderance of power vis-à-vis the Soviet Union throughout the 1950s, alarming reports of Soviet technological and military growth and assertive diplomatic and economic initiatives by the Soviets in the third world caused the U.S. to reassess the very *nature* of the Soviet threat.[15] As Robert McMahon has shown, U.S. strategists deviated from the previous administration's emphasis on military and geostrategic concerns. They emphasized instead the political, ideological, and psychological challenges posed by Soviet conciliatory tactics and initiatives in the developing world.[16] The United States continued to fret over the ramifications of the USSR's growing nuclear might to the cohesion of western alliances. But the dismantling of European empires and the emergence of dozens of newly independent states introduced a dynamic new variable into the international equation—one that threatened to alter the international *status quo* in favor of the Soviet Union. Administration officials feared that these previously "peripheral" states would gravitate *voluntarily* into the Soviet orbit, as a result of sympathy to Communist ideology, lingering hostility to European imperialists, material necessity, or admiration for Soviet industrial and technological feats.[17]

These developments led to a major reappraisal of basic national security policy, formalized in NSC 5501 and adopted in January 1955. This new report emphatically reiterated the belief that Soviet peace tactics and the approach of nuclear balance posed the foremost challenges to the United States.[18] "Greater receptivity by the allies to Soviet peace overtures" and "growing fears of atomic war on the part of the allies" threatened serious strains between the United States and its major allies. Furthermore, the growth of Soviet military only encouraged Communist expansion in the developing world, which the paper described as "a major source of weakness in the position of the free world." To meet these challenges, NSC 5501 advised, the United States should place more emphasis on building the strength and cohesion of the non-Communist world, including *both* developing nations and major industrialized allies. Solidifying principal alliances and shoring-up American leadership in the third world, then, achieved a new urgency in U.S. policy.

The Eisenhower administration responded with a greater emphasis on psychological and political activities. On the one hand, the United States needed to fortify its striking power to reassure American allies of its defense capabilities. On the other, the competition between the Soviet Union and the United States to acquire political, economic and military support from uncommitted countries meant that the United States had to present itself as the nation best-suited for leadership. Soviet initiatives challenged the United States to demonstrate its credentials as a promoter of peace and world development. As NSC 5501 stated, "The ability of the free world, over the long pull, to meet the challenge and competition of the Communist world will depend in large measure on the capacity to demonstrate progress toward meeting the basic needs and aspirations of its people." This meant encouraging modernization across the globe, fostering international trade, moderating disputes within the free world, and developing a sound economy. As the administration fine-tuned and modified this policy in the years between NSC 5501 and Sputnik, it also required countering Soviet technological feats with American successes. At the same time, Soviet tactics designed to contrast the belligerence of the United States with the peaceful intentions of the Soviet Union mandated that the U.S. conduct its Cold War activities in a way that ensured their reception as peaceful and nonaggressive.

PRELUDE TO LIFTOFF

While policymakers deliberated on the repercussions of nuclear parity and the Soviet "peace offensive" to national security, scientists, and military personnel investigated the feasibility of launching a "world-circling spaceship." These studies led to a scientific satellite program geared towards countering Soviet psychological gains. Early satellite proposals concentrated primarily on the scientific, reconnaissance, and military utility of satellite vehicles, but they also presciently noted the political and psychological ramifications of satellites. The Air Force think-tank RAND forewarned as early as 1946 that the achievement of a satellite craft "would inflame the imagination of mankind, and would probably produce repercussions in the world comparable to the explosion of the atomic bomb."[19] This study had little appreciable impact, but a subsequent report by Manhattan Project scientist Aristid V. Grosse brought the potential propaganda consequences of a Soviet first launch directly to the top levels of the government.[20] Grosse's far-sighted report, presented to Eisenhower's Assistant Secretary of Defense for Research and Development Donald Quarles in August 1953, warned that because the Soviet Union had trailed the United States in the development of atomic and hydrogen warheads, it might attempt to take the lead in the development of a satellite. Noting that the satellite "would have the enormous advantage of influencing the minds of millions of people the world over," Grosse accurately predicted that the Soviets might forego complicated instrumentation in favor of putting the satellite into orbit before the United States. "If the Soviet Union should accomplish this ahead of us," he warned, "it would be a serious blow to the technical and engineering prestige of America the world over. It would be used by Soviet propaganda for all it's worth."[21]

A number of space flight recommendations by prominent scientists and military leaders followed Grosse's report. Of special significance was a study delivered by the Technological Capabilities Panel (TCP) in February 1955. James Killian, president of the Massachusetts Institute of Technology, coordinated this influential two-volume inquiry into the problem of surprise attack.[22] In addition to its important recommendations for accelerating ballistic missile programs and for providing better strategic warning, the TCP called for the use of artificial satellites for intelligence purposes.[23] All three services proposed, or had already in progress, satellite programs. These included the Air Force WS-117L reconnaissance satellite program (later CORONA under CIA auspices), the Navy's scientific satellite proposal (later renamed *Vanguard*), and the Army Project Orbiter (later renamed Explorer). Additionally, civilian scientists urged the United States to contribute a scientific satellite as a contribution to the International Geophysical Year (IGY). In May 1955 the issue of U.S. government support for a scientific satellite came before the National Security Council, which agreed to launch such a satellite as a contribution to the IGY.[24]

In analyzing the NSC's decision to launch a scientific satellite, historians have stressed that the President agreed to the proposal only half-heartedly. Walter McDougall and Robert Divine describe this early program as a relatively unimportant, low-priority facade, designed to divert attention away from the more important military and reconnaissance programs and to establish the legal principle of "freedom of space." In McDougall's words, the NSC gave "indubitable primacy" to the protection of the military and reconnaissance programs while the prescient findings of Aristid Grosse "vanished into White House files."[25] This interpretation is consistent with the argument that Eisenhower, because of his dogmatic commitment to fiscal responsibility, only belatedly—and under considerable public and congressional pressure—pursued space spectaculars based on shoring-up American prestige. Eisenhower either did not recognize the political ramifications of being first in space, this view suggests, or consciously chose to ignore them. Consequently, historians have found in Eisenhower's aversion to prestige-oriented space stunts an easy explanation for the American failure to beat the Soviets into space.

This interpretation, however, accords too much weight to Eisenhower's post-missile gap recollections and exaggerates his personal role in directing U.S. space efforts. As pundits and congressional leaders called the President to account for failing to beat the Soviets into space, Eisenhower defended his actions by claiming the United States was not racing the Soviet Union to begin with.[26] Yet as Bulkeley has shown, the "bogus" distinction he drew between the allegedly peaceful, scientific, and disinterested American program, and the militaristic Russian one was merely a "damage-limiting public-relations exercise."[27] Eisenhower's largely self-serving memoirs too should be treated with care. Written after the missile gap episode and after Kennedy's victory in the early 1960s, Eisenhower's memoirs were clearly slanted to resurrect his damaged image, to highlight his restraint in light of JFK's free-spending policies.[28] While Eisenhower revisionists have correctly refuted the image of Eisenhower as an inactive president who followed public opinion rather than leading it, some have tended to overestimate the extent to which Eisenhower involved himself in policy making. Admittedly, Eisenhower's balanced budget mandate did limit the scope and expense of national security programs and he did devote his personal attention overwhelmingly to missile and reconnaissance programs. Yet in space policy especially, Eisenhower only established broad guidelines regarding cost and priority. Having delegated the authority elsewhere, he left the matter to the executing agencies. The "hidden hand" President clearly deferred to the advice and judgment of his advisors in this case.[29]

The traditional interpretation also overstates the importance of the "freedom of space" issue to the scientific satellite program.[30] Because the TCP recommended that government agencies re-examine international law to determine if artificial satellites violated air-space agreements—and because the IGY provided a convenient opportunity to establish the legal precedent for later reconnaissance satellites—this argument holds that the United States participated in the IGY primarily as "cover" for its intelligence operations. But the Eisenhower administration was less concerned with the "freedom of space" principle than it appears. The June 1955 interdepartmental progress report on the status of TCP recommendations, NSC 5522, supports this view.[31] Submitted *after* the authorization of the IGY satellite, NSC 5522 reported that the Departments of State, Treasury, Defense, and Justice all concurred that the launching of an artificial satellite was permissible under international law. The Departments reported that "by customary law every State has exclusive sovereignty 'over the space above its territory'. However, *air space ends with the atmosphere.* There has been no recognition that sovereignty extends into the airless space beyond the atmosphere."[32] The Departments expressed little concern for U.S. vulnerability to criticism on that front. The IGY satellite, according to Defense Department comments, was intended "for propaganda and scientific purposes." Its connection to the freedom of space principle provided a supplementary motive, but did not drive early U.S. outer space policy.

What about intelligence? To be sure, intelligence applications undoubtedly provided the primary impetus for satellite research and development. Not coincidentally, the CIA generously contributed funds to the IGY program, a sign of its importance to the intelligence community.[33] However, CIA funding of the scientific satellite does not by itself prove the Agency's overriding concern for the legality of reconnaissance overflight. Instead, the evidence suggests that Central Intelligence conceived of the IGY satellite in terms of its psychological significance to U.S. leadership. CIA comments on the TCP satellite recommendation devoted five lengthy paragraphs to discussion of its "psychological warfare value," only one short paragraph to intelligence applications, and not one paragraph to freedom of space. As the Agency commented, the Soviets undoubtedly endeavored to "further her influence over neutralist states and to shake the confidence of states allied with the United States." Dwelling on the psychological ramifications of a Soviet first launch, the CIA argued that should the Soviets launch a satellite before the United States, the American reputation as the scientific and industrial leader of the world would be called into question—a Soviet first launch would provide Soviet propaganda with "sensational and convincing evidence of Soviet superiority" to neutral and allied states. Repeating Grosse's earlier prediction, the CIA advised: "The nation that first accomplishes this feat will gain incalculable prestige and recognition throughout the world."[34]

That this view reached the highest levels is reflected in CIA director Allen Dulles's illuminating remarks at a May 1957 NSC meeting. Having briefed the Council on the limited intelligence the CIA had acquired on the Soviet project, Dulles remarked on the propaganda consequences of a U.S. abandonment of the satellite program. "If the Soviets succeeded in orbiting a scientific satellite and the United States did not even try to, the USSR would have achieved a propaganda weapon which they could use to boast the superiority of Soviet scientists. In the premises, the Soviets would also emphasize the propaganda theme that our abandonment of this peaceful scientific program meant that we were devoting the resources of our scientists to warlike preparations instead of peaceful programs."[35] Dulles suggested that the United States should continue the program not because it was necessary for intelligence applications, nor because it was necessary to establish the principle of freedom of space. Instead, as Assistant Secretary of State Christian Herter commented in reply to Dulles's analysis, the United States should launch a scientific satellite "because of the prestige it would confer on the United States."[36]

As the comments from the Central Intelligence Agency, State Department, and Defense Department suggest, the Eisenhower administration was not insensitive to the relationship between American technology and U.S. prestige. They hardly failed to appreciate the psychological significance of space flight. The IGY satellite program, it seems, was from the outset a defensive rear-guard action designed to protect the United States from a Soviet propaganda broadside. The United States had to get a satellite into space before— or at least not much later than—the Soviets, and it had to design its program in such a way that would deflect potentially damaging Soviet propaganda.

The timing of the decision itself reflects this view. Scientists had been debating the merits of an Earth satellite for several years and the TCP recommended a reconnaissance satellite in early 1955, but not until the Soviets declared their intention to launch a satellite did proposals in the United States receive pressing urgency. When, on April 16, 1955, the Soviet Union announced the creation of a space flight commission charged with orbiting a space laboratory, American scientists and policymakers reacted swiftly.[37] Joseph Kaplan of the National Academy of Scientists wrote to Alan T. Waterman, Director of the National Science Foundation, calling for swift executive action on the IGY proposal. "I should like at this time to dwell briefly on the urgency of this matter," Kaplan wrote after the Soviet announcement, reminding Waterman that if funds were not forthcoming by July, it would be virtually impossible for the satellite launch to take place before the end of the IGY in December 1958. To make perfectly clear the importance of immediate action, Kaplan enclosed a copy of *The Washington Post* article reporting the Soviet announcement and added, "the critical shortage of time cannot be over-emphasized."[38] Kaplan's sentiment is reflected in the official actions of the U.S. National Committee for the International Geophysical Year (USNC-IGY). When the USNC-IGY gave its formal approval to the project at its May 18, 1955, meeting, the official proposal noted not only the usual scientific, technical, and budgetary considerations, but also drew attention to the importance of expediting the IGY satellite launch. As the attachment labeled "Factors Affecting USNC-IGY Schedule" mentioned, "it is of interest to note that at least one other nation has announced plans for a similar program under the direction of an extremely able physicist."[39] The meaning was clear. The Soviet announcement made American success in this field imperative.

This sentiment was not lost on Eisenhower's advisors. Nelson A. Rockefeller, who succeeded C.D. Jackson as Eisenhower's psychological warfare guru, prepared a substantial memorandum for the NSC calling for immediate action. Rockefeller placed considerable emphasis on beating the Soviets to the launching pad. Noting the psychological importance of being first to launch a satellite, Rockefeller cautioned the NSC of the "costly consequences of allowing the Russian initiative to outrun ours through an achievement that will symbolize scientific and technological advancement to peoples everywhere." He urged the NSC to act quickly in order to deny the Soviets an opportunity to deprive the United States of whatever psychological and prestige awards were to be gained. Clearly sensitive to the political ramifications of space flight, he continued emphatically, "The stake of prestige that is involved makes this a race that we cannot afford to lose."[40]

In addition to speed, Rockefeller listed other essential parameters for the program, many of them firmly rooted in political and psychological concerns. Since "vigorous propaganda will be employed to exploit all possible derogatory implications of any American success that may be achieved, it is highly important that the U.S. effort be initiated under auspices that are least vulnerable to effective criticism." Rockefeller feared that first launch by the United States of an uninstrumented satellite could be quickly discounted if the Soviets were to follow it with the launching of a more sophisticated type. Even though the United States should endeavor to be first, Rockefeller cautioned, the American satellite should also possess sophisticated instrumentation. Furthermore, the U.S. should launch a satellite under international (IGY) auspices in order to ward off any Soviet propaganda which might bill the American effort as somehow militaristic or aggressive. Finally, Rockefeller advised, the American project should share with the international community the information gleaned from the satellite in order to enhance its perception as a scientific and therefore peaceful project.[41]

Rockefeller's memorandum possesses a fundamental significance. All of its themes were presented to the National Security Council; it was attached to NSC 5520, the document authorizing the IGY project, and, most important, its suggestions were incorporated into the project's guidelines.[42] NSC 5520 emphasized in no uncertain terms the importance of the timely success of the satellite project to American prestige. The document began by reminding its readers that the Soviets were believed to be working on a satellite program and warned that "considerable prestige and psychological benefits will accrue to the nation which first is successful in launching a satellite." Just as important, NSC 5520 provisions were designed to cut to a minimum ammunition available for the Soviet propaganda machine.

Historians have looked to two of these provisions—one placing the project under international auspices and the other mandating that it not materially delay the ICBM and reconnaissance satellite programs—as proof that Eisenhower appreciated neither the importance of being first into space nor the propaganda value of space flight.[43] On the contrary, these directives were predicated on maximizing American prestige gains at the expense of the Soviets. The NSC believed that the United States needed to launch a satellite as quickly as possible to reap the prestige benefits of being first into space and to demonstrate progress in ballistic missile development. Paradoxically, however, they did not want to use an actual ICBM to do so. The Council wanted to sidestep propaganda charging the U.S. with nuclear belligerence. They also did not want it to appear as if the United States were actually racing the Soviet Union, which would happen if an uninstrumented satellite were hastily hurled into space. So the American satellite had to be as "scientific" as possible and had to provide some useful information. In short, the satellite had to be quick, but not too quick; it had to demonstrate success in ballistic missile technology without using "real" ballistic missiles; and it had to be used as a propaganda weapon without appearing to do so. In light of these confusing mandates it seems that the shock of October 1957 was preordained not by the administrations *in*sensitivity to political concerns, but rather by its very fixation on them.

The three basic principles of NSC 5520—timeliness, peaceful image, and non-interference—figured prominently in the selection of a launch vehicle for the satellite. McDougall argues that assuring the project's "civilian flavor" and noninterference with military programs governed the selection of the rocket vehicle because Eisenhower did not appreciate the political implications of being first in space. If the Eisenhower administration clearly understood the ramifications of space flight, he asks, why then did it assign the satellite to the doomed Project *Vanguard*? Was not the selection of the *Vanguard* rocket, in McDougall's words, a "disaster"? If so, it was by no means clear at the time. In fact, the *Vanguard* rocket was selected precisely because it promised to deliver the best combination of *all three* considerations.

A reexamination of the evidence suggests that timeliness figured just as prominently—if not more so—in the selection of a launch vehicle.[44] It should be kept in mind that the job of selecting a launch vehicle fell to Assistant Secretary of Defense Donald Quarles, to whom the "vanishing" Grosse report was addressed, who was fully aware of Rockefeller's memorandum, and who was fully cognizant of the aims established in NSC 5520. If Quarles disregarded the importance of a timely launch, he did so despite the policy guidance papers, not because they had mysteriously vanished. Significantly, Quarles's comments to the TCP progress report NSC 5522 noted that the IGY satellite was intended "for propaganda and scientific purposes" and mentioned nothing about "freedom of space" or intelligence applications.[45]

Furthermore, the available evidence suggests that the Stewart Committee—the subcommittee (named after its director Homer J. Stewart) Quarles appointed to recommend a launch vehicle for the satellite—assigned speed a high priority.

Confronted with promising proposals from all three services, the committee selected a launch vehicle based on the three principles of NSC 5520: timeliness, peaceful image, and non-interference with ICBM development. It first ruled out the powerful Atlas rocket, the Air Force ICBM-in-progress. Even though the committee expressed concerns that the satellite might interfere with the development of this important ICBM project, timeliness also weighed heavily on this decision. The first test launches of Atlas-B were scheduled for January, February, and March 1958, at best an uncomfortably narrow margin for the IGY.[46] The committee rejected the Atlas for its guaranteed tardiness and was then left to choose between the Army Orbiter project and the Naval Research Laboratory's (NRL) project. Contrary to McDougall's claim that there was "little doubt" that the Army proposal promised a satellite soonest,[47] half of the committee members thought Orbiter less likely to succeed than the NRL Viking rocket. Moreover, some members also believed that the bigger, heavier Orbiter rocket would cost considerably more than the Viking. Finally, the Navy's satellite promised far superior instrumentation and was designed as a research rocket, not a weapon. Committee members voted three in favor of the Viking rocket, two dissented, and one abstained on account of illness.[48]

Faced with a less than decisive vote for the NRL project, Quarles accepted vigorous Army demands to reconsider. Major General Leslie Simon of the Army Ordnance Corps protested that developmental problems cast doubt on the Navy's ability to launch a satellite within the IGY. Simon's urgent plea focused on the Army's ability— and the Navy's *in*ability—to orbit a satellite before the Russians. He promised an Army launch by January 1957 and warned, "Since this is the date by which the USSR may well be ready to launch, U.S. prestige dictates that every effort should be made to launch the first U.S. satellite at that time."[49] When the Navy got word that the Army was attempting to snatch their hard-won project, the NRL responded with assurances of its own. Earning a second hearing before the Stewart Committee, the Naval Research Laboratory reversed its earlier estimates and confidently announced that "the first satellite can be launched eighteen months from the start of the program." This revision of the earlier estimate was supported from the Glenn L. Martin Company, producer of the Viking. The company's Executive Vice President stated: "We see no reason why it should not be possible to put a satellite in being in approximately 18 months."[50] As Green and Lomask stated in their official history of the *Vanguard* project, the assurances from reputable industrial firms, particularly in regard to delivery dates, impressed the Stewart Committee. The time element, enthusiastically stressed by the Army general who spoke for Orbiter, now appeared to be about equal in both the Army and Navy propositions.[51] Thus, in accordance with NSC 5520, even though the schedule may not have been the primary consideration and was certainly not the *only* consideration, it is clear that the scientists and administration officials charged with implementing the satellite decision nevertheless appreciated its significance and its connection to American prestige and credibility.

The importance placed on timeliness in the decision to pursue the satellite program and in the selection of a launch vehicle suggests that historians must look elsewhere to explain the failure of the United States to beat the Soviets into space. While a comprehensive explanation lies beyond the scope of this essay, the notion that the United States did not *want* to be first, it seems, provides too easy an answer and contradicts the evidence presented here.[52] As this essay has already shown, part of the blame must rest with the confusing mandates issued by the National Security Council. Although they wanted to launch a satellite expeditiously, they felt they needed to launch during the IGY, rather than before, in order to advance the image that the United States was working to further international scientific cooperation.[53] The administration also feared that if the United States hastily launched an uninstrumented satellite, they would "win the race, but lose the scientific competition."[54] Moreover, one must consider the impact of bureaucratic momentum on the project. In order for the NSC to reverse the decision to use the Navy's *Vanguard* rocket, convincing evidence of its certain failure needed to present itself. Such evidence did not materialize until *after* Sputnik, when the *Vanguard* rocket exploded on live television. Prior to this, the administration received repeated assurances that the launch would occur on time, with the first attempt scheduled for October 1957.[55] A third, more speculative, explanation also exists. Simply put, most Americans assumed the superiority of American science and technology. For all the warnings about Soviet scientific advances, many Americans (both in and out of the administration) found it difficult to believe that the Soviet Union might outpace the United States in a technological competition. Coming from such a perspective—as opposed to the perspective of the underdog—it would have been difficult to rationalize exorbitant expenditures to win a race that most simply assumed the United States would win.

TECHNOLOGY, CREDIBILITY, AND THE BANNER OF PEACE

Even if one accepts this view, important questions remain unanswered. If Eisenhower believed, as his post-Sputnik and subsequent "missile gap" statements suggested, that the Strategic Air Command (SAC) possessed sufficient firepower to inflict massive and unacceptable damage in response to a Soviet first strike—if, in other words, the efficacy of the U.S. deterrent to general war remained unaffected by Soviet ballistic missile deployment—why then did the administration place such a high priority on a seemingly insignificant proposal to launch a ten-pound sphere into the atmosphere? The answer to this question brings us back to the administration's broader national security concerns. The satellite proposal should be seen in relation to the administration's efforts to counteract the Soviet's supposed two-pronged strategy of nuclear threat and the promise of peace. Because the administration believed that Soviet successes in atomic weaponry threatened to dismantle U.S. alliances and encouraged third-world neutralism, the United States attached maximum urgency to its own ballistic missile programs.

Furthermore, the satellite's self-evident relation to the ICBM made early success in this field an important objective of American foreign policy. A crucial passage in NSC 5520 stated this psychological connection explicitly: "The inference of such a demonstration of advanced technology and its unmistakable relationship to inter-continental ballistic missile technology might have important repercussions on the political determination of free world countries to resist Communist threats, especially if the USSR were to be the first to establish a satellite."[56] In short, the administration was fighting fire with fire. It sought to counter Soviet "war and peace" tactics with similar efforts of its own. To counteract Soviet gains in nuclear weapons and rocketry, it endeavored to demonstrate its own proficiency. At the same time, the Soviet peace offensive pushed the administration to do so in a manner that made the United States appear the true bearer of peace and freedom.

In the face of serious challenges to U.S. leadership, then, technology itself provided a means for demonstrating not only American military power, but also America's peaceful intentions. This became increasingly clear as Soviet technological advances coupled with Soviet peace overtures again forced a review of basic national security policy. The State Department's Policy Planning Staff (PPS) recommended such a review in October 1955. The PPS expressed an especially strong concern for the emphasis in Soviet diplomacy on "amiability and lure." These flexible tactics caused American allies to let down their guard and strengthened impulses towards neutralism and disengagement. To face this challenge, the Staff warned, the U.S. needed to double its efforts to build free world unity. This meant both maintaining the credibility of the U.S. deterrent and demonstrating American peaceful intentions. Technology would help: "We shall have to replace the cement of fear with new means of cohesion [including] common efforts to use technological advances for peaceful ends."[57] At once a symbol of national power and human progress, the PPS advised, technology should form a major component of U.S. policies to contain Communist expansion and reinforce American leadership.

This sentiment was echoed two months later by an Office of Defense Mobilization-Defense Department (ODM-Defense) working group charged with recommending to the NSC responses to Soviet technological gains. The task force, warning that the Soviet Union was several years ahead of the U.S. in important fields of weaponry and perhaps leading by two or more years in ICBMs, prescribed constant vigilance and preparation. Although the United States in 1955 still possessed superior overall firepower, Soviet advances suggested that "it will take continuous, unrelenting effort on the part of the U.S. to maintain such superiority on into the future." Dire consequences awaited: "Failure to maintain technological superiority by the U.S. could result in loss of confidence by the Free World in U.S. technology and power; accelerated Soviet expansion geographically and economically; swing of important uncommitted nations into the Soviet orbit; [and] defection of important countries now members of the Free World community." Weapons systems alone would not suffice, the report cautioned. The U.S. must reflect *overall* technological superiority including technology for peaceful purposes, inventiveness in basic research, and pools of scientific and technical personnel and institutions. Achieving and maintaining of technological superiority, the working group advised, were now indispensable elements of U.S. policies to counter Soviet expansion.[58]

With each passing year the NSC placed greater and greater stress on science and technology as central components of United States national security policy. The ODM-Defense and PPS reports were transmitted to the NSC planning board and incorporated into the Eisenhower administration's third major revision of basic national security policy, NSC 5602.[59] Reflecting a broader trend in national security planning, NSC 5602 included expanded emphasis on U.S. policies to counter Soviet flexible diplomatic initiatives and technological gains. This 1956 revision of national security policy, like its 1955 predecessor NSC 5501 and its 1957 successor NSC 5707, recommended that the United States shore-up American leadership by pursuing policies which reassured the world of American peaceful intentions and technological superiority.[60]

Why such an emphasis on scientific research and technological innovation? First and foremost, the NSC called for maintaining technological superiority out of fear that the Soviets might make a breakthrough rendering the American deterrent inadequate or obsolete. The National Security Council also saw a link between technological superiority and American leadership. The NSC felt that the United States had to demonstrate to its allies its capacity to fulfill its defense commitments. Furthermore, the NSC expressed concern that Soviet technological and economic progress might serve as "an impressive example" for peoples of the developing world, which might result in the expansion of Soviet influence.[61] If the United States fell too far behind in technological innovation, the NSC believed, it would drastically weaken the position of the free world. Industrialized allies and developing nations, lacking confidence in American abilities, would increasingly operate independently of the United States, undermining its ability to contain Soviet expansion.

The years dividing the Eisenhower administration's initial satellite decisions from Sputnik suggested to the administration that its worst fears were coming true. Anticolonialist sentiment in the developing world and strains in the Atlantic alliance seemed to verify that Soviet "war and peace" tactics were working all too well. Many intelligence analysts and policymakers in the administration warned apocalyptically that unless the U.S. moved quickly, it faced isolation from its allies and perhaps even from the rest of the world.[62] The Eisenhower administration responded to these warnings immediately. By May 1956, a year after approving the IGY satellite project, the National Security Council boosted the ICBM to the highest priority of all defense programs in the country, placed the shorter-range IRBM directly beside it, and tacked second-highest priority on the satellite program.[63]

The first of these decisions came in September 1955, when the NSC met to discuss the ballistic missile program. The decision to attach to the ICBM program the highest priority above all other defense programs again illustrates the significance the Eisenhower administration attached to psychological matters in formulating its Cold War policy. This sentiment was most clearly expressed by the State Department. Acting Secretary of State Herbert Hoover, Jr., interrupted a seemingly constipated discussion concerning the precise wording of the ICBM directive—whether the program should be pursued with "all practicable speed" or "all possible speed"—to remind the NSC of the tremendous importance of the ICBM to American foreign relations. "If the Soviets were to demonstrate to the world that they actually had an ICBM before we had such a weapon," he warned emphatically, "the result would have the most devastating effect on the foreign relations of the United States of anything that could possibly happen." Hoover pointed out that the Western coalition was held together essentially by the knowledge that the United States could protect them. "If this umbrella of protection were removed," he continued, repeating familiar logic, "neutralism would advance tremendously throughout the free world." To assure continued free world leadership, in other words, the United States had to maintain its technological edge: the U.S. had to have an ICBM as soon as possible.[64]

President Eisenhower, who was not present at this meeting, removed any doubts about his feelings on these matters. He unambiguously resolved the dispute over how best to word the NSC directive, changing the phrase "all practicable speed" to "maximum urgency." He also authorized, "in view of known Soviet progress in this field," placing the ICBM program as an R&D program of the highest priority above all others. Finally, he ordered the State Department to study the political ramifications of the *intermediate* range ballistic missile, an inquiry to determine whether early achievement of an IRBM by the United States would counter the implications of a Soviet first ICBM.[65]

Of course, the most important issue at stake here was American credibility. The administration felt it had to make clear the reliability and superiority of U.S. power. However, the threat of immediate and massive destruction posed by ballistic missiles carrying hydrogen warheads complicated matters. The administration understood that in a situation of nuclear parity its freedom of action would be sharply circumscribed by the catastrophic consequences of nuclear war. In such a situation *perceptions* of power became paramount and placed a premium on overall technological superiority. As Vice President Nixon expressed at the September 1955 NSC meeting, "The important thing is not merely the achievement of a developed weapons capability in the ICBM field, but, from the point of view of foreign relations, that the peoples of the free world *believe* that you have achieved an ICBM."[66]

In the eyes of policymakers, programs like the scientific satellite were valuable because they reinforced confidence in the capacity of the United States to resist Communist threats through the visible display of technological prowess. This sentiment was most clearly expressed by Under Secretary Hoover. He called the NSC's attention to the fact that the Earth satellite was helping them overcome some of the psychological difficulties posed by Soviet nuclear capabilities. The mere knowledge that the U.S. was pursuing the project had "gone a long way to help the free peoples of the world realize that we were forging ahead in our technical capabilities." As the National Security Council believed, such confidence was the cement necessary to maintain the cohesion of the western alliance.[67]

The administration also knew, however, that Earth-circling spaceships could only go so far to demonstrate military capabilities, especially when the intelligence community predicted that the United States lagged behind the Soviets by almost two years in ballistic missile technology.[68] Consequently the NSC moved quickly to expand its missile programs, granting the IRBM the same "maximum urgency" as its intercontinental cousin. The decision to step-up the IRBM came when the study Eisenhower commissioned to forecast the consequences of Soviet first achievement of a ballistic missile system predicted disaster. The report, prepared in the State Department by the Policy Planning Staff, emphasized the psychological ramifications of Soviet first achievement of an intermediate range missile. This Soviet "first" would "reduce the free world's confidence in U.S. technological superiority and enhance its fears as to the consequences of war." As a result, the Planning Staff warned, American allies would face increased domestic pressure to adopt independent foreign policies. They would more vigorously oppose policies carrying risk of war and would be more likely to compromise on outstanding East-West issues. Moreover, the Soviets could exacerbate these trends toward neutralism by conciliatory tactics intended to persuade the allies of the wisdom of accommodation. American achievement of an IRBM at the earliest date, the report concluded, would be necessary to strengthen confidence in American retaliatory power and to prevent the erosion of Western alliances.[69]

President Eisenhower concurred that the United States needed to possess and demonstrate an effective ballistic missile capability as soon as possible. When Secretary Dulles presented the Planning Staff report to the National Security Council in December, the President reacted in an uncharacteristically vocal and temperamental fashion. With great emphasis, Eisenhower hammered Council members for apparent inaction, bureaucratic resistance, and obstructive inter-service rivalries. In response to a comment by Air Force Assistant Secretary Trevor Gardner that the Soviets possessed a two-year lead over the U.S. in ballistic missiles, Eisenhower responded cryptically, saying that he "would like to know what had been going on since last July when he had issued his strong directive on achievement of a U.S. capability in the field of ballistic missiles." Fully subscribing to the views of the State Department as to the "profound and over-riding political and psychological importance" of such weapons,

Eisenhower warned that he was "absolutely determined not to tolerate any fooling with this thing." The NSC action record for the meeting reinforced these sentiments with equal force. Eisenhower, noting that the political and psychological impact of an effective IRBM would be so great that early U.S. achievement of such a missile would be of "critical importance" to the national security interests of the United States, assigned both the ICBM and the IRBM programs highest priority of all defense programs in the country.[70]

These actions run counter to the image of Eisenhower as a President who only belatedly recognized the importance of prestige to American foreign policy. To be sure, to note the tremendous political and psychological values the administration attached to ballistic missiles and to note that the scientific satellite program was pursued because of its obvious relationship to such weapons does not necessarily mean that the administration assigned the satellite program the same priority as the missile programs. Nor does Eisenhower's enthusiastic endorsement of the ballistic missile programs mean that the President harbored equal enthusiasm for the satellite project. Indeed the evidence suggests otherwise. The missile programs were by far the most important research and development projects in progress and were so because the President and his advisors believed they should be. The evidence also suggests that Eisenhower was not overly enthusiastic about the satellite program to begin with. He was skeptical of its spiraling costs and he harbored doubts about whether orbiting such a satellite was in fact even possible.[71]

On the other hand, the Eisenhower administration placed considerable emphasis on the project, attaching to the IGY satellite a significance not fully appreciated in the historical literature. When the NSC met in May 1956 to discuss the progress of the program, the Council reaffirmed its belief that the satellite should not interfere with ICBM or IRBM programs. However, the NSC also commanded that the satellite be given sufficient priority in relation to other R&D projects to ensure a launch during the IGY. Thus the administration ranked the urgency of the satellite program just below the missile programs and above the roughly 180 other high-priority DOD programs.[72]

Just as important, it would be incorrect to conclude that because the President voiced some skepticism toward the project that he therefore failed to understand its political and psychological significance. Eisenhower's comments at the May 1957 NSC meeting are instructive. The meeting, which took place shortly after a National Intelligence Estimate (NIE) predicted the USSR would make a major effort to be first in launching a satellite,[73] reveals that Eisenhower fully understood the political implications of the scientific satellite expressed in NSC 5520. Complaining about the spiraling costs of instrumentation for the satellite, Eisenhower charged the scientists with wasting too much time and money on gadgets instead of focusing on the main objective of getting a satellite into orbit in the first place. Betraying skepticism about the technical feasibility of the project, he confessed he was annoyed by the tendency to "gold plate" the satellite before "we had proved the basic feasibility of orbiting any kind of earth satellite." He pointedly stressed "that the element of national prestige, so strongly emphasized in NSC 5520, depended on getting a satellite into orbit, and not on the instrumentation of the scientific satellite." Eisenhower's comments suggest that he was fully aware of the prestige value of the satellite long before Sputnik and the ensuing domestic outcry.[74]

SPUTNIK RECONSIDERED

If the years preceding *Sputnik 1* saw the administration placing more and more emphasis on technology as an extension of foreign policy, it remains true that the period immediately following the Soviet satellite saw these efforts stepped up dramatically. The expansion of the space program after October 1957 has been discussed in detail elsewhere, but a few comments linking national security strategy before and after Sputnik are in order.[75] Considering that Sputnik ushered in the highly politicized "missile gap," historians have generally explained the expansion and acceleration of the space program in terms of domestic politics. While one cannot disregard the very real political pressure on Eisenhower in the wake of Sputnik, the emphasis placed on domestic politics obscures the fact that the administration's actions immediately following Sputnik were consistent with the evolution of its broader national security strategy. From 1953 to 1957, as Soviet technological gains threatened confidence in American superiority and as "peaceful coexistence" appeared to undermine U.S. leadership, the Eisenhower administration placed greater and greater stress on efforts to advertise

American peaceful intentions and to maintain American technological superiority. Sputnik merely confirmed the wisdom of such policies. A first-rate propaganda victory, it seemed to prove that the Soviet "peace and war" strategy—embodied in a scientific satellite launched by a powerful ballistic missile—was working spectacularly. Indeed, the Sputnik outcry seemed to verify the most dire predictions of intelligence officers and policymakers as confidence in American leadership appeared shaken around the world and at home.

To the NSC, *Sputnik 1* not only sapped U.S. prestige by suggesting Soviet technological superiority, but it also lent added credibility to Soviet pronouncements—particularly Khrushchev's claim a few months earlier that it had successfully tested an ICBM.[76] At a meeting of the National Security Council a few days after the launch of *Sputnik 1*, Allen Dulles interpreted the event as part of a trilogy of Soviet propaganda moves—the August ICBM test, the September hydrogen bomb tests, and the satellite launch. Together, he warned, they seemed to provide the world with convincing evidence that the Soviets possessed a substantial lead in the technology necessary to construct an operational ICBM. As Donald Quarles admitted, Sputnik revealed that the Soviets possessed even more competence in long-range rocketry and in auxiliary fields than the U.S. had given them credit.

Additionally, Council members interpreted reactions to Sputnik abroad as verifying earlier fears that Soviet technological success would undermine American leadership. As the Director of Central Intelligence warned, Sputnik was exerting a "very wide and deep impact" in Western Europe, Africa, and Asia. Under Secretary of State Christian Herter echoed Allen Dulles's assessment. He reported to the NSC that even the best allies "require assurance that we have not been surpassed scientifically and militarily by the USSR." The situation appeared even more disastrous outside the Western alliance, Herter cautioned, because the Soviet feat seemed to affirm the wisdom of neutrality. The neutral countries, he noted, "are chiefly engaged in patting themselves on the back and insisting that the Soviet feat proves the value and wisdom of the neutralism which these countries have adopted." To the NSC, then, *Sputnik 1* confirmed predictions that Soviet technological spectaculars could deal a severe blow to U.S. prestige and credibility.[77]

The administration's overall response to Sputnik is best summarized in a State Department memorandum evaluating reactions to the Soviet satellite. The document is especially important because many of its concerns came to be addressed by subsequent revisions of U.S. space policy. The memorandum outlined four major effects on world public opinion:

(1) Soviet claims of scientific and technological superiority over the West and especially the U.S. have won greatly widened acceptance.
(2) Public opinion in friendly countries shows decided concern over the possibility that the balance of power has shifted or may soon shift in favor of the USSR.
(3) The general credibility of Soviet propaganda has been greatly enhanced.
(4) American prestige is viewed as having sustained a severe blow.

Harking back to the earlier predictions of NSC 5520 and 5522, the State Department expressed the belief that Sputnik's repercussions would be greatest among the newly independent or dependent peoples, largely preoccupied with economic development. The technologically less-advanced areas of the world would be most easily "dazzled" by the feat. They were also the areas "least able to understand it" and "most vulnerable to the attractions of the Soviet system." The State Department warned that by demonstrating the ability of the Soviet system to compete on a technological level with the West, Sputnik meant that developing nations would be more likely to turn to the Soviet bloc for technical and material aid. This, the Department reasoned, would place them directly or indirectly in the Soviet camp. Because "the satellite, presented as the achievement and symbolic vindication of the Soviet system, helps to lend credence to Soviet claims," Sputnik paved the way for an intensive psychological warfare campaign.[78]

These views were reflected in the National Security Council's second major statement of outer space policy, NSC 5814. Formally adopted in August 1958 after the creation of NASA, it no longer repeated the 1953–1957 warnings that the USSR was threatening to surpass the United States in science and technology. By this time, the Soviets had orbited two other satellites, one carrying a live dog and the other nearly 100 times heavier than the biggest American satellite.[79] These successes, compounded by televised American failures, including the *Vanguard* launch—the "ignominious flop" that burst into flames before viewers around the world[80]—seemed to verify the NSC's predictions. As NSC 5814 stated, "The USSR has surpassed the United States and the free world in scientific and technological accomplishments in outer space, which captured the imagination and admiration of the world." Echoing lines of reasoning established long-before Sputnik, the Security Council warned that further Soviet demonstrations of superiority in outer space technology would "dangerously impair" confidence in overall U.S. leadership. Strength in space technology was necessary to "enhance the prestige of the United States among the people of the world and [to] create added confidence in U.S. scientific, technological, industrial, and military strength."[81]

The National Security Council dictated the overall goals and parameters of the now-adolescent space program. In addition to important military and reconnaissance applications, the NSC directed NASA to "judiciously select" projects designed to achieve a "favorable world-wide psychological impact." Of the possibilities—Earth satellites, lunar rockets, human Earth-orbital flight, planetary probes, human circumlunar flights—human space flight ranked highest because "to the layman" manned exploration represented the "true conquest" of outer space. "No unmanned experiment can substitute for manned exploration in its psychological effect on the peoples of the world." Besides prestige-oriented space spectaculars, NSC 5814 also called for international cooperation in space activities and directed the U.S. to seek a treaty banning the use of space for military purposes, both to establish the United States as leader in the use of outer space for peaceful purposes.

The Eisenhower administration's 1958 outer-space policy statement suggests that Sputnik inspired an increased emphasis on the psychological impact of U.S. policies. This included a renewed emphasis not only on space programs but on other foreign *and domestic* policies as well. As a revision of basic national security policy advised in May 1958, "The psychological impact abroad of our policies, both foreign and domestic, plays a crucial part in the overall advancement of U.S. objectives. It is essential therefore that along with the pertinent military, political, and economic considerations, the psychological factor be given due weight during the policy forming process."[82] But what if there had been no Sputnik? As this essay has argued, the expanded emphasis on psychological matters did not just derive from domestic pressure nor did it stem solely from the Sputnik challenge. It should be seen instead as part of a broader response to the changing nature of the Cold War, a by-product of "peaceful coexistence" in an era of nuclear devastation. As a 1960 NIE put it, peaceful coexistence, "a strategy to defeat the West without war," involved a political struggle to capture the support of peoples across the world. By manipulating issues of peace, disarmament, anticolonialism, and economic development, by dramatizing the growth of Soviet power, and by capturing the imagination of the world's peoples with their technical prowess, the Soviets threatened to attract the allegiance of the underdeveloped and uncommitted states against the West.[83] If an era of "total Cold War" developed after October 1957, in which science, technology, education, and the pursuit of national prestige ranked with military and economic strength as vital forces in the U.S.-Soviet struggle, it was as much a product of the changing nature of the Russian-American rivalry as it was a Sputnik product. Sputnik simply provided the defining image of a struggle already underway and a race already being run.[84]

1. The author wishes to thank Wilson Miscamble, C.S.C. of the University of Notre Dame, Dwayne A. Day of the George Washington University Space Policy Institute, Ronald Doel of Oregon State University, and the faculty and graduate student members of the Cold War History Group of the University of California, Santa Barbara, for reading earlier drafts of this paper and for providing valuable criticism.

2. Michael S. Sherry, *In the Shadow of War: The United States Since the 1930s* (New Haven, CT: Yale University Press, 1995), p. 216.

3. Robert Divine, *The Sputnik Challenge* (New York: Oxford University Press, 1993), p. 110.

4. Historians have emphasized Eisenhower's prudent response to the "Sputnik panic" and the missile gap crises which followed. In what has become a widely accepted view, Eisenhower refused to panic in the face of considerable congressional and public pressure for a massive arms build-up in reaction to Sputnik. Because of his commitment to balanced budgets and his determination to avoid an arms race, Eisenhower resisted political pressure to recklessly pursue a "Manhattan approach" to ICBM development or to pursue expensive space spectaculars based on shoring-up American prestige. According to this perspective, U-2 intelligence convinced Eisenhower that the United States deterrent remained unaffected by the Soviet success. Nonetheless, Eisenhower's refusal to divulge the source of his confidence (the U-2) meant that he could not convince the public of the prudence of his policies. For a concise summary of the traditional view, see Giles Alston "Eisenhower: Leadership in Space Policy," in Shirley Anne Warshaw, ed., *Reexamining the Eisenhower Presidency* (Westport,

CT: Greenwood Press, 1993). The standard work on the Cold War in space is Walter A. McDougall, ... *The Heavens and the Earth: A Political History of the Space Age* (New York: Basic Books, 1985). Robert Divine especially emphasizes the importance of domestic politics in propelling the space race. See Divine, *Sputnik Challenge*. For other historians who support this interpretation, see Stephen Ambrose, *Eisenhower: The President* (New York: Simon and Schuster, 1984), esp. pp. 423–61; Jack Manno, *Arming the Heavens: The Hidden Military Agenda for Space, 1945–1995* (New York: Dodd, Mead & Company, 1982); Zuoyue Wang, "American Science and the Cold War, The Rise of the U.S. President's Science Advisory Committee," Ph.D. diss., University of California, Santa Barbara, 1994, pp. 100–82; Paul B. Stares, *Space Weapons and U.S. Strategy: Origins and Development* (London: Croom Helm, 1986), pp. 38–58; and Sherry, *Shadow of War*, pp. 214–37. On Eisenhower's response to the missile gap see McGeorge Bundy, *Danger and Survival: Choices About the Bomb in the First Fifty Years* (New York: Random House, 1988), pp. 236–358; Peter J. Roman, *Eisenhower and the Missile Gap* (Ithaca, NY: Cornell University Press, 1995).

5. McDougall emphasizes that the challenge Sputnik posed to America's image and prestige transformed the conflict into "total Cold War." By presaging nuclear parity and suggesting Soviet scientific superiority, Sputnik turned the superpower rivalry into a colossal public relations contest: "The Cold War now became total, a competition for the loyalty and trust of all peoples fought out in all arenas of social achievement." See McDougall, ... *The Heavens and the Earth*, p. 8. Divine's recent study confirms McDougall's conclusion that public pressure following Soviet successes in space forced a major reorientation in space policy. Largely a study in presidential politics, Divine's book laments the reversal of Eisenhower's prudent and restrained economic policies and blames the reversal on opportunistic politicians and pundits. Political realism, Divine emphasizes, dictated that Eisenhower reorient American space policy towards capturing the lead in space exploration. It was not until the last two years of Eisenhower's presidency that "Ike would give as much weight to intangible factors such as world opinion and prestige as to missiles and space craft." See Divine, *Sputnik Challenge*, esp. pp. 183, 205.

6. On the U.S. spy satellite program see Jeffrey T. Richelson, *America's Secret Eyes in Space: The U.S. Keyhole Spy Satellite Program* (New York: Harper & Row, 1990), pp. 1–123; and Kenneth E. Greer, "Corona," in Kevin C. Ruffner, ed., *Corona: America's First Satellite Program* (Washington, DC: History Staff, Center for the Study of Intelligence, Central Intelligence Agency, 1995.)

7. Two historians who challenge this view, in different contexts, are Robert J. McMahon and Rip Bulkeley. McMahon suggests that image, prestige, credibility and other psychological considerations always exerted a powerful influence on Eisenhower's foreign policy decisions. See Robert J. McMahon, "Credibility and World Power: Exploring the Psychological Dimension in Postwar American Diplomacy," *Diplomatic History* 15 (Fall 1991): 455–71. Bulkeley's critique of space historiography also notes the importance Eisenhower attached to psychological matters. He checks the Eisenhower administration's post-Sputnik rhetoric against its actions prior to the Soviet launch and finds that contrary to the administration's public statements, its space efforts were "not part of a disinterested policy of support for pure science." Rather, the administration was concerned from the start that the satellite project "should be conducted and publicly presented in such a way as to secure the maximum benefit for the prestige and influence of the United States in the propaganda competition with the Soviet Union." See Rip Bulkeley, *The Sputniks Crises and Early United States Space Policy: A Critique of the Historiography of Space* (Bloomington: Indiana University Press, 1991), p. 162.

8. McMahon defines prestige as an "elusive concept" that connoted "a blend of resolve, reliability, believability and decisiveness; equally important, it has served as a code word for America's image and reputation." See McMahon, "Credibility," p. 455. John Lewis Gaddis suggests that by as early as 1950, when the Truman administration adopted the important

policy paper NSC-68, American security "had come to depend as much on *perceptions* of the balance of power as on what that balance actually was." Consequently, "judgments based on such traditional criteria as geography, economic capacity, or military potential now had to be balanced against considerations of image, prestige, and credibility." Like the authors of NSC-68, Gaddis contends, Eisenhower and his advisors attached great importance to perceptions of power. John Lewis Gaddis, *Strategies of Containment: A Critical Appraisal of Postwar American National Security Policy* (New York: Oxford University Press, 1982), pp. 92 and 144. For analysis of credibility and deterrence see Robert Jervis, *Perception and Misperception in International Politics* (Princeton, NJ: Princeton University Press, 1976) and Richard Lebow, *Between Peace and War: The Nature of International Crises* (Baltimore, MD: John Hopkins University Press, 1981). On prestige and the space race see William H. Schauer, *The Politics of Space: A Comparison of the Soviet and American Space Programs* (New York: Holmes & Meier, 1976), esp. pp. 91–105.

9.   On the Soviet Union and nuclear weapons see David Holloway, *Stalin and the Bomb: The Soviet Union and Atomic Energy, 1939–1956* (New Haven, CT: Yale University Press, 1994) and Richard Rhodes, *Dark Sun: The Making of the Hydrogen Bomb* (New York: Simon & Schuster, 1995). On Stalin's successors (especially Khrushchev) and technology, see Vladislov Zubok and Constantine Pleshakov, *Inside the Kremlin's Cold War: From Stalin to Khrushchev* (Cambridge, MA: Harvard University Press, 1996), esp. pp. 138–209.

10. On the "new look" see Gaddis, *Strategies*, pp. 127–97; Bundy, *Danger*, pp. 246–60; Robert A. Divine, *Eisenhower and the Cold War* (New York: Oxford University Press, 1981), pp. 33–70. For extensive and detailed discussion of the strategy see Saki Dockrill, *Eisenhower's New-Look National Security Policy, 1953–1961* (New York: St. Martin's Press, 1996). On John Foster Dulles and massive retaliation see John Lewis Gaddis, *The United States and the End of the Cold War: Implications, Reconsiderations, Provocations* (New York: Oxford University Press, 1992).

11. See for example NIE 11-4-54, August 27, 1954, *Declassified Documents Catalog*, 1981, document 283A [hereafter *DDC* with year/document number].

12. On Soviet peace initiatives after Stalin see James Richter, *Khrushchev's Double Bind: International Pressures and Domestic Coalition Politics* (Baltimore, MD: Johns Hopkins University Press, 1994); Voltech Mastny, *The Cold War and Soviet Insecurity: The Stalin Years* (New York: Oxford University Press, 1996), pp. 171–98; Adam Ulam, *Expansion and Coexistence: Soviet Foreign Policy, 1917–1973* (New York: Praeger, 1974), pp. 496–694; Zubok and Pleshakov, *Kremlin's Cold War*, pp. 138–235.

13. NSC 162/2, October 30, 1953, Dwight D. Eisenhower Library [DDEL], White House Office [WHO], Office of the Special Assistant for National Security Affairs [OSANSA], NSC series, Policy Papers subseries, box 6, folder NSC 162/2.

14. National Security Council Planning Board, "Tentative Guidelines Under NSC 162/2 for FY 1956," June 14, 1954, DDEL, WHO, OSANSA, NSC series, Policy Papers subseries, box 11, folder NSC 5422.

15. See JCS to Wilson, June 23, 1954, *FRUS 1952–1954*, 2:680–681; John Dulles to the NSC, November 15, 1954, *FRUS 1952–1954*, 2:772–776; Streibert to the NSC, November 19, 1954, *FRUS 1952–1954*, 2:784; and Allen Dulles to the NSC, November 18, 1954, *DDC*, 81/415B. Their concerns were likely based, in part, on intelligence estimates warning of Soviet advances in guided missiles technology. The first estimate to treat the subject was NIE 11-6-54, November 5, 1954. See Steury, "Missile Gap," *Intentions and Capabilities*, p. 55. See also NIE 11-4-54, August 24, 1954, *DDC*, 81/233A.

16. Robert J. McMahon, "The Illusion of Vulnerability: American Reassessments of the Soviet Threat, 1955–1956," *International History Review* 18 (August 1996): 591–619. On the Truman administration's geostrategy see Melvyn P. Leffler, *Preponderance of Power: National Security, the Truman Administration, and the Cold War* (Stanford, CA: Stanford University Press, 1992).

17. McMahon, "Illusion," pp. 591–619.

18. NSC 5501, January 7, 1955, DDEL, WHO, OSANSA, NSC series, Policy Papers subseries, box 14, folder NCS 5501. For similar sentiments from the intelligence community, see NIE 11–3–55, May 17, 1955, DDC, 78/22B.

19. Douglas Aircraft Company, Inc., "Preliminary Design of an Experimental World-Circling Spaceship," Report No. SM-11827, May 2, 1946 in John M. Logsdon, gen. ed., *Exploring the Unknown: Selected Documents in the History of the U.S. Civil Space Program,* vol. 1 (Washington, D.C.: NASA SP-4407, 1995), pp. 236–44. For more on RAND see Merton E. Davies and William R. Harris, *RAND's Role in the Evolution of Balloon and Satellite Observation Systems and Related U.S. Space Technology* (Santa Monica, CA: RAND Corp., 1988).

20. R. Cargill Hall, "Origins of U.S. Space Policy: Eisenhower, Open Skies, and Freedom of Space," in Logsdon, *Exploring,* pp. 213–29.

21. Grosse to Quarles, "Report on the Present Status of the Satellite Problem," August 25, 1953, in Logsdon, *Exploring,* pp. 267–69.

22. McGeorge Bundy described the Killian report as one of the "most influential in the history of American nuclear policy" for its impact on the development of ballistic missile systems, intelligence collection recommendations, and efforts to provide better strategic warning to the threat of surprise attack. See Bundy, *Danger and Survival,* p. 325. See also James Killian's comprehensive memoir, *Sputnik, Scientists, and Eisenhower: A Memoir of the First Special Assistant to the President for Science and Technology* (Cambridge, MA: MIT Press, 1978).

23. Portions of the report remain classified, although a general summary is in *FRUS 1955–1957,* 19:41–56. The declassified sections of the TCP report and related documents may be found on *DDC* 93/2972, 93/3111, 96/2701, and 96/2778. White House Office, Office of the Special Assistant for National Security Affairs: Records, 1952–61, NSC Series, Subject Subseries, box 11, Technological Capabilities Panel. See also Bundy, *Danger and Survival,* pp. 325–28; and McDougall, ... *The Heavens and the Earth,* pp. 115–18.

24. Hill, "Origins," p. 221.

25. McDougall, ... *The Heavens and the Earth,* pp. 119–21.

26. See for example, Eisenhower's news conferences, October 3, 1957, and October 9, 1957, *Public Papers of the Presidents of the United States: Dwight D. Eisenhower,* pp. 707–709, 719–32.

27. See Bulkeley, *Sputniks Crisis,* pp. 120, 154–60.

28. John Robert Greene, "Bibliographic Essay: Eisenhower Revisionism, 1952–1992, A Reappraisal" in Shirley Anne Warshaw, ed., *Reexamining the Eisenhower Presidency* (Westport, CT: Greenwood Press, 1993), pp. 209–19.

29. For the classic revisionist account, see Fred I. Greenstein, *The Hidden-Hand Presidency: Eisenhower as Leader* (New York: Basic Books, 1982). For historiographical discussion see Warshaw, ed., *Reexamining the Eisenhower Presidency* and Roman, *Missile Gap,* pp. 9–18.

30. McDougall, ... *The Heavens and the Earth,* pp. 119–21, 134. For an alternative view, see Dwayne A. Day, "A Strategy for Reconnaissance: Dwight D. Eisenhower and Freedom of Space," unpublished manuscript from his forthcoming book.

31. NSC 5522, June 8, 1955, DDC 96/2811.

32. NSC 5522, June 8, 1955. Emphasis added.

33. Day, "Strategy for Reconnaissance."

36. NSC 5522, June 8, 1955. Interestingly, the NSC directed the CIA to discredit Soviet prestige and the prestige of International Communism through (unspecified) covert means. See NSC 5412/2, December 28, 1955, DDEL, WHO, OSANSA, Special Assistant Series, Presidential Papers subseries, folder 1955 (1).

35. NSC meeting, May 11, 1957, DDEL, Ann Whitman file, NSC series, box 8, folder 322nd meeting of NSC.

36. NSC meeting, May 11, 1957.

37. "Interplanetary Commission Created," *The Washington Post*, April 17, 1955, in Logsdon, *Exploring the Unknown*, p. 308.

38. Kaplan to Alan Waterman, National Science Foundation, May 6, 1955, in *Ibid.*, pp. 302–303. This was Kaplan's second letter on the subject and contained an urgent tone not present in the first. See Kaplan to Detlev Bronk, President, National Academy of Sciences, March 14, 1955, in *Ibid.*, pp. 301–302.

39. USNC-IGY, "Minutes of the Eighth Meeting," May 18, 1955, in *Ibid.*, p. 295–308.

40. Rockefeller to Lay, May 17, 1955, DDEL, Ann Whitman file, Administration series, box 31, folder Rockefeller, Nelson 1952–1955.

41. Rockefeller to Lay, May 17, 1955.

42. NSC 5520, May 20, 1955, DDEL, WHO, OSANSA, NSC series, Policy Papers subseries, box 16, folder NSC 5520.

43. Alston, "Eisenhower: Leadership in Space Policy," pp. 104–106; McDougall, ... *The Heavens and the Earth*, pp. 132–34; Divine, *Sputnik Challenge*, pp. 102–105.

44. Most historians argue that the Eisenhower administration attached very little importance to speed because of his overriding concern for ballistic missile programs and his determination to maintain the American project's civilian character. McDougall, ... *The Heavens and the Earth*, p. 122; Divine, *Sputnik Challenge*, pp. 102–10; Alston, "Eisenhower: Leadership in Space Policy," pp. 104–105. Bulkeley, on the other hand, provides convincing evidence to the contrary. He writes, "the historical perception of ... Eisenhower's early space policy has been coloured by the fact that after *Sputnik 1* the administration staged a damage-limiting public-relations exercise ... which drew a bogus but meretricious distinction between its own peaceful, scientific and allegedly disinterested satellite project and the somehow more sinister and less noble, if more effective, Soviet one." See Bulkeley, *Sputniks Crisis*, pp. 156–62.

45. NSC 5522, June 8, 1955.

46. It should be noted here that McDougall also cites the official *Vanguard* history by Green and Lomask for his interpretation of these events. Constance McLaughlin Green and Milton Lomask, *Vanguard: A History* (Washington, DC: Smithsonian Institution Press, 1970), p. 41.

47. McDougall, ... *The Heavens and the Earth*, p. 123.

48. Green and Lomask, *Vanguard*, pp. 45–52.

49. Quoted in *Ibid.*, pp. 52–53.

50. Quoted in *Ibid.*, p. 54.

51. Quoted in *Ibid.*, p. 55.

52. For a different explanation from an author who concurs that the U.S. accorded the schedule a high-priority, see Bulkeley, *Sputniks Crisis*.

53. NSC meeting, May 3, 1956, DDEL, Ann Whitman file, NSC series, box 7, folder 283rd Meeting of NSC; NSC meeting May 10, 1957, Ann Whitman file, NSC series, box 8, folder 322nd Meeting of NSC.

54. See Day, "Strategy for Reconnaissance."

55. NSC meeting, January 24, 1957, DDEL, Ann Whitman file, NSC series, box 8, folder 310th Meeting of NSC; see also NSC meetings of May 3, 1956, and May 10, 1957.

56. NSC 5520, May 20, 1955.

57. Department of State, "General Comments on NSC 5501," October 3, 1955, *FRUS 1955–1957*, 19:123–25.

58. "Report by the ODM-Defense Working Group," December 20, 1955, *FRUS 1955–1957*, 19:173–177.

59. NSC 5602/1, March 15, 1956, *FRUS 1955–1957*, 19:242–268.

KENNETH A. OSGOOD

60. NSC 5707/8, June 3, 1957, DDEL, WHO, OSANSA, NSC series, Policy Papers subseries, box 20, folder NSC 5707/8. See also NSC meeting, February 28, 1957, *DDC*, 96/1053.

61. NSC 5602/1, 263. See also NSC meeting, May 31, 1956, Ann Whitman file, NSC series, box 7, folder 286th Meeting of NSC.

62. See for example NIE 100-5-55, June 14, 1955, *FRUS 1955–1957*, 19:85–87; NSC 5525, August 31, 1955, *FRUS 1955–1957*, 9:529–31; DOS comments on NSC 5501, October 3, 1955, *FRUS 1955–1957*, 19:123–25; and NIE 100-7-55, November 1, 1955, *DDC* 88/3158.

63. For official DOS chronologies of important developments in the ICBM and satellite programs see *DDC* 94/2000, *DDC* 94/2000 and *DDC* 81/221A.

64. NSC meeting, September 8, 1955, DDEL, Ann Whitman file, NSC series, box 7, folder 258th Meeting of NSC.

65. Ibid.

66. Ibid. Emphasis added.

67. Ibid.

68. NIE 100-7-55, November 1, 1955, *DDC*, 88/3158.

69. DOS memo, undated, *FRUS 1955–1957*, 19:154–61.

70. NSC meeting, October 1, 1955, *FRUS 1955–1957*, 19:166–70. See also footnote no. 9, ibid., 170.

71. These sentiments are well documented by Divine and McDougall and are also illustrated by Eisenhower's comments at the NSC meeting, May 3, 1956.

72. NSC meeting, May 3, 1956.

73. NIE 11-5-57, March 13, 1957, in Steury, *Intentions*, pp. 59–62.

74. NSC meeting, May 10, 1957.

75. In addition to the works cited in note 4 above, see Roger D. Launius, *NASA: A History of the U.S. Civil Space Program*, (Malabar, FL: Krieger Publishing Co., 1994) and John Logsdon, "Opportunities for Policy Historians: The Evolution of the U.S. Civilian Space Program," in Alex Roland, ed., *A Spacefaring People: Perspectives on Early Space Flight* (Washington, D.C.: NASA SP-4405, 1985), pp. 81–107.

76. In a memorandum to the President, White House Press Secretary James Hagerty cautioned that the satellite does tend to corroborate the Soviet ICBM claim of August 27, 1957. Hagerty to Eisenhower, October 7, 1957, *DDC*, 87/126.

77. NSC meeting, October 10, 1957, DDEL, Ann Whitman file, NSC series, box 9, folder 339th Meeting of NSC.

78. DOS, "Reaction to the Soviet Satellite," October 16, 1957, *DDC*, 81/373A. For similar expressions see also: USIA Circular Telegram to All Missions, October 16, 1957, *FRUS 1955–1957*, 24:167–68; CIA Consultants to Allen Dulles, October 23, 1957, *DDC*, 95/3241; NIE 11-4-57, November 23, 1957, *FRUS 1955–1957*, 19:665–72; Arneson to John Dulles, November 14, 1957, *FRUS 1955–1957*, 11:768–69; Soviet Embassy to DOS, November 16, 1957, *FRUS 1955–1957*, 24:185–87; USIA Report, DDEL, WHO, Office of the Special Assistant for National Security Affairs: Records, 1952–61, OCB Series, Subject Subseries, Folder: Space Satellites, Rockets, etc. (1); USIA survey, December 1957, *DDC*, 86/2225; USIA memo, December 17, 1957, *FRUS 1955–1957*, 11:779–82.

79. After the launch of *Sputnik 3*, weighing 2,925 pounds, Khrushchev remarked to Arab leader Gamel Abdel Nasser that the United States "will need very many satellites the size of oranges to catch up." Lester A. Sobel, ed., *Space: From Sputnik to Gemini* (New York: Facts on File, 1965), p. 40.

80. Killian, *Sputnik, Scientists, and Eisenhower*, p. 119.

81. NSC 5814/1, August 18, 1958, DDEL, WHO, OSANSA, NSC series, Policy Papers subseries, box 25, folder NSC 5814/1.

82. NSC 5810/1, "Psychological Aspects of U.S. Policies," May 5, 1958, DDEL, WHO, OSANSA, NSC series, Policy Papers subseries, box 25, folder NSC 5810-Basic National Security Policy.

83. NIE 11–4–60, undated, in Steury, *Intentions*, pp. 141–46.

84. Wang, "American Science," p. vii.

CHAPTER 8

Orbiter, Overflight, and the First Satellite: New Light on the Vanguard Decision

Michael J. Neufeld

The August 1955 decision to choose the Navy's Vanguard over the Army's Orbiter as the first U.S. scientific satellite project remains one of the most controversial episodes in the history of the United States space program. There seems little doubt that it was the critical decision on the American side that led to the Soviet Union's Sputnik surprise of October 1957. In the wake of that shock, the success of the Army's reactivated project, now called Explorer, in putting up a satellite after an embarrassing Vanguard failure, confirmed the opinion of many that a terrible mistake had been made in 1955.

It was only in the 1960s, after the immediate recriminations had passed and more information became available, that R. Cargill Hall and official Vanguard historians Constance Green and Milton Lomask formulated more nuanced and balanced explanations of that decision. Among the factors they noted were: 1) the superior electronics, scientific experiment capacity and growth potential of the Vanguard proposal of the Naval Research Laboratory (NRL); 2) the likelihood that its launch vehicle, based on the Viking and Aerobee sounding rockets, would interfere less with high-priority military missile programs; 3) the greater ease of declassification of aspects of the program as a result, in keeping with the International Geophysical Year (IGY) of 1957–58, under which the satellite was to be launched; and 4) the possibility of anti-German prejudice in the selection committee against the Orbiter's design team, led by Wernher von Braun at Redstone Arsenal in Huntsville, Alabama. Hall, Green and Lomask also deepened our knowledge of how close and contested that decision had been in the so-called Stewart Committee—named for its chairman, Homer Joe Stewart of the Jet Propulsion Laboratory, who ironically had led the opposition to the decision of his own committee.[1]

The publication of Walter A. McDougall's Pulitzer-prize-winning *The Heavens and the Earth* in 1985 launched a new historiographical phase. McDougall was able to show, based on declassified National Security Council (NSC) documents, that there was a hidden agenda behind the Eisenhower Administration's May 1955 decision to support the concept of an IGY satellite. Establishing the precedent of "freedom of space" with a peaceful scientific satellite would smooth the way to overflying the Soviet Union with military reconnaissance satellites. How important "overflight" was in the decision is still somewhat debatable—as evidenced by the contributions of Kenneth Osgood and Dwayne Day to this volume—but McDougall was able to show that it had considerable influence at least from the level of Donald A. Quarles, Assistant Secretary of Defense for Research and Development, upwards to the President himself. This has been confirmed by further research by Cargill Hall and by Day.[2]

As a corollary of this thesis, McDougall speculated that the decision of the Stewart Committee was heavily influenced and perhaps pre-determined by this hidden agenda. The choice of the NRL's sounding-rocket-based Vanguard assured "the strongest civilian flavor" in the first U.S. satellite, in contrast to the Army's Orbiter, which used the Redstone ballistic missile as the first stage and Army solid rockets for the upper stages. Going further out on a limb, McDougall also speculated that establishing "freedom of space" was so important to the Eisenhower Administration that it consciously decided to risk the USSR launching first, which would be another way to establish the principle of overflight. This possibility was "less desirable, but it was not worth taking every measure to prevent."[3]

While the latter hypothesis has found few backers because of a dearth of positive evidence, the idea that Vanguard was virtually certain to win the support of the Stewart Committee because it looked "more civilian" for overflight purposes has been influential.[4] But newly declassified documents, combined with research into neglected sources from the Stewart Committee, reveal that overflight had little or no influence on the Vanguard decision; moreover, that decision was even closer than previous accounts suggest. Recently declassified minutes of Quarles' Research and Development Policy Committee show that this key advocate of the hidden agenda of overflight almost overturned the decision on August 16, 1955, by arguing for an interservice program in which the Army launch vehicle would inject NRL's satellite into orbit—a fact revealed here for the first time.

In order to better explain why the heavily favored Orbiter actually lost, I will also closely examine the history of that proposal, which arose months before Vanguard, and a third, weaker competitor from the U.S. Air Force. The Army project, which was in fact an interservice collaboration with the Office of Naval Research (ONR), helped bring forward the concept of a satellite in the scientific community and the secret counsels of the Administration. Yet it was ultimately undermined by its very origins as a low-cost, "quick and dirty" way to beat the Soviets into space.

ORBITER'S ORIGINS

As is well known, Project Orbiter began in June 1954, when Lt. Commander George Hoover of ONR's Air Branch initiated a practical study of how a minimal satellite could be injected into orbit around the Earth. The background for Hoover's initiative was the growing advocacy for military and scientific satellite concepts, both in the classified world and in the public arena, although it must be remembered how outlandish spaceflight still appeared to many only a few years before Sputnik.[5] The advocates were a heterogeneous group, including hard-core space enthusiasts like rocket engineer Wernher von Braun and International Astronautical Federation (IAF) President Frederick C. Durant III, upper-atmosphere scientists like astronomer Fred L. Whipple and physicist S. Fred Singer, and far-sighted officers like George Hoover. Given the Cold War interpenetration of American science, engineering and the military in the national security state, the communities of space advocates, scientists and officers actually overlapped a great deal.

A prime example is Fred Durant, who played a crucial role in Orbiter's origins. In mid-1954, Naval Reserve officer and rocket engineer Durant was not only IAF President and Past President of the American Rocket Society (ARS), he was also secretly working for

```

``````

``````

the scientific intelligence unit of the CIA. Until the CIA declassifies its documents on the early satellite projects, we cannot get much past informed speculation, but this much is clear: Durant was not only advocating space travel in the public arena through ARS and IAF, he may also have been using the IAF to gather intelligence on foreign rocket programs for the CIA at a time when concern over Soviet missile development was rising rapidly. Durant was apparently the middleman who brought Hoover together with Wernher von Braun, for a meeting on June 25 in Washington that also included Fred Whipple, Fred Singer and a few others. Durant continued to coordinate with von Braun in the months ahead, and was a central figure on the informal Orbiter committee from its origins until the end, which indicates that the CIA was interested in a scientific satellite months earlier than has been previously recognized. Confirming that supposition is this statement from a January 1955 Army-Navy proposal to the Secretary of Defense: "The Central Intelligence Agency has shown intense interest in Project ORBITER. The Agency apparently thinks considerable psychological warfare value and scientific prestige will accrue to the United States if we launch the first artificial satellite."[6]

Von Braun came to Hoover and Durant's meeting armed with a ready-made proposal for a minimal satellite launcher based on the 200-mile-range Redstone missile and three upper-stages of clustered, unguided Loki anti-aircraft rockets. Loki clusters had actually already been built by William Bollay's Aerophysics Development Corporation for the Air Force's Hypersonic Test Vehicle program. After an August 3 visit by Hoover and a superior officer to Huntsville to further confer on a joint program, von Braun and his associates, notably Gerhard Heller, produced a formal proposal dated September 15, 1954. In it, they stated: "The establishment of a man-made satellite, no matter how humble, would be a scientific achievement of tremendous impact ... . It would be a blow to U.S. prestige if we did not do it first." This remark reveals von Braun's Cold War concerns about the Soviet Union, but undoubtedly also those of Durant and others interested in the "psychological warfare value" of a satellite.[7]

Von Braun's proposal soon acquired the informal name "Project SLUG" because he contemplated putting merely an inert 5 lb. body into
``````

orbit using the Redstone/Loki combination. (The formal codename came only in January, after a brief interlude as "Project ORBIT.") The primary reasons for choosing this bare minimum approach to a satellite appear to be: 1) beating the Russians; and 2) keeping the budget to a minimum in view of the Eisenhower Administration's parsimonious approach to defense spending, strategic nuclear forces aside. In missilery, only the USAF intercontinental ballistic and cruise missile programs carried maximum priority in 1954, in part because of the growing Soviet capability in missiles. Thus von Braun's September proposal mentioned an added cost of only $100,000 for the next fiscal year, to be financed by ONR, and took for granted considerable built-in expenditures by the Redstone missile program. It is noteworthy that von Braun did not consider developing a more powerful upper stage like the upgraded Aerobee the NRL later used, as that did not fit with the minimum development approach.[8] Contributory reasons for this approach likely included the difficulty of interservice projects—ironic in view of ONR's central role in Orbiter—and the relative isolation of the Huntsville Germans from the upper atmospheric rocket community, the users of Aerobee and Viking.

Von Braun and company apparently assumed that a 5 lb. minimal satellite was too small to carry even a radio beacon, a decision that would later prove damaging. Tracking the body, using its orbit to measure the Earth's gravitational field and extreme outer atmosphere—even proving it went into orbit—thus became a central concern, since only optical means could detect it. Von Braun's group proposed that the satellite be a balloon or a mechanically unfolding sphere about twenty inches in diameter, and there was also discussion of the use of flares and other means of increasing visibility. Considerable effort was expended from fall 1954 to July 1955 to show that such a body could be spotted and tracked, since it still would be at the limit of naked-eye detection. Centrally involved were Fred Whipple, the Harvard astronomer and meteor expert, and Clyde Tombaugh, the discoverer of Pluto. Tombaugh, who was employed in optical missile tracking at the Army's White Sands Missile Range in New Mexico, was convinced that optical determination of the satellite's orbit would be much easier if that orbit was nearly equatorial. Huntsville wanted an equatorial launch too in order to achieve the maximum velocity from the rotation of the Earth, since the rocket's payload capacity was so marginal. But this approach meant that the launching would have to be done from a remote Pacific island chain, or from the Navy's experimental missile ship *USS Norton Sound*, which had earlier fired a Viking.[9]

While "SLUG" was being defined and refined over the fall and winter of 1954–55, it periodically intersected a second track of satellite advocacy. Key members of the U.S. scientific community, notably Lloyd Berkner, Joseph Kaplan and Fred Singer, had set out to engineer the international scientific unions' endorsement of the idea of launching satellites during the IGY, with the ultimate objective of gaining U.S. government support. The complex details of that campaign are adequately treated elsewhere, but it is noteworthy that after an Orbiter meeting in Washington on September 6, Fred Whipple, Gerhard Heller of Redstone Arsenal and J.B. Kendrick of Aerophysics (the Loki cluster contractor) attended a meeting of the Upper Atmosphere Rocket Research Panel on September 8–9, also in Washington. The first day was classified at the SECRET level, the same level as Orbiter, so formally or informally they must have briefed the space scientists on the project.[10]

As the Orbiter proposal moved forward to the Secretaries of the Army and Navy in December, and to Defense Secretary Charles E. Wilson in January, those individuals in the scientific community with security clearances and knowledge of Orbiter, notably Whipple, no doubt used that information to back the feasibility of a satellite in discussions of an IGY satellite in the National Academy of Sciences (NAS) and National Science Foundation (NSF). A second channel was the CIA, which was supportive of Orbiter and was receiving reports from Durant, and then endorsed the IGY satellite proposal of the NAS and NSF in spring 1955 from the very top, in the person of Director Allen Dulles. The CIA's primary rationale was again the psychological warfare value and prestige of a satellite if the U.S. was first—and the damage it would cause if the USSR did it instead. A third channel was the enormously influential "surprise attack study" of the Technological Capabilities Panel (TCP) led by James Killian of MIT, which delivered its report on Valentine's Day 1955. Among its recommendations was the launching of a scientific satellite specifically because of its value as a precedent for later reconnaissance vehicles, i.e. overflight.[11] We do not know what impact Orbiter might have had on that Panel, but as the only practical satellite proposal on the table in the winter of 1954/55, it was certainly known to TCP members. It at least made the case that a satellite was feasible.

COMPETITION EMERGES

In Orbiter documents from that period, the assumption is plain that it would soon become the official and only interservice satellite project. From the beginning the two main sponsors, the Army Ordnance Corps and the Office of Naval Research, had attempted to bring the Air Force into the project to sew up a tri-service basis for it. Von Braun wrote to Durant on September 10, 1954: "Airforce [sic] cooperation would be highly desirable, but [Gen. Leslie] Simon [Chief of Ordnance R&D] said that he would go along even [if] the Airforce doesn't feel like joining us." When Orbiter was pushed up to the highest levels in December-January, similar offers were always mentioned. They may, of course, have been mere political gestures. Since World War II, inter-service conflict over "roles and missions" in the guided missile field had already led to great bitterness. USAF suspicion of the Army's motives in the Korean War era "verged on paranoia," according to the Air Force's own history of its ballistic missile program: "Everywhere, it seemed, there was evidence of a conspiracy." Yet Orbiter advocates were doubtlessly honest in their desire to bolster their proposal with Air Force participation.[12]

It was to no avail. The Air Force representative tried to undermine the joint Army-Navy Orbiter proposal in the meeting of Assistant Secretary Quarles' Committee on General Sciences on January 20, 1955, in part by objecting to the cost. Four days later, a RAND Corporation scientist told an Army representative that "Rand has recently advised the Air Force that a minimal satellite should not be attempted at this time for the reason that it would alert the Russians to the possibility of a military satellite useful for reconnaissance"—ironically almost in exact contradiction to the overflight logic that the TCP Report and Quarles would soon use to justify the IGY satellite. And at the end of March, the head of the USAF ballistic missile program, Maj. Gen. Bernard A.

Schriever, advised his superior, Gen. Thomas Power, that he wanted nothing to do with the scientific satellite, which he saw as deliberately underrated in difficulty by its advocates, not militarily useful, a hindrance to the crash ICBM program, a management headache if a tri-service program, and (by implication) aid to the enemy—i.e., the Army missile team in Huntsville, which was also working on a concept for a 1500 mile missile that he wanted to thwart.[13]

Schriever's attitude meant that Air Force advocates who wished to jump on the scientific satellite bandwagon found themselves without adequate support even within their own service. The "World Series" proposal first floated in April actually came out of the USAF research center at Holloman Air Force Base, New Mexico. Interestingly, von Braun's former guidance chief in Germany, Ernst Steinhoff, was one of the principal architects of the Holloman proposal, which planned to mount a modified Aerobee on top of an Atlas to launch a much bigger payload than that of Orbiter or Viking/Vanguard. But it received only lukewarm support from Air Research and Development Command because of Schriever's opposition to any interference in the ICBM program. Reflecting USAF priorities, on March 16 that service did however formalize its requirement for a military reconnaissance satellite under the rubric Weapons System 117L.[14]

April also saw the emergence of what would become the real threat to Orbiter, the proposal of the Naval Research Laboratory's Milton Rosen, chief engineer of Viking. He had been present at the September 1954 Upper Atmosphere Rocket Research Panel meeting and had become involved in the satellite studies of the American Rocket Society and U.S. National Committee for the IGY. Rosen thus had intimate knowledge of Orbiter and it weaknesses, and decided that he could create his own plan for a more efficient system that was also more closely suited to the scientific community's IGY research interests. Since NRL was responsible to ONR, his project also had the effect of undermining the Navy's backing for Orbiter.[15]

The sudden surfacing of two rival proposals in April 1955 ended any possibility that the all-important Quarles would quickly give his blessing to Orbiter as the official U.S. IGY satellite. He had been officially informed about the project on December 21, 1954, when E. R. Piore, Chief Scientist of ONR, notified him of $68,000 in study contracts it would soon let. Army Ordnance's Los Angeles representative, James B. Edson, reported a month later that:

> Dr. Quarles is favorably disposed and a briefing for him was scheduled 19 January 1955. The key item in obtaining budget justification was determination of the density of the upper atmosphere for use in intercontinental ballistic missile trajectory calculations. [This argument is not present in other documents, but is implicit in the interest in upper atmosphere science.] Prestige and propaganda aspects were also recognized.[16]

Faced, however, with multiple proposals, Quarles threw the matter back into his Committee on General Sciences for further study, while also putting the entire satellite matter on the agenda of his highest advisory body, the tri-service R&D Policy Council.

At the April 18 meeting of the latter, Quarles told the members that the request of the National Science Foundation and the U.S. National Committee for the IGY for federal government support for a satellite would proceed: "If the NSC [National Security Council] feels there is justification for supporting [this project] (and there are certain persons in high places in our Government who think it would be disastrous if another Government did this first) and asks Defense to do so—Defense will undertake the venture on a tri-service basis." Quarles apparently did not mention the highly secret agenda of overflight as a precedent for space reconnaissance, which other documents show was much on his mind because of the TCP report, but seized rather on the Soviet threat. Only two days earlier, Moscow Radio had announced the existence of a Soviet spaceflight commission. That announcement doubtlessly aided Quarles in his swift engineering of NSC resolution 5520 of May 20, 1955, approved six days later, which made the satellite official policy—a process described elsewhere by McDougall and Hall, and in this volume by Osgood and Day.[17]

Meanwhile, Quarles' Committee on General Sciences had reported on May 4; the results of its deliberations did nothing to discourage Orbiter's advocates from believing that the official blessing of their project was inevitable. The Committee Chairman noted that the Secretary of the Navy had formally sent an Orbiter proposal document to Quarles on March 23, estimating a satellite launch by fall 1957 for $8.5 million, followed on April 15 by a further Navy proposal for a Viking-based system (i.e. Vanguard) for $7.5 million, and an Air Force proposal for Atlas-Aerobee with a larger payload, but at a cost of $50–100 million at a later date. Intriguingly, the Navy presented the second project as a "backup for ORBITER I and a possible second phase of a scientific satellite program"—not as a competitor. The Committee recommended that all three be pursued— Orbiter as the program for $6 million, Viking/Vanguard as a $5.5 million backup with "procurement of selected long lead-time items," and the Air Force project only as a design study for $1.25 million. The Technical Appendix to NSC 5520 of May 20, which presumably was prepared by Quarles' staff in the intervening weeks, shows the strong influence of Orbiter on the satellite concept, although the Financial Appendix mentions both Orbiter and "Viking" as viable candidates, but in a way that could be consistent with the Committee recommendation.[18]

THE STEWART COMMITTEE

Quarles, however, did not feel he could justify the relatively extravagant approach of parallel programs, perhaps because he rightly did not believe the low cost estimates he was being given. By June 1 he and his staff had formulated a plan to create an Advisory Group on Special Capabilities (the term Ad Hoc was soon added, but then removed again in the fall), another body with a deliberately opaque name. As Chairman he chose Homer Joe Stewart, a leading engineer-administrator at the Jet Propulsion Laboratory in Pasadena. Stewart sat on the Air Force Scientific Advisory Board, but his institution, although a part of the California Institute of Technology, was virtually an Army arsenal. He had been drawn into Orbiter planning by April, when he had completed a study of von Braun's September 1954 proposal for JPL Director W. H. Pickering, who was looking for a way into the Huntsville-ONR project. Stewart thus was not an unbiased observer, but at the time virtually all rocket engineers or upper-atmosphere scientists had a service affiliation or connection to one of the proposals, so small was the community and so intertwined was it with the military.[19]

The Stewart Committee was to have eight or nine members, with two being nominated by each service and two by Quarles' office. In the end there were eight; but one, astronomer-engineer Robert C. McMath, scarcely participated because of illness. The available documents still do not allow us to know who was appointed by whom, but reasonable guesses can be made for most members; this issue is important because service orientations appear to have contributed to the Committee's ultimate split in favor of Vanguard. Stewart and chemical engineer Clifford C. Furnas, Chancellor of the University of Buffalo, both had strong Army affiliations, and would form the ultimate pro-Orbiter minority. Charles C. Lauritsen, a physicist and colleague of Stewart at Caltech, had been intimately associated with Navy rocket development since 1940. Orbital mechanics specialist George Clement worked for the RAND Corporation, a USAF think tank, and J. Barkley Rosser, a Cornell mathematician, was affiliated with RAND. UCLA physicist Joseph Kaplan was Chairman of the U.S. National Committee for the IGY, and was thus a likely appointee of Quarles to oversee the important issue of which proposal would become the IGY satellite. That leaves two, one of which must have been a Navy nominee: either McMath, who ultimately cast his vote for Orbiter in a letter too late to affect the deliberations, or General Electric rocket engineer Richard W. Porter.[20]

Porter may be the most interesting case. He had been the head of the Army-sponsored GE Hermes rocket project begun in 1944, had played a central role in bringing the core of von Braun's "rocket team" from Germany, and had been associated with the Germans when they were based in the Texas/New Mexico desert until 1950. But the Army had cancelled Hermes in 1954, in part because the GE leadership was uninterested in being a military missile contractor. Whatever his Army connections, Porter had a vested interest in the NRL project: Milton Rosen proposed to use the Hermes GE X-400 27,000 lb. thrust liquid-fuel engine in the upgraded Viking that was the first stage of his booster. In addition, Porter had participated in the American Rocket Society panel, chaired by Rosen, that led up to the scientific community's approach to the Administration in the spring.[21]

Whether Porter actually leaned toward Viking/Vanguard before the Committee met is unknown, but we do know that Kaplan had a definite preference for the NRL project. In his May 6 letter to the Director of the National Academy of Sciences, Kaplan gave political rather than scientific reasons, although the larger scientific payload (10–40 lb.) Rosen promised must have influenced him. Open sponsorship by the U.S. National Committee plus the "Viking and Aerobee" combination would, he stated, "clearly establish ... the civilian character of the endeavor." In addition, the NRL system would create "no security classification considerations"—an assertion that was certainly exaggerated. The competitor Redstone missile Kaplan referred to only obliquely as "German V-2 developments," indicating that he thought the German connection of the Orbiter project would be a liability in the international arena. Viking/Aerobee, on the other hand, presented the best face for multi-national science and American foreign policy. Further evidence that bias against the Huntsville Germans might have influenced the vote of a Stewart Committee member is lacking, but it is noteworthy that Fred Singer wrote to Wernher von Braun on August 24 expressing his dismay at the "rather antagonistic feelings" about von Braun and Redstone Arsenal he had encountered recently among "some people," including "members of the IGY group."[22]

Could the hidden agenda of overflight also have influenced Kaplan's preference? It cannot be ruled out, given his high-level connections to the decision-making process as Chairman of the U.S. National Committee for the IGY, something no other member had. This access might have allowed him insight into the thinking of Quarles and others, but Kaplan had many other compelling reasons to prefer the "more civilian" Vanguard because of his role as U.S. representative to an international enterprise, his interest in science, and his possible anti-German bias.

The question remains whether Quarles' mandate to the Stewart Committee tilted the process towards Vanguard specifically for overflight purposes. Stewart stated in a recent telephone interview that at an initial briefing for his Committee, one of Quarles' staffers did mention that Vanguard "might have a more civilian aspect," which could indicate that the Assistant Secretary leaned slightly to the NRL at this stage. At some point, although possibly not in this meeting, the overflight issue was mentioned, Stewart stated; it was not a new issue as he had himself discussed satellites and the legal limits of air space in a 1946 Army meeting. "Freedom of space" was a known issue in the rocket community, which is not the same thing as saying that anyone knew it was important to the Administration's rationale for a scientific satellite. Stewart also asserted that overflight "had no bearing on the Committee's discussions" and was never mentioned there, which is consistent with the surviving documents.[23] Of more consequence was Quarles' specification that the satellite project not interfere with existing missile programs, although Orbiter's advocates could plausibly claim that the impact of their project on Redstone would be virtually nil.[24]

Why then did Orbiter find itself on the losing side when the Committee met in Washington July 6–9 and in Pasadena July 20–23? To understand this we must look not only at the known positive attributes of Vanguard—innovative, miniaturized electronics, a larger scientific payload, and a more efficiently designed launch vehicle—but also at Orbiter's deficiencies and why they were not much remedied before the presentations to the Stewart Committee. The hindsight induced by the Explorer success of 1958 has blinded many to the fact that Orbiter was not Explorer: it was inferior.

Most fundamental was the limited capacity of the basic Orbiter launch vehicle only a 5 lb. payload, whereas the later Explorer I had a 17 lb. payload attached to a 14 lb. burnt-out fourth stage. Orbiter had a lengthened Redstone 1st stage; Explorer had the same, but used a special hypergolic (self-igniting) fuel combination that boosted first-stage thrust from 75,000 to 83,000 lb. The basic Orbiter proposal had 37 Loki solid rockets in three stages (30/6/1) spun up before launch for stability; the Explorer Jupiter-C vehicle used 14 scaled-down six-inch-diameter Sergeants (11/3/1) in a similar arrangement. It is true that the proposal Wernher von Braun and Col. John Nickerson presented to the Stewart Committee on July 7 and 9 did discuss optional, as yet untested seven inch Sergeant rocket stages differently arranged (7/1/1) than the later Explorer. Homer Joe Stewart had first proposed the scaled-down Sergeants to Pickering in April, and JPL pitched the idea to von Braun at a Pasadena meeting shortly thereafter. Yet attacks by opponents against the Lokis had a lingering effect. The Loki I (the version in production) did not have a great reliability record, the larger the number of rockets the greater the odds that one would fail, and even one failure could prevent the satellite from reaching orbit. Critics in the Air Force and elsewhere seized on this argument, and in spite of reassurances by the manufacturer, Orbiter advocates could never convince the skeptics that its vehicle was highly reliable. In the end the Stewart Committee preferred the Sergeant option and scarcely discussed the reliability issue in its final report, but the damage had been done: the Orbiter launch vehicle entered the competition without the reputation of being much more reliable than the development-intensive Viking/upper-stage combinations.[25]

The 5 lb. inert sphere was also the subject of criticism, as we have seen, notably from scientists who wanted some real capacity for IGY experiments, and from opponents in the other camps who doubted that the body could be optically tracked. The imaginative use of a transistorized light-weight transmitter in the NRL satellite, combined with the proposed "Minitrack" tracking system, in the end forced von Braun and company to concede the idea in their July 1955 proposal that the NRL radio system could be carried in its satellite, or alternately, it could carry a radar reflector designed by another Army laboratory that supposedly could have a small power source that would allow data transfer by modulating the reflected signal. In mid-July von Braun also became very interested in a new lightweight JPL transmitter, part of the later "Microlock" system. Yet the scientific return from the Orbiter satellite remained confined to tracking its orbit to measure the Earth's gravitational field and extreme outer atmosphere, as long as the payload was only 5 lb. In contrast, the 17 lb. Explorer payload included both Minitrack and Microlock transmitters, plus James Van Allen's cosmic ray experiment that led to the discovery of the radiation belts.[26]

Why did von Braun and the Army stick to the minimal approach, with Loki clusters and an uninstrumented or barely instrumented satellite as their first choices, in face of criticism even before the Stewart Committee met? (Summer 1955 documents indeed treat it solely as an Army project; George Hoover of ONR had been forced to pull back because his superiors had switched Navy backing to Milton Rosen's project.)[27] First and foremost, Orbiter's advocates apparently continued to believe that the satellite's primary purpose was to beat the Soviets into space as quickly and cheaply as possible. In reality, the situation had become much more complicated: the scientists' wanted good IGY science, which was compatible with the agendas of Quarles and his superiors to have the satellite set an overflight precedent and generate international prestige, while not slowing down ballistic missile projects deemed more crucial to the Cold War. Everyone wanted the United States to be first, but nobody wanted to compromise their other agendas in the process.

The conviction of von Braun and Army Ordnance that their program concept was correct contributed to an overconfidence that bordered on arrogance. At a July 16 meeting in Los Angeles, one week after the first-round presentations, von Braun talked as if the go-ahead order would be received shortly. And on July 22, James B. Edson, who had been transferred from Pasadena to the Pentagon, stated that "OCO [Office of the Chief of Ordnance] personnel expect, because of the excellence and soundness of the Ordnance presentations, that the Ordnance system will be recommended."[28] The roots of this overconfidence included the self-confidence of the Huntsville Germans, who saw themselves as the world's most experienced rocket group, plus the fact that Orbiter had been the dominant player in the field for so long that it advocates found it hard to take the other projects seriously. Homer Joe Stewart described one of his most "vivid memories of that period" was Clifford Furnas' statement in a JPL parking lot after one of the meetings: "'You know, these Vanguard people are serious with their proposal.' It just hadn't occurred to him up to that point that they really thought of this as something that might be done. This was kind of a shock to me too, although I'd recognized it a little earlier."[29]

By the time that Furnas and Stewart realized the seriousness of the Vanguard proposal, the fate of Orbiter already hung in the balance. Kaplan, Porter and Lauritsen all leaned toward the Navy, leaving the two RAND-affiliated and presumably Air-Force nominated members, Clement and Rosser, with the potential deciding votes. In letters summarizing their position after the first round of meetings, neither Clement nor Rosser was enthusiastic about Orbiter or Vanguard, which they saw as marginal in payload and unlikely to meet the IGY deadline of a successful launch by the end of 1958. Clement leaned slightly toward the Air Force "World Series" proposal, but in a position consistent with RAND advice to the Air Force months before, was unenthusiastic about any small scientific satellite program. Rosser advocated using the rocket booster of the Navaho cruise missile together with an Aerobee-Hi second stage and unspecified third stage, an idea that came out of Air Force and North American Aviation people in the Los Angeles area.[30]

This position was odd and irrelevant, and Clement's view had scarcely more impact, given that the other members of the Committee wanted an IGY satellite, but did not see "World Series" as a serious contender for launching it, although they were very interested in an Atlas-based "Phase II" satellite as a follow-on. For the IGY, however, the time factor for Atlas was even more doubtful, the USAF gave the proposal little support, and it was a certainty that it would interfere in the ICBM program. It might be the case, as is traditionally asserted, that Clement and Rosser fell into line with the majority by pleading lack of expertise in rocket engineering, but it is striking that both ended up in a position consistent with Air Force attitudes— namely, that in the missile business the Army was a bigger threat to USAF "roles and missions" than the Navy; the Army also was a competitor for some of the same components, notably rocket engines.[31] In general, it is again noteworthy that it was the two most identifiably Army oriented Committee members, Stewart and Furnas, who formed the losing side. Service loyalties and interservice rivalry clearly cast their shadow over the outcome.

The hard core of pro-Viking/Vanguard support was thus Kaplan, Porter and Lauritsen. Kaplan's motives have already been described, and Lauritsen's career was closely tied to the Navy. Porter, however, did explain his position in a July 14 letter, and it is very consistent with the majority position in the Committee draft report of July 22, and in the official report of August 4.[32] A lengthy explication is unnecessary because the available evidence confirms the original interpretations of Hall, Green and Lomask, as against McDougall and others: the key factors were the majority's belief that the NRL Viking-based vehicle would require little more development than Orbiter, would be a more efficient design with one less stage (and hence would be more reliable), would carry a larger and more scientifically useful payload, and would be more easily declassified for international cooperation purposes. The good track record of NRL's Viking program was another plus. The fact that Orbiter would interfere in the Redstone missile project was barely mentioned in the majority's rationale because the argument was so weak. Beyond what was best for IGY, the Committee did not address political concerns at all, although at least Kaplan was predisposed to believe that Vanguard's "civilian" appearance would be better for prestige and propaganda purposes than a military rocket with German roots. The newly available and declassified documents show no evidence that the Stewart Committee had a predetermined agenda to pick Vanguard, or that overflight entered into its deliberations. What the new evidence does more clearly reveal is the impact of service loyalties on the Committee split.

A NEAR RUN THING

The Committee's report shocked a complacent Army Ordnance, and galvanized its service leadership to try to get Assistant Secretary Quarles to overturn the recommendation. Before the R&D Policy Council meeting on August 16, Ordnance's Assistant Chief for R&D, Maj. Gen. Leslie Simon, wrote a counter-proposal based on a week of feverish calculations in Huntsville, Pasadena and Washington. This memorandum, dated August 15, attempted to sway the decision with renewed evidence of Soviet competition. Just over two weeks earlier, in an unexpected turn of events, President Eisenhower had announced the IGY satellite project to the nation, possibly because of intelligence reports that the Soviets might announce first. Uncoincidentally, the Sixth International Astronautical Federation Congress was to convene on August 2 in Copenhagen, and Soviet delegates were to appear for the first time. Immediately after the opening, which was chaired by IAF President Fred Durant, the Congress passed a Swiss resolution to send a cable of congratulations to Eisenhower. Not to be outdone, Soviet delegate Academician Leonid Sedov held a news conference at his embassy soon thereafter, announcing a Soviet program to launch a satellite. Some media reported that he said that the launch would be in eighteen months (i.e., six months before the IGY began on July 1, 1957), and although this was allegedly withdrawn the next day, it had an impact in Washington.[33]

Simon's August 15 memo offered two possibilities: an upgraded Redstone first stage using North American Aviation's 135,000 lb. thrust engine in place of its 75,000 lb. one, making a satellite of 162 lb. possible, or a reduction in the weight of the original long-tank version of the Redstone by 1700 lbs., leading to a payload of up to 18 lbs. "The first orbital flight for this [latter] configuration can be scheduled for January 1957 if an immediate approval is granted. Since this is the date by which the U.S.S.R. may well be ready to launch, U.S. prestige dictates that every effort should be made to launch ... at this time." He also asserted that the project would have no impact on Redstone missile timetables, expressed doubt about NRL's ability to complete its projected rocket development on time, and indicated a willingness to let NRL build the heavier satellites.[34]

The next day the Donald Quarles' R&D Policy Council met at the Pentagon with Stewart present. After the Air Force and Navy gave their preference for the NRL proposal and the Army delegate summarized Gen. Simon's arguments, there was an extensive discussion. Quarles put forth a surprising tri-service compromise that essentially accepted Simon's offer and overturned the Stewart Committee recommendation: the Army would be given $20 million to run the program and would provide the booster, the NRL would provide the satellite and tracking, the Navy would fund the development of the Aerobee-Hi as an alternate second stage, and the Air Force would work toward a backup based on its military satellite (i.e., WS-117L) and provide launch support. Perhaps by sheer force of personality, Quarles got the Council to approve, conditional upon the sanction of the Stewart Committee, which was to report in a week on the practicality of the weight reduction, the feasibility of December 1956 as first launch date, and the impact on other missile programs of diverting 135,000 lb. engines. At the end of the meeting, Quarles announced what everybody knew, namely that it would be his last meeting, as he had just been sworn in as Secretary of the Air Force.[35]

Along with desire for interservice collaboration, which Quarles explicitly stated, it is apparent that beating the Russians must have led him to support the Orbiter booster. However, nowhere in the minutes or in the supporting documents did he or anyone else discuss the overflight factor—so closely held was the secret. Even key actors were apparently excluded; Army Ordnance documents show no awareness that it was a crucial rationale for the IGY satellite. One thing is clear: if Quarles had had a strong preference for Vanguard because it looked "more civilian" for overflight purposes, or for propaganda purposes, or even because it interfered less in missile programs, he never would have made his (hitherto unknown) August 16 proposal. He could simply have endorsed the Stewart Committee report, backed as it was by two out of three services.[36]

The August 16 meeting produced another week of feverish studies. Huntsville and Pasadena, with renewed hope, worked out calculations for the various combinations of Quarles' concept, and the NRL sought countervailing arguments to stop the threat to its recent, and even to its staff members, surprising victory. The Committee met again in Washington on August 23 to hear new presentations by von Braun and Rosen, among others, but Quarles' concept could not break the existing deadlock. The vote was four to two (we do not know who was missing), the two presumably being Stewart and Furnas. The Committee concluded that the weight savings were feasible, but only with extensive redesign, and that the December 1956 goal was doubtful and could be met "only with extraordinary effort and unusual organizational arrangements." Even the minority expressed this skepticism while supporting the revised, Army dominated proposal. Moreover, both sides had little faith in the ability of such a project to overcome interservice barriers.[37]

As for the upgraded Redstone, the Air Force stated that it could supply 135,000 lb. engines in mid-1957 "without interference with priority military programs," but the majority already had discussed the idea in the August 4 report, and was unimpressed because it added development time and effort that cancelled out any apparent advantage for the Orbiter vehicle. The August 23 majority of four also restated its preference for the more efficient staging and the guided second stage of the Viking-Aerobee combination. Equally important was the ultimately illusory promise made by the prime Viking and Vanguard contractor, Glenn L. Martin Company of Baltimore, to deliver the vehicle in eighteen months—the same as the Army. When the R&D Policy Council convened the next day, chaired by Quarles' former deputy, the compromise plan was dead. The battle was over: Vanguard had won. On September 9, 1955, Deputy Defense Secretary Reuben Robertson issued the formal instruction to the services.[38]

CONCLUSIONS

The near-reversal of the Vanguard decision in the R&D Policy Council brings home again how close and contingent that decision was. If Quarles had still chaired the Council on August 23, would it have made a difference? If Robert McMath, the third pro-Orbiter member of the Stewart Committee, had participated in July, could the three-three tie have influenced the two "fence-sitters," as Green and Lomask have speculated? If one or two members of those originally picked had been different, would Orbiter have won? We can never know. But the newly discovered and declassified records do show fairly conclusively that Quarles did not set up the Committee with a pre-determined agenda to choose the NRL proposal—whether because of Vanguard's "more civilian" appearance for overflight purposes, as McDougall has argued, or because it would not have interfered in military missile programs, as others have argued. It is true that the pro-Vanguard majority used the non-interference and ease of declassification issues to bolster their arguments. Yet the core of that majority—Kaplan, Porter and Lauritsen—also had personal agendas that inclined them to the NRL proposal, were genuinely impressed with some of the superior aspects of the Vanguard launch vehicle and satellite as compared to Orbiter, and perhaps as a result, managed to convince themselves that the development effort would not be much greater.

With hindsight we know that the latter judgement was wrong, that von Braun's Huntsville group probably could have made a satellite launch attempt by the end of 1956. But hindsight has also obscured the fact that Orbiter was not Explorer, that the minimal development and payload approach chosen by von Braun, and supported by Army Ordnance, ONR and others, contributed to its demise. If they had promised the instrumented payload carried by the later Explorer things might have turned out differently. Only after losing in the Stewart Committee did they change their conservative strategy, but it was too late.

In both the original and revised strategies for Orbiter, as well as in the Administration's satellite decision of April 1955, the Soviet factor played a critical role; it is perhaps in this area that the new documents yield the most interesting insights. Beating the Russians for prestige and psychological warfare reasons was the CIA's primary reason for supporting Orbiter and then NSC 5520—and it was a concern that the space enthusiasts like Wernher von Braun shared. Making sure that the U.S. put up the first satellite in a fiscally restrained environment led logically to von Braun's mimimum-development approach. But the scientific community's parallel advocacy effort complicated matters, because it created a new, powerful constituency for an instrumented IGY satellite, and unwittingly fed into the hidden agenda of overflight that emerged out of the TCP Report.

In this environment, the Soviet announcements of April and August 1955 had interesting effects. They set off shock waves in the classified deliberations of the Administration, reminding everyone that the Soviets could be first. Quarles, for one, seems to have reacted both times by underlining the need to prevent that possibility. Yet the net effect was never to overturn all the other agendas at work in the satellite decision-making process, which included interservice rivalry, overflight, science, international cooperation, prestige, and the desire to create a rational, fiscally conservative program. Only after Sputnik, in hindsight, did the overriding importance of beating the Russians at any cost seem obvious; beforehand American elites may have had a hard time believing that the U.S. might lose, no matter how often they were reminded of it. And in any case, the Stewart Committee majority saw the Vanguard project as delivering a better satellite on a similar timetable to Orbiter; thus the latter would not have been the automatic choice even if being first into orbit had been the sole priority.

In sum, all monocausal and deterministic arguments regarding the Vanguard decision fail. Multiple, often conflicting factors led to this chain of events, which in turn did much to set up the Sputnik shock of 1957. It is a powerful reminder of the contingent nature of history.

I wish to thank John Bluth of the JPL Archives, Jacob Neufeld of the Air Force History Support Office, and Dwayne Day of George Washington University, for their helpful assistance in procuring documents, and Dwayne Day, Fred Durant, Matt Bille, Dill Hunley of NASA Dryden, and Cargill Hall of NRO, for their comments on earlier drafts. Responsibility for all interpretations and errors of fact remains my own.

1.   R. Cargill Hall, "Origins and Development of the Vanguard and Explorer Satellite Programs," *Airpower Historian* 11 (Oct. 1964), pp. 101–12; Constance McLaughlin Green and Milton Lomask, *Vanguard: A History* (Washington, DC: NASA SP-4202, 1970), Chap. 3; Charles Lindbergh's Introduction to the latter book also pithily summarizes these factors

on p. vi. For purposes of convenience, I will use the Vanguard label for the NRL project even though it was not officially so named until fall 1955.

2. Walter A. McDougall, ...*The Heavens and the Earth: A Political History of the Space Age* (New York: Basic Books, 1985), chap. 5; R. Cargill Hall, "The Eisenhower Administration and the Cold War: Framing American Astronautics to Serve National Security, *Prologue* 27 (Spring 1995): 58–72, and "Origins of U.S. Space Policy: Eisenhower, Open Skies, and Freedom of Space," in *Exploring the Unknown: Selected Documents in the History of the U.S. Civil Space Program*, edited by John M. Logsdon, et al. (Washington, DC: NASA SP-4407, 1995), 1:213–29; Dwayne A. Day, "A Strategy for Space: Donald Quarles, the CIA and the Scientific Satellite Programme," *Spaceflight* 38 (September 1996): 308–312, and "A Strategy for Reconnaissance: Dwight D. Eisenhower and Freedom of Space," in *Eye in the Sky: The Story of the Corona Spy Satellites*, edited by Dwayne A. Day, John M. Logsdon and Brian Lattell (Washington, DC: Smithsonian Institution Press, 1998), pp. 119–42.

3. McDougall, *Heavens and the Earth*, pp. 123–24.

4. Notably, three recent popular histories take the thesis for granted: Curtis Peebles, *The Corona Project: America's First Spy Satellites* (Annapolis, Md.: Naval Institute Press, 1997), pp. 23–24; Helen Gavaghan, *Something New Under the Sun: Satellites and the Beginning of the Space Age* (New York: Copernicus, 1998), p. 27; Thomas A. Heppenheimer, *Countdown* (New York: John Wiley & Sons, 1998), pp. 99–100. Robert A. Divine's *The Sputnik Challenge: Eisenhower's Response to the Soviet Satellite* (New York: Oxford University Press, 1993), pp. 4–5, 8, asserts that a combination of overflight and non-interference in military missiles predetermined the outcome. Dwayne Day has offered various explanations, but his contribution to this volume seems to accept the non-interference factor alone as decisive.

5. Hall, "Origins and Development," pp. 101–105; Green and Lomask, *Vanguard*, chap. 1, and for a broader view, RIP Bulkeley, *The Sputniks Crisis and Early United States Space Policy* (Bloomington: Indiana University Press, 1991), esp. chaps. 4–9.

6. That Durant was working for the CIA either directly, or as a consultant from Arthur D. Little, Inc., has been made known to me by a number of independent sources, and is confirmed by one document, Durant's memo to the CIA Assistant Director of Scientific Intelligence regarding a meeting on UFO's in January 1953; see Logsdon, ed., *Exploring the Unknown*, 1:201–11. For Durant's role in Orbiter, see the sources cited in the previous note, plus Durant to von Braun, September 4, 1954, and reply, September 10, 1954, in the Wernher von Braun (hereinafter WvB) Papers, Durant correspondence file, U.S. Space and Rocket Center, Huntsville, Alabama (hereinafter SRCH), and pictures of the Orbiter meetings in Washington in March 1955 and in Huntsville and Cape Canaveral in May: Green and Lomask, *Vanguard*, p. 20, and "ORBITER (File #1)," in National Archives College Park (hereinafter NACP), RG 156, E. 1039A, Box 91. The January 1955 statement comes from "Proposal to the Secretary of Defense for the approval of *PROJECT ORBITER*," attached to Maj.Gen. Leslie Simon to Chief of Naval Research, January 27, 1955(?) in the file just cited. Cargill Hall and Dwayne Day have shown that the CIA supported the idea of satellite in the spring of 1955 and later put money into Vanguard when it was running into trouble. See the articles cited above in fn. 2.

7. Von Braun notes on call to Hoover and preparation for meeting, June 17, 1954, in WvB desk calendar, January 4-November 2, 1954, in WvB Papers, SRCH; typewritten agenda and handwritten notes to the June 25, 1954 meeting in Durant "Orbiter" folder, Durant personal papers (copies made with the permission of Fred Durant); 1972 oral history interview (hereinafter OHI) of Homer Joe Stewart by James H. Wilson, Jet Propulsion Laboratory (JPL) Archives, part 11; WvB, "A Minimum Satellite Vehicle: Based on Components Available from Missile Developments of the Army Ordnance Corps," September 15, 1954, copy in NASA History Office, WvB Bio. file, text reprinted in Logsdon, ed., *Exploring the Unknown*, 1:274–81.

8. Von Braun's desk calendar notes for June 17, 1954, show that he did consider other rocket engines, including the GE motor later used in Vanguard, and upgrading the Redstone from 75,000 lb. to 130,000 lb. thrust, an idea that would resurface a year later. See desk calendar of January 4-November 2, 1954, WvB Papers, SRCH. For "SLUG" see 1954-55 documents in "ORBITER (File #1)," NACP, RG 156, E. 1039A, Box 91. On the Air Force ICBM program after 1954, see Jacob Neufeld, *The Development of Ballistic Missiles in the United States Air Force 1945-1960* (Washington, DC: Office of Air Force History, 1990), chap. 4.

9. WvB, "A Minimum Satellite Vehicle," September 15, 1954, in Logsdon, *Exploring the Unknown*, 1:279-280; "Feasibility of Observing and Tracking a Small Satellite Object" by Varo Manufacturing Company, Inc. (Davis, Whipple and Zirker), June 25, 1955, NACP, RG156, E.1039A, Box 88; Orbiter reports of J. B. Edson, Jan.-Mar. 1955, "OUTLINE OF CONFERENCE IN THE PENTAGON 5 MAY 1955 ON OBSERVATION OF PROPOSED ARTIFICIAL SATELLITE—PROJECT ORBITER," Tombaugh trip report of June 2, 1955, and "OUTLINE OF CONFERENCE IN LOS ANGELES 16 JULY 1955 ON PROPOSED ARTIFICIAL SATELLITE PROGRAM—PROJECT ORBITER," NACP, RG156, E.1039A, Box 91, "ORBITER (File #1)." See also the folder on Tombaugh's project "Search for Natural Satellites of the Earth," 1952-55, in the latter box. The only ambiguous reference to a transponder being too heavy ("& too expensive") is contained in Fred Durant's notes to the September 6, 1954, Orbiter meeting in the Durant "Orbiter file," but it appears implicit in all of the documents cited above. The only other alternative—passive detection by radar—was viewed as infeasible given the satellite's small size.

10. Because the previous Panel meeting on April 29 had indicated that a classified meeting would be held on September 8, it is likely that a covert campaign had been launched among the scientists to put satellites on the agenda of the international unions' meetings in the fall. Lt. Cdr. Hoover's initiative in June to start a practical study may well have been a reaction to that campaign. See "Minutes of Meeting of the Upper Atmosphere Rocket Research Panel," April 26 and September 9, 1954 (September 9 is Part II only, Part I for September 8 classified and not yet found), copies from Homer Newell, National Air and Space Museum (NASM) Archives; WvB notes, August 3, 1954, in WvB desk calendar, January 4-November 2, 1954, and Durant to WvB, September 4, 1954, Durant Corr. file, both in SRCH, WvB Papers; Durant notes, September 6, 1954, Orbiter meeting, Durant "Orbiter" file. Durant's letter states that it was Whipple who pointed out the September 8 meeting on June 25. For the history of the scientists' campaign, see Bulkeley, *The Sputniks Crisis*, pp. 95-99, 125-33; McDougall, *Heavens and the Earth*, pp. 119-21; Allan Needell, *Cold War Science and the American State: Lloyd Berkner and the Balance of Professional Ideals* (forthcoming, Harwood Academic Publishers), chap. 12; and the previously cited contributions of Hall and Day.

11. On the CIA and the TCP, see NSC 5522, June 8, 1955, Dwight D. Eisenhower Library, White House Office, Office of the Special Assistant for National Security Affairs: Records, 1952-61, NSC Policy Papers, Box 16; McDougall, *Heavens and the Earth*, pp. 115-20; Hall, "Origins of U.S. Space Policy," in Logsdon, *Exploring the Unknown*, 1:218-20, and the contributions of Osgood and Day in this volume.

12. Orbiter documents, Dec. 1954-Mar. 1955, in NACP, RG156, E.1039A, Box 91, "ORBITER (File #1)"; WvB to Durant, September 10, 1954, in WvB Papers, SRCH; Neufeld, *Development of Ballistic Missiles*, pp. 86, 88, 91-92.

13. Minutes of the January 20, 1955, meeting of the CGS of the OASD(R&D), January 24, 1955, attached as an appendix to the R&D Policy Council minutes of April 18, 1955, in NACP, RG319, E.39, Records of the Office of the Chief of [Army] R&D, Records Relating to the R&D Policy Council, Box 2; J.B. Edson to Lt.Col. Nickerson, January 25, 1955, transmitting "Memorandum Report RAND Information on Project SLUG— By H. Morris," in NACP, RG156, E.1039A, Box 91, "ORBITER (File #1)"; Schriever memo to Gen. Power, March 30, 1955, "SUBJECT: Redstone—Scientific Satellite,"

in Schriever Papers, microfilm roll 3524, frames 407–408, Air Force History Support Office, Bolling AFB, DC.

14. The only substantive document from the "World Series" proposal so far available is Appendix B to the "Report of the Ad Hoc Advisory Group on Special Capabilities," RD 263/9, August 4, 1955 (hereinafter Stewart Comm. report), copy in NACP, RG156, E.1039A, Box 95, and microfilm copy in JPL Archives, roll 10–3. See also Hall, "Origins and Development," p. 107; McDougall, *Heavens and the Earth*, p. 121. Cargill Hall believes that Quarles not only orchestrated the whole IGY satellite decision, but also forced the Air Force to submit a proposal over the objection of its leadership (private communication with the author). Presumably Quarles' intent was to ensure that the competition had a tri-service character.

15. "Minutes of Meeting of the Upper Atmosphere Rocket Research Panel," September 9, 1954, copies from Homer Newell in NASM Archives; Technical Panel on Rocketry documents, January—March 1955, in Logsdon, ed., *Exploring the Unknown*, 1:297–301; Green and Lomask, *Vanguard*, pp. 26–27; Naval Research Laboratory (NRL) Rocket Development Branch, "A Scientific Satellite Program," April 13, 1955, copy supplied by J. Tugman, NRL; Milton Rosen oral history interview with the author, transcription in process.

16. Piore to Quarles, December 21, 1954, attached to the R&D Policy Council minutes of April 18, 1955, in NACP, RG319, E.39, Records of the Office of the Chief of R&D, Records Relating to the R&D Policy Council, Box 2; quotation from J.B. Edson, "*Fourth Status Report Week Ending 21 January 1955 PROJECT SLUG*", in NACP, RG156, E.1039A, Box 91, "ORBITER (File #1)."

17. R&D Policy Council minutes of April 18, 1955, as cited in the previous note; extract from "Scientific Intelligence Digest OSI 55–10, 9 May 55, CIA," on the April 16 Moscow Radio announcement, and Stewart Alsop, "Debate on The Satellite," May 26, 1955, *Durham Morning Herald* (clipping), in NACP, RG156, E.1039A, Box 91, "*ORBITER* (File #1)"; *Washington Post* article, April 17, 1955, and NSC 5520 in Logsdon, ed., *Exploring the Unknown*, 1:308–13; McDougall, *Heavens and the Earth*, pp. 120–21; Hall and Day as cited in fn. 2; and in this volume, Day and Osgood, and for the Soviets, Siddiqi and Gorin.

18. Robert W. Cairns, Chairman, Coordinating Committee on General Sciences, memo to ASD (R&D) [Quarles] on "Scientific Satellite Program for the Department of Defense," May 4, 1955, in NACP, RG156, E.1039A, Box 91, "ORBITER (File #1)"; NSC 5520 in Logsdon, ed., *Exploring the Unknown*, 1:310–311. In this and a March document attached to the R&D Policy Committee April 18 minutes, there is mention of an "ORBITER II" with a bigger, instrumented payload, but it was not proposed for funding even by Orbiter's Army/Navy backers, thus the designation "ORBITER I."

19. Attachments, June 1, to the R&D Policy Council Minutes of meeting June 7, 1955, in NACP, RG319, E.39, Records of the Chief of R&D, Records Relating to the R&D Policy Council, Box 2; Stewart memo to Pickering, April 19, 1955, in JPL roll 10–3, frames 547–561. This microfilm roll is Stewart's file on the Stewart Committee, 1955–58, a declassified photocopy of which has been deposited in its entirety in the NASM Archives. The frame numbering is from the photocopy. The microfilm was originally cited in Clayton R. Koppes, *JPL and the American Space Program* (New Haven: Yale University Press, 1982), which descibes JPL's relationship with the Army. For the origins of the intertwining of the military and upper atmosphere rocketry, see David H. DeVorkin, *Science With a Vengeance* (New York: Springer Verlag, 1992).

20. Athelstan Spilhaus of the University of Minnesota was also named as alternate to Kaplan, but did not participate. Green and Lomask, *Vanguard*, pp. 35–36, 48; attachments to the R&D Policy Council Minutes of meeting June 7, 1955, in NACP, RG319, E.39, Records of the Chief of R&D, Records Relating to the R&D Policy Council, Box 2; Furnas notes on

February 28, 1955, meeting of the Army Ordnance Advisory Committee, in Dwight D. Eisenhower Library, Furnas Papers, Box 1, "Army. Ordnance Advisory Committee, 1955"; Albert B. Christman, *Sailors, Scientists, and Rockets* (Washington, DC: Naval History Division, 1971), pp. 86ff.; Droessler to Stewart, June 27, 1955, with attached draft charter and Committee list for nine members, official charter, July 13, Rosser to Smith, July 12, Rosser to Stewart, July 25 (both from RAND), and McMath to Stewart, August 1, 1955, in JPL roll 10-3, fr. 19–22, 72–74, 565–566, 576–577, 683–684; Appendix C to Vanguard report, in U.S. Congress, House of Representatives, Committee on Appropriations, Subcommittee on Department of Defense Appropriations, *Hearings on Department of Defense Appropriations for 1960*, part 6, 86th Congress, 1st Session, April 14, 1959, pp. 81–82.

21. Porter OHI by David DeVorkin, 1984, in the NASM Archives; ARS report, November 24, 1954, in Logsdon, *Exploring the Unknown*, 1:281–83.
22. Kaplan to Waterman, May 6, 1955, in Logsdon, ed., *Exploring the Unknown*, 1:302–303; NRL, "A Scientific Satellite Program," April 13, 1955, copy from J. Tugman, NRL; Singer to WvB, August 24, 1955, copy in Ordway Collection, folder "Project SLUG/ORBITER: AMBA Documentation, 1954–1955," SRCH. For the anti-German issue see also Green and Lomask, *Vanguard*, p. 48. It should be noted that von Braun and over a hundred Peenemünde veterans and family members had just become American citizens in a highly publicized ceremony in Huntsville on April 15.
23. Stewart telephone interview by M.J. Neufeld, December 18, 1997, notes on file; Stewart's files on the Committee on JPL roll 10-3.
24. Supplement, June 14, 1955, minutes to the R&D Policy Council meeting of June 7, in NACP, RG319, E.39, Records of the Office of the Chief of R&D, Records Relating to the R&D Policy Council, Box 2; Stewart Comm. report, C-10.
25. Hoover memo to Fortune/ONR, March 28, 1955, about the March 17, 1955 Orbiter meeting, Durant personal papers; Stewart memo to Pickering, April 19, Thackwell/Grand Central Rocket Co. to Los Angeles Ord. District, May 11, Grand Central Rocket Co., "Reliability History of Loki Rockets Manufactured by JPL and the Grand Central Rocket Co.," June 27, Stewart to Smith/OASD(R&D), July 13, 1955, in JPL roll 10-3, 23ff., 30ff., 547–561, 568–574; J.B. Edson reports of January 21 and March 28 (latter incl. USAF criticism) and Maj. Williams' notes on April 25 meeting at JPL, in NACP, RG156, E.1039A, Box 91, "ORBITER (File #1)"; JPL Publication No. 47, "A Feasibility Study of the High-Velocity Stages of a Minimum Orbiting Missile," July 15, 1955, History Collection Doc. 3-593, JPL Archives; Stewart Comm. report, 10, 14, C-6, C-8; Stewart OHI by Wilson, JPL, part 11; for the Explorer Jupiter-C, see WvB, "The Redstone, Jupiter and Juno," in Eugene M. Emme, ed., *The History of Rocket Technology* (Detroit: Wayne State University Press, 1964), pp. 107–21, esp. 111–13.
26. Stewart Comm. report, A-9 to A-11, C-12; Green and Lomask, *Vanguard*, pp. 43–48; "OUTLINE OF CONFERENCE IN LOS ANGELES 16 JULY 1955 ON PROPOSED ARTIFICIAL SATELLITE PROGRAM—PROJECT ORBITER," NACP, RG156, E.1039A, Box 91, "ORBITER (File #1)." Von Braun and his group also remained fixated on the equatorial launch concept, and discussed using the Galapagos or Gilbert Islands. Unbeknownst to them, this ran against the grain of NSC 5520 itself, which stated a preference for a launch from Cape Canaveral into an orbit of thirty-five-degrees inclination to minimize launch costs, but presumably also to increase the number of foreign countries that would be overflown, without directly overflying the USSR. Since both sides in the Stewart Committee rejected the equatorial launch expedition, however, for reasons of cost, better geophysical coverage and tracking assistance from allies, it is not clear that it had any effect on the decision. NSC 5520 Technical Appendix in Logsdon, ed., *Exploring the Unknown*, 1:279–80, 312; Stewart Comm. report, 8, C-11.
27. Milton Rosen oral history interview by the author, July 1998, transcription in process.

28. "OUTLINE OF CONFERENCE IN LOS ANGELES 16 JULY 1955 ON PROPOSED ARTIFICIAL SATELLITE PROGRAM—PROJECT ORBITER," NACP, RG156, E.1039A, Box 91, "ORBITER (File #1)"; and Edson to Hirshhorn/White Sands, July 22, 1955, in same box, file "Search for Natural Satellites of the Earth."

29. Stewart OHI by Wilson, JPL, part 11. Note that we cannot take the words he puts in Furnas' mouth too literally, because the term Vanguard is anachronistic.

30. Green and Lomask, *Vanguard*, p. 48; Rosser to Smith, July 12, and Clement to Smith, July 13, in JPL roll 10–3, fr. 576–577, 586–587. For the North American proposal see also Albert E. Lombard, Jr., Scientific Advisor, Directorate of Research and Development, Deputy Chief of Staff, Development, USAF, to Stewart, July 14, 1955, in JPL roll 10–3, fr. 47. C. C. Furnas discussed the split vote of the Committee in "Why Did U.S. Lose the Race? Critics Speak Up," in *Life* (October 21, 1957), pp. 22–23, but mistakenly increased the number of members from seven to nine and the "fence-sitters" from two to three. A 1969 Furnas account of the decision was published obscurely and posthumously as "Birthpangs of the First Satellite," *Research Trends* (Spring 1970): 15–18. It is even more inaccurate, but gives a distorted version of Quarles' near-reversal of the decision in August—something I have found nowhere else in the published literature.

31. Porter to Smith, July 13, 1955, in JPL roll 10–3, 579–584; Stewart Comm. report, 3; Hall, "Origins and Development," pp. 107–108; Green and Lomask, *Vanguard*, pp. 41, 48–49. For six months after the Vanguard decision there was serious, but hitherto virtually unknown classified discussions in the Stewart Committee and the R&D Policy Committee about a USAF-launched "Phase II" heavy scientific satellite. Ultimately the concept died in spring 1956 because of lack of Air Force and Administration interest. See the September 1955—February 1956 materials in JPL roll 10–3; the Air Force proposal of January 1956 in NACP, RG156, E.1039A, Box 88; "Policy Council Meeting 15 Dec 1955," in RG 319, E.39, Records of the Chief of R&D, Records Relating to the R&D Policy Council, Box 2.

32. Stewart to Smith, July 13, draft report documents, July 22, in JPL roll 10–3, fr. 568–574, and 50–70, and Stewart Comm. report, August 4, 1955.

33. Untitled 8 Aug. document by Stewart(?) in JPL roll 10–3, fr. 87; Simon to Stewart, August 15, 1955, quoted (with redactions) in Vanguard report, *Hearings... for 1960* (see fn. 20), 58–59; Green and Lomask, *Vanguard*, pp. 36–39, 52–53; Frederick C. Durant, III, "Impressions of the Sixth Astronautics Congress," *Jet Propulsion* (December 1955): 738–39, copy in Durant bio. file, NASM Archives; contribution of Asif Siddiqi in this volume. Cargill Hall has pointed out that Eisenhower's announcement of the IGY satellite immediately followed the President's return from the Geneva summit where the Soviets had rejected his "Open Skies" proposal. He may also have decided to announce suddenly because the overflight logic of the program—a scientific satellite as a stalking horse for a reconnaissance satellite—seemed more urgent than ever. R. Cargill Hall, private communication with the author. Documentary evidence as to why the President made the announcement decision is, however, still lacking.

34. Simon to Stewart, August 15, 1955, quoted in Vanguard report, *Hearings... for 1960* (see fn. 20), pp. 58–59, and in Green and Lomask, *Vanguard*, pp. 52–53. I have not yet been able to find a declassified original.

35. Minutes of R&D Policy meeting of August 16, 1955, in NACP, RG319, E.39, Records of the Office of the Chief of R&D, Records Relating to the R&D Policy Council, Box 2.

36. In fairness, it must be pointed out that the Air Force delegate, Lt. Gen. Donald Putt, expressed only a mild preference for the NRL proposal, based on the common sources for rocket engines and components that the Army and USAF drew on for their missile programs; he was more interested in the Atlas-based "Phase II" satellite. See ibid.

37.   WvB desk calendar entries, August 15, 19 and 22, 1955, in WvB desk calendar, 1954–56, SRCH, WvB Papers; Nickerson teletype to Stewart, August 19, and Stewart to Chairman, R&D Policy Council (J.B. Macauley), August 24, 1955, in JPL roll 10–3, 86–87, 90–95; Furnas, "Why Did U.S. Lose the Race?" pp. 22–23; Hall, "Origins and Development," pp. 108–109; Green and Lomask, *Vanguard*, pp. 53–54.
38.   Stewart to Macauley, August 24, 1955, in JPL roll 10–3, 86–87; R&D Policy Council minutes of August 24, 1955, in RG319, E.39, Record of the Office of the Chief of R&D, Records Relating to the R&D Policy Council, Box 2; Stewart Comm. report, pp. 9–14; Green and Lomask, *Vanguard*, pp. 53–57.

Part 3
Ramifications and Reactions

Ramifications and Reactions

John M. Logsdon

To a significant degree, the lasting impacts of the launch of *Sputnik 1* and the reactions around the world to that achievement were a product of the fact that the launch came as a surprise—a surprise because it was the Soviet Union, not the United States, that was first to achieve this technological milestone. In October 1957, very few were prepared, at least on the basis of prior thinking, to identify the significance of the Soviet demonstration of technological capability in terms of its impact on politics and policy, military security, international relations, science and technology, popular culture, or indeed of its relevance to every-day life. The fact that the Soviet first did have widespread and long term impacts testifies both to the inherent importance of the shift in human affairs it signified—what McDougall has called a "saltation"[1]—and to the opportunity it provided to define the event in ways that advanced pre-existing agendas.

Two key individuals were *not* surprised by the launch of the satellite itself. What was surprising to them was the intensely excited reaction around the world, and particularly in the United States, to the event. They were of course Soviet leader Nikita Khrushchev and U.S. President Dwight Eisenhower—Khrushchev, because just over two months earlier he had given the Soviet "Chief Designer" Sergei Korolev permission to develop a small satellite for launch as soon as possible, and Eisenhower, because U.S. intelligence warned him of the impending launch.

As his son Sergei reports in the following essay, Khrushchev in the immediate aftermath of the *Sputnik 1* launch treated the achievement as significant, but not one with major political significance. However, the "sensational headlines" from the West "quickly corrected Moscow's position." Khrushchev "recognized his mistake and seized the initiative. From now on space and satellites became his propaganda hobby-horse." Having the flexibility and political insight to capitalize on Soviet space successes was one of Nikita Khrushchev's most impressive leadership achievements.

Eisenhower was slower to react; of necessity, his reaction had to be defensive. There had been some in the Eisenhower administration who had all along been warning that the nation first to launch a satellite would reap large propaganda benefits. As one example, in 1955 Eisenhower's Special Assistant Nelson Rockefeller in a memorandum to the National Security Council as it debated whether to authorize a U.S. scientific satellite had said that he was "impressed by the psychological ... advantages of having the first successful endeavor in this field result from the initiative of the United States, and the costly consequences of allowing the Russian initiative to outrun ours through an achievement that will symbolize scientific and technological achievement to peoples everywhere." Rockefeller added that "the stake of prestige that is involved makes this a race we cannot afford to lose."[2] The President had been well aware of such warnings; he made a judgment that they were not correct. This may have been one of the more profound misjudgments in his career.

Thus there was virtually no advance preparation for how the White House should respond, if the Soviets indeed were first to launch a satellite. It took Eisenhower over a month to develop a strategy of attempting publicly to minimize the significance of the Soviet satellites, while at the same time taking the steps within his administration to accelerate the U.S. space effort. (The much larger *Sputnik 2* had been launched at November 3, close to the annual Soviet celebration of the Bolshevik revolution. Sergei Khrushchev's account suggests that the timing was his father's initiative.)

By the time he did act, Eisenhower had lost for the White House much of the initiative in shaping the U.S. response to *Sputnik*. As Eilene Galloway suggests in her essay, Senate Democratic Majority Leader Lyndon B. Johnson was quick to recognize the importance of space leadership to the U.S. national interest, and inserted the Congress into the process of shaping an appropriate U.S. response to the Soviet achievement. Over the succeeding months, it was the interaction between the White House and Congress that defined the organization and character of the U.S. response in space (and elsewhere) to *Sputnik*. Neither branch of government had the upper hand. In the days immediately after the *Sputnik 1* launch, Johnson and his political associates also recognized that his perceived lack of leadership on the space issue also could make President Eisenhower and the Republican party vulnerable in the 1958 Congressional and 1960 Presidential elections.[3] Balancing considerations of partisan politics and the national interest between late 1957 and the 1960 presidential election was difficult, but the historical record shows that both Eisenhower and Johnson met the challenge well.

Galloway makes a provocative claim in her essay, one that deserves scholarly debate with respect to whether it reflects primarily Galloway's career-long internationalist bent or is also an accurate historical assessment. She suggests that "*Sputnik 1* was launched at exactly the right time ... . *Sputnik* had a unifying effect, drawing together all the societal elements needed for finding ways to establish conditions to ensure peace and prevent war in this pristine environment." She notes that if the United States had been first to launch a satellite, the reaction in the United States and around the world would have been much different, since America was expected to be first. Instead, "the fear of orbiting weapons ... galvanized political decisionmakers to take immediate action for achieving U.S. preeminence in outer space." The notion that one of the most important impacts of *Sputnik 1* was tilting the balance toward making space a weapons-free environment, with the United States in the lead in keeping it so, is certainly intriguing.

One must note that Galloway equates enlightened leadership in this new sector of human activity with "U.S. preeminence." That is not exactly the ways things looked to those European countries, particularly the United Kingdom and France, who were considering their own future in space in the years immediately after *Sputnik*. John Krige reports that the first inclination in the United Kingdom was to develop an autonomous space capability, including both a launcher and satellite-building capacity, and that elsewhere in Europe scientists were taking the lead in shaping a European space effort with scientific objectives at its core. By early 1959, however, the United States had set up a new civilian space organization, NASA, charged both with assuring that the United States would be "a leader" in space activities and with carrying out a program of international cooperation. Those in charge of NASA saw international cooperation as one instrument for achieving U.S. leadership in the emerging space field, and crafted offers of U.S. assistance that were seen from the U.S. perspective as very generous. Those offers also, reports Krige, had the effect of "systematically cutting the ground away beneath an autonomous space effort in Europe." European space aspirations became enmeshed in the Cold War rivalry between the United States and the Soviet Union; as Krige notes, "*Sputnik* turned space from a realm of dreams and fantasies into a locus of superpower rivalry, a major battlefield in the technological cold war." In this setting, Europe had very little room for independent action.

As mentioned earlier, the fact that the Eisenhower White House was not well prepared to react to the Soviet Union being first in space made it possible for various interests within the United States to define an appropriate U.S. response in terms favorable to their objectives. Perhaps foremost of those interests was the U.S. academic community, both in terms of its educational and research functions. As John Douglass observes in his piece, "*Sputnik* did not determine the evolving role of the federal government in higher education. But it did alter the way that Americans viewed advanced training and research," and it made U.S. research universities into "a national defense tool." By the end of his administration, Dwight D. Eisenhower not only warned the nation of the dangers of concentrating power in a military-industrial complex, but also in a scientific and technological elite. It was the reaction to *Sputnik* that elevated that elite to national influence in the late 1950s, as Gretchen Van Dyke notes in her essay.

The essays that follow raise two intriguing "what if" questions that cast the impact of *Sputnik* into sharp relief. Galloway asserts that if Vanguard had been orbited before *Sputnik 1*, "we [the United States] would not have been able to establish a coordinated national and international framework for developing the beneficial uses of outer space as quickly and successfully as we did." Indeed, if the United States had been first into space (as it almost certainly could have been, if the White House had allowed the von Braun team to launch a satellite as soon as it could), it is probable that the whole evolution of the space sector would have been notably different. The world *expected* the United States to be first. It was the uncertainty—indeed fear—of the implications of the Soviet surprise of being first that made U.S. decision makers conclude that space had to be an arena for Cold War competition—peaceful competition, but competition nonetheless.

The bias toward space initiatives that allowed the United States to demonstrate that in space it was technologically at least the equal of the Soviet Union was already evident in the way the United States formulated its initial space policies and programs during the remainder of the Eisenhower administration. It peaked, of course, with the May 1961 decision by President John F. Kennedy to challenge the Soviet Union to a race to be first to send people to the moon.[4]

Here, Sergey Khrushchev's essay is particularly fascinating. He corroborates what is becoming increasingly clear from the U.S. historical record—that President Kennedy would have preferred to cooperate in space with the Soviet Union rather than up the stakes in the space race, and thus made repeated cooperative overtures to the Soviet leadership. When Kennedy, a week after proposing the U.S. moon project to the U.S. Congress, suggested to Nikita Khrushchev at their first meeting in Vienna that the United States and the Soviet Union cooperate in a moon project, his son reports that Khrushchev was "unprepared" to respond. In June 1961, there had been no decision to undertake, and only limited discussion about, a Soviet lunar landing program. Thus Khrushchev, fearing that intimate space cooperation with the United States would reveal Soviet weaknesses in its missile program and faced with the opposition to close cooperation with the United States on the part of both his military advisers and space leader Sergey Korolev, did not respond. To Sergey Khrushchev, this was "a definite mistake. A combined lunar project ... might have been a turning point in the relations between our two countries."

There was a second chance. President Kennedy repeated his offer of U.S.-Soviet cooperation in 1963. This time, according to his son, Nikita Khrushchev's reaction was positive. The Soviet Union had developed an ICBM arsenal, "and from the combined work the Americans could learn about our strength and not our weakness." In addition, their joint efforts to solve the Cuban missile crisis had led Khrushchev "to trust Kennedy more."

With Kennedy's assassination in November 1963 and Khrushchev's removal from power in October 1964, the opportunity for transforming what started as a race with the launch of *Sputnik 1* on October 4, 1957, into a joint effort to send the first humans to the moon disappeared. What if things had been different? Sergey Khrushchev suggests that "the first men on the Moon would have been Neil Armstrong and Yuri Gagarin, and the Cold War might have ended fifteen years earlier." Perhaps, realistically, the Cold War impacts of *Sputnik* and the events that followed might have made this outcome politically infeasible. But space has always been a realm for dreamers; it is thus legitimate to dream of what might have been.

1.   Walter A. McDougall, ... *The Heavens and the Earth: A Political History of the Space Age* (New York: Basic Books, 1985), p. 6. However, McDougall is no longer convinced that the term is correctly applied to *Sputnik's* impact; see the Introduction to this book for his current thinking on the issue.

2.   The memorandum is included in John M. Logsdon *et al.*, *Exploring the Unknown: Selected Documents in the History of the U.S. Civil Space Program*, Vol. I (Washington: NASA SP-4407, 1995), pp. 312–13.

3.   For a discussion of this point and of Eisenhower's overall strategy for coping with the *Sputnik surprise*, see the essay by David Callahan and Fred I. Greenstein, "The Reluctant Racer: Eisenhower and U.S. Space Policy," in Roger D. Launius and Howard E. McCurdy, eds., *Spaceflight and the Myth of Presidential Leadership* (Urbana: University of Illinois Press, 1997).

4.   John M. Logsdon, *The Decision to Go to the Moon: Project Apollo and the National Interest* (Cambridge: MIT Press, 1970).

CHAPTER 9

The First Earth Satellite: A Retrospective View from the Future

Sergey Khrushchev

Forty years have passed since the launch of the first Earth satellite, almost a whole lifetime, but it seems that it all happened, if not yesterday, only last week—quite recently. This launch has been enthusiastic round-the clock work, not done under constraint, not because it was necessary, but because it was of interest, stemming from the naive confidence that you are equal to anything, that you can touch the stars with your hand, that the Moon is only a stopping station, the first step, and that the next stop is Mars, and after this anywhere. For me this was the best and the brightest time of my life. I am sure that this is true not only for me, but also for my contemporaries not only in the Soviet Union, but in America too. The scientists of both countries rushed forward like men possessed, not wishing to grant their rival a single inch, but in contrast to all the other races of those years this was a race not to the death, but to immortality. It pleased fate—and in this was a sign of Providence—to distribute the prizes among the contestants evenly, without giving offense.

I did not have occasion to participate directly in the development of the first satellite. I worked not with Sergey Korolev, but with his competitor Vladimir Chelomey. However on the strength of my service and family position I was able to be in close touch with the events of these years.

It must be said that the idea itself of flights into space did not seem to us, the schoolboys of yesterday, as something fantastic. Beginning with H. G. Wells' Martian invasion of Earth, and also "Aelita" and the books of Aleksey Tolstoy about flight from the Earth to Mars, we had also accumulated genuine knowledge from books of popular science by Perelman and Shternberg, and towards the middle of the 1950s we were waiting for the time when the engineers would turn the dream into reality. For this reason, when on the 26th of February 1956, as a 21-year-old student at the Moscow Institute of Power Engineering, my father took me to the firm of Korolev, where the happy manager showed us, among other wonders, his design for the first Earth satellite in the history of mankind, I was breathless with delight. My reaction was: now at last they have decided to do what should have been done long ago. Of the incredible complexity and boldness of what they had conceived at that time I had no inkling. Understanding came later when I myself had the good fortune to design and launch space vehicles.

My father also did not seem particularly delighted, all the more so since already a month before then he had signed the government decree on the development of the satellite. He was more concerned about the carrier-rocket of the satellite, the intercontinental ballistic missile R-7, which had given us the hope that with its appearance the threat of the nuclear destruction of our country through a preemptive strike by the United States would become history. From intelligence reports he knew that various plans for our total destruction were being seriously considered by American military men—and not only by the military.

For this reason, after hearing the report of the chief designer, my father was primarily interested in the question of whether the launch of the satellite would not distract the group involved in the project from the main task at hand. And having received assurances to the contrary, his mind was put at rest.

During those days neither my father nor Korolev imagined that they were discussing an event which would truly shake and overturn the world, and which would begin the countdown of a new space era in the development of humankind. No, they did not remain indifferent. My father had dreamed not of a political career, but that of an engineer. With a neophyte's enthusiasm, he was delighted at being associated with these technical achievements, and together with all Soviet people dreamed of overtaking the Americans, overtaking them in their ability to work, to grow corn, to turn out automobiles, fountain pens, and all the rest. The launch of the satellite was offering an opportunity, rare in those times, to demonstrate our technical superiority. Naturally, if we were able to forestall the Americans. Korolev warned my father that in the U.S. they were not asleep, and that if we did not hurry we could find ourselves in the second, that is, in last place. My father shared the concern of the chief designer, but, as I have already mentioned, not to the detriment of the main task—the defense of the country.

For Korolev the launch of the satellite was the realization of a long-held dream, the dream of a young man who had become an adult, of a man who was strong in spirit, and who strove to be a leader everywhere and in everything.

For both of them, my father and Korolev, the launch of the satellite signified the affirmation of our, and their own personal, leadership in competition with the most highly developed country in the world. And with such events as the flight of the ANT-25 over the North Pole in the U.S. in 1937, or winning in football against the English in 1949, there could scarcely be anything greater.

On October 4, 1957, my father stopped in Kiev on his way home from a vacation on the shore of the Black Sea, and I accompanied him. My father loved Kiev with its hills looming over the Dnieper, and its streets shaded by giant chestnut trees—the same streets in which he had positioned so many forces after the devastating German onslaught. My father naturally knew, as did I, that the launch of the satellite was scheduled for the evening of that day. However, the hero of the beginning of that day was not the satellite, but a tank, or rather tanks, which were capable of crossing the river over the bottom. The military men intended to demonstrate them to their Commander-in-Chief particularly on that warm sunny, autumn day, and particularly in Kiev. In October-November 1943 the armies in which my father was serving, under withering German fire, had forced a crossing of the Dnieper with the loss of many thousands of lives. Contemporary technology had made it possible to carry out the same feat while, although not eliminating losses completely, at least reducing them substantially.

My father returned from the maneuvers late in the evening thoroughly satisfied with what he had seen. After dining he settled down in one of the halls of the Marinskiy Palace with local Ukrainian officials. The palace had been built by Rastrelli on the occasion of a visit to Little Russia by Catherine II. Over the centuries the palace had become fairly dilapidated. Towards evening it grew cold on the street, and the wind blew unmercifully through the cracks, but absorbed in their conversation those who were in conference did not notice the drafts. What specifically the conversation was about, I do not remember. I remember only that the local officials were trying to squeeze more money for investments out of my father, and that he was desperately resisting this. All of this took place in an exceptionally friendly manner, for you see, my father had worked in this city with these people for more than a decade and they thoroughly understood each other. During the conversations night was approaching. My father began to glance at the door, behind which the duty adjutant sat behind a battery of telephones. I was the only person in that room who understood what these glances meant. My father was waiting for THE NEWS.

Sometime around or after 11:00 p.m.—not attaching the proper importance to the time, I had not glanced at my watch—the door creaked slightly, and an assistant pushed himself through the opened crack and whispered something in my father's ear. He nodded to the gathering, saying: "I'll be back shortly," and went into the next room. I alone was able to guess that at this moment they were reporting to my father from the testing ground about victory or … .

My father was not gone long. When after several minutes he appeared in the doorway, smiling broadly, a weight was lifted from my heart. If it was Korolev who had just called, then the launch had proceeded normally.

My father silently went to his seat and sat down, but he was in no hurry to continue the conversation. With an unhurried glance he looked at those present, and his face was simply beaming. "I am able to report to you a very pleasing and important piece of news," he finally began. "Korolev has just called"—(here he assumed a secretive expression)—"He is the designer of our rockets. Bear in mind, it is not necessary to remember his name—it is a secret. So here it is, Korolev has reported that this evening an artificial earth satellite has just been launched."

My father included all of those present in his triumphant gaze, and every one smiled politely, not understanding what it was that had happened.

Forgetting the subject of the preceding conversation, my father switched over to rockets. He began to tell them about the radical

change in the balance of power in the world which had been caused by
the appearance of the ballistic rocket. Those present listened in silence.
Their faces remained indifferent. They had grown accustomed to
paying attention to my father regardless of what he was saying. The
Kievans were hearing about rockets for the first time and could not
clearly picture what sort of thing they were. But since my father was
saying that it was an important matter, then it must be.

My father referred to the satellite as a very important and prestigious achievement, which had been derived from the intercontinental
rocket. And he stressed the point that here we had succeeded in overtaking America.

"The Americans have been proclaiming to the whole world that
they are preparing to launch an earth satellite. It is only the size of an
orange," continued my father. "We didn't say anything, and now
our satellite is no midget but weighs 80 kilograms, and is circling the
planet."

The assistant appeared again, and reported that shortly they would
be transmitting radio signals from the satellite. A radio receiver
standing in the corner was switched on. Everyone listened with
curiosity to the intermittent beep of the first communication session
from space.

It was not for nothing that Korolev had hurried to be the first to
launch a satellite. Now had come the moment of his anonymous
worldwide triumph—he was ahead of the whole planet. On his
return to Moscow my father immediately got in touch with Korolev.
He very much wanted to ask him about details, and to learn how it
had all come about.

Who first began the conversation about the possibility of launching a second satellite on November 7, on the anniversary of the
October Revolution, is now impossible to say. The historians of
rocket technology attribute this to my father, and indeed in a cate-
gorical and officious way. To me it seems problematical. My father
understood that he was not competent in technical questions, and
that here everything depended on the chief designer—and even not
on him alone, but also on the state of readiness of what they had
been working. Issuing orders could only give rise to haste, which
might result in harm to the project. For this reason, my father usually
stated his wishes in the form of questions. He may have asked Sergey

Pavlovich: "And would it not be possible to gladden the Soviet people with still one more launch, preferably on the holiday?" Korolev immediately seized on this idea which my father had let fall. Within several days he called my father and said that the launch was a reality, and that for the first time in history a living creature, a dog, would fly into space. After the second satellite other launches followed.

This kind of pressure put my father on his guard. He was concerned that such a turn of events might be detrimental to the tests of the military rocket. Propaganda was propaganda, and science was science, but national defense came first. Korolev assured him this would in no way happen. The tests could be shelved for a year, or possibly a little longer. Whether he had miscalculated his capabilities, had been mistaken, or had practiced a deceit, he was able to add the "seven" to the arsenal in only slightly more than three years, in the beginning of 1960.

Work on the second satellite was carried on day and night. To meet the deadline required accomplishing something which was almost impossible, for you see, only a month remained until the 7th of November.

To launch the second satellite by the day of Red October, and to again demonstrate to the world his leadership, even though it was anonymous, was something that Korolev himself wanted a hundred times more than any one else. However, every one in the design office knew that they were working without sparing themselves, in order to carry out the personal directive of "Himself," and the chief designer reported to the Kremlin every day on the progress of his work. This simple device was used not only by Sergey Pavlovich. When I began work, our chief designer more than once referred to the directives of my father and to the fact that Nikita Sergeyevich was personally observing how things were going. It must be said that these kinds of tactics were successful. People were imbued with the significance of their work, worked tirelessly, and accomplished "the impossible."

At least the second satellite was finished on time. On November 3 the dog Layka was put into orbit in a satellite weighing 508 kilograms. My father was in ecstasy. I was too, although I could not put aside the thought of what an evil fate had been prepared for the likable little mutt. I tried to speak about this to my father, who for his whole life had been a great animal lover, but he only waved away my questions, saying: "The scientists know what they're doing." A memorial to the dog-cosmonaut was the "Layka" cigarettes with the attractive, shaggy and smiling muzzle on the package.

Korolev was beginning to enjoy himself and suggested the launch on May 1 of a further, still heavier satellite, with a weight of 1,327 kilograms.

The launch was scheduled for April 27, 1958, but ended in an accident and the satellite was lost. A second, reserve model of the satellite was put into orbit on May 15.

My father drew his own conclusions, and when piloted flights began he became adamant. He rejected and forbade any suggestions for setting a date for the launch of a cosmonaut. His argument for his position was simple: "The joy of success can in a single hour turn into a mournful funeral march. Risk and losses in a new undertaking are inevitable. For this reason one should not tempt fate." And he added with a smile: "Hurry and people will laugh at you."

Leaf through the calendar, and you will see that until 1964 no piloted flights were carried out on holidays.

My father's distaste for stunts in the air and later in space, stunts which threatened people with destruction, had its roots in the period of the 1930s. It grew out of the tragedy of the *Maksim Gorkiy*. This was the name of the largest of our aircraft constructed by Tupolev. It was built as a single model and claimed to have no equal in the world. The people were proud of the *Maksim Gorkiy*. In the consciousness of the people the aircraft was identified with the power of the country and with the indisputable merit of socialism. They showed it off at every suitable opportunity. On one May Day celebration, they conceived the idea of sending the most outstanding products of the aircraft industry on an excursion into the skies over Moscow. And for the purpose of comparison they sent with the *Maksim Gorkiy* Polikarpov's high-speed I-5 fighter plane. Those who arranged the event thought that against the background of the fighter plane the giant aircraft would appear even more grandiose. The fighter plane circled around the *Maksim Gorkiy* like a wasp around a honey jar. Miscalculating a turn, however, the fighter pilot sliced into the wing of the larger plane. Both of the aircraft crashed to earth. All of the passengers and crew were killed.

The tragedy had a shattering effect on my father, who in those years was secretary of the Moscow party committee. People had been killed ... in addition to this, Stalin accused my father and the chairman of the Moscow Soviet, Bulganin, of indulging in a dangerous escapade. The festive, bravura music was changed to a funeral march. My father and Bulganin, along with others, carried the urns containing what they had managed to scrape together from the pieces.

The tests of the "seven" continued along with the launching of satellites. The successes were followed by a whole series of explosions, crashes, failures. It is not my task to write about them. But there is one incident which I consider it necessary to mention.

On that particular day my father had come home not exactly upset, but deeply troubled. A nightmarish incident had taken place. During the launch of the "seven" the instrument controlling the flight distance failed. The rocket flew over Kamchatka, passed beyond the borders of our territory, and hurtled in the direction of the U.S. Fortunately it did not reach the neighboring continent. The fuel ran out, and the last stage together with the warhead fell into the Pacific Ocean. No one knew whether the Americans had located the flight or not. My father was worried. You see in this way a war could be *provoked.*

The launch of the first satellite was reported by the Soviet press on the next day, October 5, with discretion—which matched our understanding of what had happened. They said it was an achievement, but one which was not of the common type. The "other side of the world," bursting with sensational headlines, quickly corrected Moscow's position. My father realized his mistake and seized the initiative. From now on space and satellites became his propaganda hobby-horse. And it was not only a matter of propaganda, for my father no less than Korolev was attracted by his association with great technical achievements. He loved to talk with Korolev, not in his Kremlin office but at his dacha on the Crimean shore of the Black Sea, and to the gentle splashing of the waves on the shingle to question him in a leisurely way about details, and to listen to his fantastic plans for flight into the far reaches beyond the atmosphere. True, enthusiasm was enthusiasm, but he was always mindful of government expenditure. Space was space, but for my father the main goal in his competition with the U.S. was not here, but in food production. His main slogan was: "We will overtake the U.S. in the production of meat, milk, and butter for per capita of the population." And of course also in housing—people were still living in basements, and several families were sometimes living in one room. But new homes were expensive, oh how expensive! This is where the money should primarily be spent, and only secondarily on defense and on space. The needs of space exploration should be made commensurate with the resources of the country, for "you should cut your coat according to the cloth."

And so with the launch of the satellite began the first stage of the Soviet space epic, the space race with the United States, even though no mention had been made of a race. The Soviet Union, my father, and Korolev were at the start when in 1954 they set the goal of creating a ballistic missile which could drop a hydrogen-nuclear warhead on American territory at a range of 8,000 kilometers, a warhead whose weight was estimated by its creator, Andrey Sakharov, to be 5.5 tons. The intercontinental ballistic rocket R-7 created in Korolev's design office was not used as a weapon, because its military use turned out to be too expensive, but it served well as a means of confirming the priority of the Soviet Union in not only the mastery of circumterrestrial space.

On September 14, 1959, a pennant-bearing sphere sent by Korolev reached the Moon and sprayed its surface with a multitude of five-cornered fragments accurately stamped with the seal of the Soviet Union. The date coincided with my father's official visit to the U.S., and he wanted to give the American president a copy of the pennant—ours, of course—at the moment of the arrival of the plane at Andrews Air Force Base. His assistants persuaded him to observe diplomatic decorum and wait to present the gift during his first visit to the White House.

Later came the triumph of the launch of the first man into space on April 12, 1961. The reception of the cosmonaut Major Yuri Gagarin in Moscow was perhaps comparable to the victory celebrations at the end of World War II.

My father was glad to support the initiatives of Korolev, who had squeezed out of the R-7 everything conceivable and inconceivable: the prolonged flight of a man in orbit, the flight of a woman cosmonaut, and finally the flight of three cosmonauts. For the time being this could be achieved if not without expenditures, at least at the cost of not really serious losses.

Korolev finally abandoned the military aspects of rocket technology. His new status as a pioneer, even though an anonymous one, evoked the deadly envy of other rocket scientists, especially Yangel and Chelomey. Wisecrackers even circulated a joke: "Korolev is working for TASS" (the telegraph agency of the Soviet Union, which circulated over the whole world information about the succession of achievements in space).

Space as a source of military intelligence in those years was not given any special priority, and was on the same level as other projects being worked on by the orders of the Ministry of Defense. My father, unlike the Americans, did not suffer from excessive curiosity. American nuclear power did not cause him uncertainty, and where their rockets were stationed was not so important to him—more important were the political agreements which would prohibit their military use.

Peaceful, working space—communications, meteorology, and other fields in general—had third priority. Various dilettantes, who were still unknown, had begun to work on them. But all things come to an end. Gradually the R-7 exhausted its potentialities, and was transformed from a record-holder into a space workhorse. And the Americans, now seriously worked up, decided to accept the challenge and show who was really first in the rocket-space race.

May of 1961, and the announcement by President John F. Kennedy of the beginning of the "Apollo" project and plans for landing a man on the Moon, marked the beginning of the second stage in the development of Soviet cosmonautics. My father was faced with the decision of whether to accept the challenge and be prepared to spend billions for the sake of keeping the palms of victory, or whether he should step aside and allow his undoubtedly richer competitor to get ahead of him. A competitor for whom the expenditure of twenty billion dollars or so for the sake of maintaining his ascendancy was not a matter of great concern. My father was not prepared to answer this question. And what is especially surprising, neither was Korolev.

In this context some explanation is necessary. By nature Korolev was not a pioneer or an originator of technical ideas. He was a brilliant organizer and manager who possessed a unique ability to pull together and win over ordinary workers, cause them to work selflessly in order to realize his idea, and rallied his comrades around himself in a tight-knit group subject to a single will, and inspired with his enthusiasm those in high places on whom the financing of his projects and the support of his work depended. But at the same time he was in constant need of external nourishment from fresh ideas. In separating from among them the idea which was most fruitful, Korolev possessed phenomenal intuition and would throw himself into its realization. If such ideas and their originators were not near him, the whole huge, detached mechanism began to spin its wheels. Korolev did well with the R-7. Mikhail Tikhonravov, a man of brilliant intellect and imaginative scope, suggested a package of rockets be used for the launch of the satellite. What could not be done by one rocket could be accomplished by tying rockets together. Tikhonravov was not capable of realizing his idea himself since he was totally devoid of organizational talents. Korolev however seized on the idea which had been given him, developed it, and incorporated it into his design. Korolev cannot be blamed for the fact that everything that was developed in Experimental Design Office-1 (OKB-1) was attributed to him. In spite of his ambitious nature he did not appropriate other men's ideas, although he could have easily done so. On the contrary, he determinedly supported and nourished his talented associates, promoting their progress. Beside him, in his design office, a whole galaxy of Academicians was formed, still another confirmation of his organizational ability. Such a thing was never seen in other design offices. The directors of these offices, scientists for the most part, could not tolerate potential competitors working alongside them. But for Korolev they were not competitors but the source of his strength, a life-giving source. Then why do all the projects developed in the OKB-1 bear Korolev's name? Design organizations in the Soviet Union, as well as in most companies in the West, most frequently were formed around their organizers, the originators of ideas and inventions. However, companies in the West, according to the rate of their growth, logically separated their functions into management and design, and in the USSR these two functions, even in enterprises with many thousands of workers, were centered in one person. The manager of the enterprise and the chief designer were one and the same person, the originator of everything that took place within the walls of his organization. According to this scheme, it would not be Kelly Johnson, but rather the President of the Lockheed Corporation who would be considered the originator of the high-altitude U-2 aircraft.

All that we have said above does not detract from the genius of Korolev, but only puts it in its proper framework. And it explains the logic of the events of the second stage of the development of the Soviet space program. At the beginning of the lunar race, Tikhonravov was not working with Korolev. He was working at Korolev's design office and was occupied solely with space vehicles and not with the questions of placing them in orbit. It was there that he was realizing his talent, and among the "rocket men" Korolev had none better than him. The new heavy space carrier rocket N-1, in contrast to the military rockets—it even received the new index "N"—originated in the collective mind. In other words, it was a combination of various ideas current at the time, and was not the product of the "mad" schemes of a prophetic inventor.

In the period of 1957–64, the appearance of the N-1 was constantly changing. In the beginning the rocket was oriented towards nuclear engines. This fashion passed and nuclear engines were replaced by oxygen-hydrogen engines, and so on ad infinitum. In addition, the weight which was being carried into orbit "jumped around" and was changed, so that the purpose of what they were developing was constantly changing. At the beginning there was the super-heavy Earth satellite, which was later replaced by the lunar ship. However the lunar project was not considered to be a goal in and of itself, but only a stage in the flight to Mars. Having begun with 40 tons injected into orbit, by 1961 "the collective mind" had stopped at 70–80 tons, which, as the saying goes, was "neither fish nor fowl." For an orbital station it was a bit too much and a bit too expensive, and for a lunar ship clearly not enough. However the decision was to try and squeeze the moon rocket into this narrow Procrustean bed.

One of Korolev's closest associates, Boris Chertok, remarked: "In developing the flight to the Moon they were not put off by the question of what the lunar module should be like and what kind of rocket should be designed to go with it, but they adapted the space ship to the carrier rocket, limiting themselves to the latter's capacities—75, later 90, and finally, after exhausting all reserves, 95 tons of payload in circumterrestrial orbit. Clearly not much for a full-scale expedition to the Moon."

In the meantime, Korolev the manager was possessed by one very important but completely non-technical concern: How to maintain his monopoly in space and not let his competitors thrive—Chelomey with his UR-500 (Proton) and Yangel with the R-56 were knocking at the door with ever greater persistence.

The last draft design of the N-1 was to a considerable degree aimed at the resolution of this very task: the creation of a universal carrier rocket, "cutting off oxygen," which would be up-to-date and competitive. The models of 1961–63—which were represented in the N-1—and those of the following years still put into orbit the same 70–75 tons. But at the same time a quasi-rocket, built on the basis of the second and third stages of the N-1, covered the same tasks as did the UR–500 of Chelomey—which according to Korolev's way of thinking made the development of the latter superfluous. His pursuit of universality made Korolev reject the package of "sausages," the hinged fuel tanks on the first stage, and in principle also reject the generally accepted construction of the body of a rocket fuel tank. Korolev thought—he told me about this himself in 1963—that such a design had no future, since the strength of aluminum alloys would limit the size of the carrier rocket to a launching weight of 4–5 thousand tons. This leap into the future predetermined the choice of spherical fuel tanks and the decision to locate an assembly factory at the launch site. It would in fact have been impossible to transport spherical fuel tanks with a diameter of 10 meters by train.

Not only the striving for universality, but the creation of a carrier rocket which would be suitable for all eventualities put Korolev in a spot. A no less important and even fateful event was his break with his old friend, his engine-maker competitor, Valentin Glushko. Glushko envied Korolev for the glory he had acquired more than did the others, and felt that he himself had been unjustly passed over. In his opinion it was the engines, and not the fuel tanks and everything else that Korolev was making, which ensured that all these pieces of metal were put into space. And all the glory was going to Korolev; he only stood in Korolev's shadow. A rebellion was imminent.

Korolev was not able to keep his "empire" intact much longer. Already in 1959, in spite of desperate counter-measures by the chief designer, Glushko "turned traitor." He supported Yangel in the creation of the "acid" intercontinental ballistic rocket R-16, by which he deprived Korolev forever of his monopoly in this field. This befitted the tyrant, and Korolev decided to punish the "apostate," and to stop his cooperation with Glushko at the first opportunity. In actual fact he had deprived himself of his chief partner. Glushko went over to Yangel and Chelomey, but he continued to work with Korolev under constraint and without enthusiasm, only because the government directives obliged him to do so. And there was no one to replace him. With difficulty Korolev agreed that Nikolay Kuznetsov—a designer sent by God, but a specialist in turbo-reactive engines—would make the engine for the N-1. In this new field, in spite of all his talents, he was condemned to being an apprentice.

And there was still another thing. Having opposed the lunar program, Korolev decided to skip a highly important step—the stage of scrupulous tests on the ground. At that time, it had become customary to blame the authorities for this, implying that they begrudged the money for ground-based testing facilities. I believe that this explanation is a result of hindsight, since Korolev, and only Korolev, determined the technical component of the project—on him also depended the stages of the testing.

The government might in general refuse to carry out a program because of its cost, but neither the Council of Ministers, nor the Military-Industrial Commission, nor the Ministry of Defense would run the risk of interfering in technical decisions if the Chief Designer firmly stood his ground—especially a chief designer of Korolev's caliber. This was not the case. Korolev worked a design step-by-step, which had become an established principle in aeronautics, but in the case of rocket construction was a relic of the past, an anachronism. And this is not a supposition. Mark Gurevich, a designer of the school of Semyon Lavochkin, and formerly the leading developer of the intercontinental winged rocket *Burya*, during the time he was working with Chelomey did not have a single conflict with Korolev. Gurevich directed the designing of the marine radar reconnaissance satellite, and the task of putting this satellite into orbit fell to Korolev's R-7. In the process of coordinating the technical parameters, Gurevich more than once had occasion to talk with Korolev—or more exactly the teacher instructed the "aircraft man" in the subtleties of rocket technology. Korolev believed that separating the tests into stages, the prolonged and scrupulous elaboration on the ground, only delayed the project. In his opinion, flight tests could reveal any defects much more quickly and productively and save valuable time. Gurevich did not raise objections to Korolev, but in his planning preferred to act in accordance with aeronautical traditions. Korolev's idea worked somehow as long as rockets were still relatively simple, but in the development of the N-1 it played a bad joke on its originator.

After Korolev's death in 1966 a major negative was added to all of these factors. It was not possible for the Chief Organizer, the man who could move mountains and brook all obstacles, to turn the impossible into the possible. His successor, Vasiliy Mishin, was excellent at calculating trajectories, but did not have the slightest idea how to cope with the groups of many thousands of people, the management of whom had been loaded onto his shoulders, nor to make the huge, irreversible government machine work for him. It was the death of Korolev which put the final period on the Moon race. All succeeding events resembled prolonged agony, the agony of a huge organism which does not want to accept the fact that its hour of death has come.

Whatever the main error in the Moon race, in the contest for the dubious privilege of leaving one's footprints in the lunar dust, it was not in the field of technology but in politics. It had never been my father's plan to spend substantial sums in order to support our priority in space. However during the four years after October 1957 he had become so accustomed to our position of leadership that he had not imagined anything else was possible. Initially my father did not attach much importance to the challenge of John F. Kennedy, and Korolev assured him of the unshakeability of his leadership. The fact was that he did not speak about, or more probably did not know himself, how much the maintenance of this leadership would cost. The proposal which was made by President Kennedy in Vienna in 1961, that the efforts of our two countries be united in a single lunar project, found my father unprepared. The military men came out against this proposal—they wanted to protect their secrets. Korolev was also against it, since he did not like the idea of sharing the palms of leadership with any one. My father himself was also doubtful—in the area of intercontinental rockets we were still quite weak—and the Americans might find out about this. In general he kept silent, and from the perspective of today this was a definite mistake. A combined lunar project would not only have saved face for us and saved a bundle of money, but it also might have been a turning point in the relations between our two countries. It could have been, but it did not happen.

Having rejected cooperation with the Americans my father was not in a hurry to get involved in the race, not to speak of an announcement of the priority of the Soviet Moon program on a national level. Korolev also did not press, and limited himself to the issuing of a government resolution in June 1961 about developing space research. Such resolutions—appeals were easily signed, for they contained no specific tasks—were not taken seriously and they were soon forgotten.

The uncertainty lasted until February 1962 and a session of the Council of Defense which was devoted to matters of rockets. By this time Korolev had revised his N-1, and he obtained the approval of the Council of Defense to begin a draft design for the lunar complex N-1-L-3. Again this was without any kind of calculation of the expenditures necessary for the whole program. Thus passed the years 1962, 1963, and 1964.

In the middle of 1963 President Kennedy repeated his proposal to unite the efforts of the two countries in a lunar program. Initially he spoke about this at a meeting with Soviet ambassador Anatoliy Dobrynin on August 26, 1963, and later he repeated his proposal officially in a speech from the podium of the U.N., in September of the same year. This time my father reacted positively, for by now we had a sufficient number of the R-16 missile, and from the combined work the Americans could learn about our strength and not our weakness. On the other hand he was attracted by the perspective of sharing expenses and in this way economize his own resources. Also the political climate had changed after the Cuban missile crisis, and my father began to trust Kennedy more. Naturally there could be no question of American subsidizing of Soviet developmental work, as is occurring now—such a thought never entered his head. And so to divide up the whole complex, so that each partner would do his share fairly independently and with good will, was absolutely realistic. My father told me that autumn of his positive reaction to the proposal of the American president.

But fate decided otherwise. First, in November 1963 Kennedy was killed. Lyndon Johnson, the new American president, did not return to the theme of the cooperation of the two countries in space. Then in October 1964 my father was removed from power. The chance for reversing the course of events was never realized. Otherwise the first men on the Moon would have been Neil Armstrong and Yuri Gagarin, and the Cold War might have ended fifteen years earlier.

In general, it seems to me that if my father had remained in power longer, our participation in the Moon race would have ended in and of itself as soon as he realized what expenditures it was costing the country. Why do I believe this? It is still the same thing: in my father's eyes the first priority was by no means the Moon, but the solution of the problem of food supply and the construction of housing. In this context I recall the following example. During the 1960s Moscow won the right to be host to the World's Fair. My father wholeheartedly supported this very prestigious undertaking. He supported it until he learned that the organization of the Fair would cost the Soviet budget billions of dollars. He immediately called for a review of the already signed agreements, for under the conditions of those years he did not believe in hypothetical revenue from tourism. The Agreement was reviewed. The World Exhibition was held in Brussels or Montreal (I don't remember for sure). The Moon project could have shared the fate of the World Exhibition but didn't. Leonid Brezhnev succeeded my father in the Kremlin and money was no longer counted.

There was one other point: the Moon race that took place *between* Korolev and Chelomey rather than between Russians and Americans. This episode in itself was not a major event in the history of space exploration, but aroused a lot of speculations. Korolev's heirs even claim that they lost because they were forced to scatter their efforts. Was that the case? I once had the occasion to say that in the fervor of their strife, both Chelomey and Korolev would rather see Americans succeed than their respective rival. And this is the truth. But in fact the competition of the two chief designers ended without ever having started. In September 1964, at a routine review of military technology at Baykonur, Chelomey proposed making the 4,500 ton UR-700 launch vehicle and delivering two cosmonauts to the Moon with the aid of a 145 ton Earth orbiter. Leonid Smirnov, chairman of the Military-Industrial Commission, was tasked with reviewing the proposal and drafting an opinion. For this purpose, the Afanasyev (who headed the newly established "rocket" Ministry of General Machine Building)-Keldysh (who was then touted in the newspapers as the "theoretician of cosmonautics," the President of the USSR Academy of Sciences) Commission was established. It is hard to say whether my father would have supported Chelomey or prefer the well-reputed Korolev to a beginner. However, after my father was forced to resign, the Commission's decision was pre-determined. Korolev was backed by the omnipotent Dmitriy Ustinov, who had a huge influence over Brezhnev who had come to power. In November 1965, Chelomey lost a race that had never even begun. He only managed to "produce" a few volumes of a pre-conceptual design. The money he spent came only from the OKB-52's funds intended for advanced research, and that was all. As the N-1 failed repeatedly, the concept of the UR-700 was revived on several occasions, but never finalized to be ready for a serious job. I will not comment on Chelomey's project of the flight around the Moon, since it was pretty insignificant within the totality of the lunar project.

Why was the N-1 "closed" in 1974? This can be explained in numerous ways, but no one knows the truth. I, too, have a version ready. Ustinov, who was behind the OKB-1 and shared the responsibility, was aware of the adventurous nature of launching a ship to the Moon with only 95 tons of Earth-orbiting spacecraft (5 tons without any weight reserves). If the lunar spacecraft got even a bit heavier, everything was ruined. And Ustinov knew from his abundant experience that structures tend to become heavier but never lighter. So he decided to take no more risks and closed the entire research under the pretext of an accident. Again, this is only my version, but I was well familiar with both Ustinov and the works of the Military-Industrial Commission.

In the meantime, in 1967 Chelomey, who had to forego the Moon, started designing the multiton *Almaz* inhabited orbital station to be launched on the UR-500. Thus, he ushered in a third page in the history of Soviet cosmonautics. This turned out to be one of the most successful. Many pioneering concepts were implemented on *Almaz*: it was the first to become stabilized by flywheel mass, it was the first to have an 11 meter picture camera/telescope on board, it was the first to utilize a radar capable of reconnaissance under cloudy conditions with a good resolution both on land and on sea. But Chelomey was not meant to enjoy the fruits of his labor. At the same time as the resolution closing the N-1 and the lunar spacecraft project was being drafted, Ustinov ordered Chelomey to hand any available materials on *Almaz* over to Vasiliy Mishin, Korolev's successor. The OKB-1 installed some equipment of its own on *Almaz* and in 1971 launched *Salyut*, the world's first orbital station, the first in a series of space stations.

Around that time, the Americans also launched a station of their own. A new space race began to take shape but never happened. The Americans spent some time working on a space station but decided, for whatever reason, that future prospects in astronautics lay elsewhere, and focused their attention on building a winged spacecraft.

The Soviet space program decided to follow both routes, since in Brezhnev's times, unlike Khrushchev's times, money was being spent with ease, and both space stations and *Buran*, the Soviet Shuttle were funded generously.

It is funny, how some aspects of the Soviet space program depended on decisions made in Washington. Early in the sixties, Chelomey's design bureau was designing its own winged spacecraft/bomber, the *Raketoplan*. They even manufactured two winged models, MP-1 and MP-2 and launched them using the R-12 and R-14 missiles to check the thermal conditions of atmospheric entry. At the same time, a similar preliminary project, the *Dyna-Soar*, was underway in the U.S. The project went stop-and-go, at times it went forward, at times its funding was reduced. The Soviet General Staff was tracking the Americans very closely, as if its supervisors were in Washington rather than in Moscow. Whenever the Americans intensified work on the *Dyna-Soar*, the Soviet General Staff would send reports to the government in support of *Raketoplan*. Whenever the Americans scaled their project down, the General Staff would immediately suggest to close down *Raketoplan*.

Regardless of the experience Chelomey had, Ustinov did not put Chelomey in charge of the *Buran* winged spacecraft project. Instead he established a new design bureau, literally out of nowhere, under Lozino-Lozinskiy. Thus the third stage of Soviet cosmonautics went on, without either significant achievements or losses for that matter. Invaluable experience was gained in long-term orbital flights. It went on during the stagnation in the early 1990s when the country's economic and technical infrastructure collapsed along with the collapse of the Soviet Union, thus destroying the foundation that supported space programs.

Around 1992, the fourth, and perhaps the final development stage of the Soviet cosmonautics set in, the period of dying, of agony. Dying—unless some emergency external assistance or resuscitation comes in. When a manufacturing, primarily high-tech industry disintegrates, along with its supporting scientific infrastructure, and switches course to sponging on the cheap sale of the country's natural resources, the survival of a single branch, even though it be the most prestigious one, is not to be hoped for. Sure, the agony can be extended by artificial feeding, but that is all. In the foreseeable future, one cannot hope to see a turn for the better in Russia, a return to a modern economy based on the manufacturing of an end product. In the meantime, any experience we may have accumulated will have been lost, forcing us to start from scratch again.

It is very embarrassing to lose the experience and knowledge accumulated over the past forty years, especially the unique knowledge in the field of long-term human space flights. Saving whatever can be saved could only be accomplished by connecting surviving teams to a healthy economy directly, bypassing the corrupt Russian authorities, similar to what is done to a heart in fatal condition. How do we get this done? By having purchase orders funded directly, the way it is done in building elements for the International Space Station at the KB Salyut once headed by Chelomey, or at Khrunichev Production Association, or by setting up joint ventures of the Sea-Launch type, an entity that joined Korolev's Energiya and Yangel's Yuzhmash, one-time deadly rivals, under its auspices? Or are there any other ways? I don't know, I am not in a position to make decisions, but something has to be done, and quickly.

This is a sad ending to an article on the long-term repercussions of launching the Earth's first artificial satellite. And I find little comfort in the adage about everything having a beginning and an end. Because the end, the dying away of Russian cosmonautics, is not a result of natural life processes, but instead a consequence of blunders committed by those who had a highly irresponsible attitude to governing Russia. But that is another story.

Whatever current world developments may be, October 4, 1957, opened another glorious page in the book of the Earth's history, and we are currently living in a different epoch, a space epoch. And this is the most important and the most long-term consequence of this epoch-making event that occurred forty years ago. What Korolev and his comrades-at-arms did was transform the faraway cosmos into a part of reality of our life here on the Earth. Humanity will never forget Korolev or his accomplishments.

Building a Third Space Power: Western European Reactions to Sputnik at the Dawn of the Space Age

John Krige

The launch of Sputnik by the Soviet Union on October 4, 1957, consolidated space as a domain of international competition and national security. From the mid-1950s onwards the Eisenhower administration had been taking steps, under the cover of the scientific objectives of the International Geophysical Year, to secure freedom of access to space for reconnaissance satellites.

The development of Intercontinental Ballistic Missiles had qualitatively transformed the vulnerability of the U.S. and the Soviet Union to nuclear attack and new methods of surveillance and of rapid warning were imperative to enhance security. By being the first to penetrate the "new high ground," Sputnik severely undermined that security and dealt a major political and ideological blow to presumed U.S. leadership. It created the illusion that there was a significant technological gap with the Soviet Union, and it spawned a climate of national anxiety which was translated into a massive acceleration and expansion of the civilian and military space programs. Sputnik was not simply a Soviet scientific and technological first. It was also a warning that "the evil empire" would soon be technologically capable of striking American targets directly from Soviet launching pads. It was seen too as a sign of the economic strength and even political superiority of state socialism, the triumph of rational planning over laissez-faire liberal democracy. Space became synonymous with superpower rivalry, and a major battlefield in the technological cold war.[1]

The intimate association of space with national security and technological dominance which we find in the United States and the Soviet Union is not present in Europe in the 1950s (or even in the decade thereafter). Certainly in the latter half of the 1950s first Britain and later France took steps to defend themselves using Intermediate Range Ballistic Missiles (IRBMs). Six months before Sputnik the British Minister of Defence, Duncan Sandys, had confirmed that the country would develop its own nuclear deterrent based on the V-bomber force equipped with hydrogen weapons and an IRBM called *Blue Streak*.[2] And at the turn of the decade de Gaulle began to put in place the elements of his independent nuclear capability, the *force de frappe* that included powerful rockets in its panoply of arms.[3] But these programs were far from maturity: Britain's *Blue Streak* and France's *Diamant* were planned to be operational in the mid-1960s. In the interim European security was assured through NATO while London took steps of its own to develop bilateral dual-key defensive measures with Washington.[4] Sheltering under the American nuclear umbrella Europe felt far less vulnerable than the United States to a new Soviet threat, and anyway it simply did not have the technological resources nor the infrastructure at its disposal to deal with any such threat directly.

Along with military dependence went political resentment. In the 1950s many senior British government officials were irritated by what they saw as the arrogance of U.S. power. They defined their role as providing a sensible counterweight to American overreaction, a mature voice in world affairs whose accomplishments were not appreciated across the Atlantic and whose task it was to instill some realism into the strategic thinking of their dangerously inexperienced ally. The American leadership was seen to combine military power with political naiveté, and the Sputnik shock was just the kind of thing that it needed to make the country face up to its limitations. The Americans, one internal Whitehall minute noted patronizingly in December 1957, "need Pearl Harbors from time to time, and perhaps it is just as well that the Russians are capable of delivering them in this relatively harmless way."[5] Britain could only benefit from the new situation. For a decade the British had been smarting under what they felt was the cruel injustice of the McMahon Act, adopted by Congress in July 1946. This bill prohibited the passing of nuclear information to foreign nationals, so effectively putting a halt to Anglo-American nuclear collaboration, which had been so valuable to both partners since 1941. Minor modifications were made in 1954, when the U.S. authorities were permitted to supply information on the external characteristics of atomic weapons, though sharing information on warhead design was still prohibited.[6] Sputnik changed all that. A few days after Sputnik II was launched, Duncan Sandys told the House of Commons that "the Sputniks [...] have helped to precipitate closer collaboration with the United States [...] the new impetus towards unrestricted collaboration [...] offers new prospects which we dared not hope for a few months ago."[7] And indeed in 1958 the McMahon Act was revised to permit the exchange of information about the design and production of nuclear warheads and the transfer of fissile materials to countries that "had already made substantial progress in the development of atomic weapons," meaning Britain above all. There were further relaxations in 1959.[8]

Sputnik then did not send shock waves of anxiety through U.K. government circles. On the contrary it was seen as being to Britain's advantage, as showing the Americans that they were not invulnerable and that they needed their allies, so breathing new life into the faltering Anglo-American "special relationship" in the nuclear field just when the U.K. was developing new technologies for its independent deterrent.

The Sputnik that was tracked by hundreds of amateur radio enthusiasts and by the giant radiotelescope at Jodrell Bank as it crossed British soil was seen in the U.K. as a scientific and techno-logical marvel above all.[9] When doubts and criticism did come, and they came quickly, it was not over a potential military threat but over the orbiting of Laika. It was the switchboards and postbags of the Royal Society for the Protection of Cruelty to Animals, and not those of the Ministry of Defence, which were overwhelmed by anxious and furious citizens in November 1957. In France too the satellite was quickly stripped of any immediate security implica-tions. "The victory of the Soviet engineers ('techniciens')," wrote Le Monde a couple of days after the news of 'le Spoutnik' broke, "can be seen to be purely scientific." The editorial recognized that missiles and satellite launchers originated in the same technology of course, but it insisted on maintaining a strict divide between the objectives of the two. Sputnik was a small, temporary star that would soon disappear like an inoffensive meteorite and "it would need the extrapolation of a pessimistic novelist to furnish it already with a thermonuclear explosive or with the eye of a mechanical spy." The Soviet satellite was before all else a symbol of scientific and technological progress: "There is no cause for alarm for the moment," Le Monde reassured its readers, who were anyway far more concerned about the grueling war in Algeria than about any potential invasion from space.[10]

In 1957, then, Sputnik and space meant very different things on opposite sides of the Atlantic. In the United States they were charged with layers of meaning, a new focal point in the titanic confrontation between the two world systems. Their significance in Europe was quite other, far thinner, far more restricted. My aim in this paper is to explore that other, to show how space was perceived in western Europe, in contrast to the United States, at the end of the 1950s and early 1960s. More specifically I want to trace how, for scientific and political elites alike, the meaning of space changed in the years immediately following the launch of Sputnik. My analysis will concentrate mostly, though not exclusively, on Britain, not only because she was the leading technological power in Europe at the time, but also because at present we know far more about the situation in the UK than in her neighbors. I shall show how space, originally seen primarily as a domain for conducting scientific research and gaining national prestige, gradually came to assume new functions, functions more congruent with those that it had in the U.S. (and the USSR), yet functions which were expressed in a form specific to a Europe of medium-sized powers. It was the launcher above all (though not only) that sparked this change. It was a change that extended space from a realm controlled by scientists seeking knowledge and prestige into being also a political platform for governments to unite against the communist threat, to build a European aeronautical industry and to create the foundations of a "third space power" alongside the United States and the Soviet Union.

1958: FROM SOUNDING ROCKETS TO SATELLITES

If the launch of Sputnik created a sense of "euphoria" in parts of the British scientific community it was not simply because of the scientific and technological achievement that it represented. It was also because they could, literally, hardly believe their eyes or their ears. When Eisenhower announced in July 1955 that the U.S.A. planned to launch an earth satellite in the framework of the International Geophysical Year (IGY), to be held in 1957–1958, the U.K. had not yet officially embarked on a sounding rocket program for upper atmosphere research. Indeed, meeting in April 1956, the British members of the IGY National Committee "were very skeptical about earth satellites ever appearing and if they did, that anything of real interest would result." Massey, the leader of the British scientific community in this field, was thus as "surprised" and "greatly amazed" as everyone else to hear of the launch of Sputnik at the reception which he and three other British delegates (the only representatives from Western Europe) were attending at the Soviet Embassy in Washington D.C. in October 1957. Six weeks later on November 13, 1957, after eight months of tests, Britain's first scientific experiments began using their sounding rocket *Skylark*.[11]

The British sounding rocket program, like that in other European countries and in the United States, benefited enormously from military interest and support.[12] Indeed, according to Massey, the initiative for launching an upper atmosphere research program came not from the scientists themselves but from the Ministry of Supply, which was responsible for the U.K.'s guided missile R & D program.[13] Approached in May 1953 by an official of the Ministry, Massey was reputedly asked whether he would be interested in using rockets available from the Ministry for scientific research.[14] He of course agreed. Subsequent negotiations led to a request to the Treasury for £100,000 for rocket research in the upper atmosphere to be spread over four years. Half of this money was to be paid to the Royal Society who would distribute it to university groups for building scientific instruments. The remainder would be paid to the Ministry of Supply to cover the cost of the rockets and other facilities.[15] In the absence of a safe launching ground in a small country like Britain, this meant sending the *Skylarks* aloft from the Angle-Australian rocket range in Woomera, near Adelaide, with the rather perverse effect that the British experiments measured conditions in the Southern Hemisphere and observed the southern sky.[16] The Treasury agreed to this request in 1955 telling the Ministry of Supply that its principal reason for doing so was not because the Royal Society thought it to be "a good piece of pure research" but because there was "a defence interest in the widest sense."[17]

With the sounding rocket program getting into its stride in 1958, and with the conviction that the satellite data that U.K. scientists needed could be obtained from the U.S. program, there was little enthusiasm at senior levels in the Royal Society or in the Ministry of Supply for a British space program. A number of factors led to a change in perceptions by the end of the year.

First, there was pressure from some scientists, notably Bernard Lovell, the director of Jodrell Bank, and some government departments, notably the Foreign Office, who insisted that Britain would be "classed as an underdeveloped country" if she did not launch her own satellite.[18] This appeal to national prestige and world status touched a sore nerve in a country trying to adjust to its diminishing influence in a world increasingly dominated by the superpowers. As Britain's representative to the UN, Sir Pierson Dixon put it, after the Suez debacle "... I remember feeling very strongly that we had by our action reduced ourselves from a first-class to a third-class power."[19] In short it seemed that any government which wanted to have some weight in international affairs had to have a space program.

Second, there was pressure from the United States for Britain to embark on a satellite program. In September 1958 U.S. officials suggested launching British instruments on American-built satellites from Woomera more or less free of charge. A month later, in October, the U.S. government released a report praising British achievements and extolling the virtues of international collaboration in space. Behind this encouragement was the State Department's wish that the next nation to enter space be from the "Free World", not the communist bloc. Britain was then the only allied power in Europe potentially capable of playing this role and State hoped, far too optimistically as it turned out, that a British payload could be ready within about six months.[20] In this way another dramatic ideological victory for the Soviet Union and its satellite states could be averted.

Finally, and crucially, there were renewed prospects of using Britain's IRBM *Blue Streak* as the first stage of a satellite launcher. This is a complex issue that requires some elaboration.

Blue Streak was a liquid-fuelled rocket based on the American *Atlas* intercontinental ballistic missile. It was powered by two Rolls Royce engines built under license from North American Aviation's Rocketdyne division and its range was specified as 1,500 miles initially but capable of development to 2,500 miles.[21] Embarked upon in 1954 with strong American support and help, the British missile was soon overtaken by the parallel development in the U.S. of the *Thor* and *Jupiter* IRBMs intended for deployment in the European theatre. In March 1957 British and American negotiators meeting in Bermuda agreed in principle to install *Thors* on British soil (*Jupiter* was destined for Italy and for Turkey) under a dual-key arrangement, a decision which immediately raised fears of duplication and doubts about the need to continue with a costly and slower indigenous program.

Blue Streak weathered this storm predominantly because the Air Ministry, supported by the Minister of Defence Sandys and his Chief Scientist Brundrett were emphatic that the missile was the cornerstone of Britain's independent deterrent and the necessary symbol of great power status. As one memo put it, "unless we are to become a second-class power we must make an independent contribution to the deterrent; the ballistic rocket is the only weapon likely to last and therefore we must have it."[22] But *Blue Streak* was never safe. Parts of the Air Ministry interpreted the claim that *Blue Streak* was the only durable weapon system as a threat to the V-bomber force. The Admiralty saw it, and rightly so, as the major obstacle to the development of a Polaris-type submarine nuclear deterrent. And they had a strong argument in their favor: the question of vulnerability. *Blue Streak* was a fixed-site land-based missile and very expensive measures would have to be taken to protect the missile from destruction by incoming enemy nuclear weapons by housing it in suitably reinforced underground silos. Submarine-launched weapons were not, of course, subject to the same danger.

Sputnik, indirectly at least, changed the terms of this debate. As we saw earlier it quickly led to amendments to the McMahon Act and to important relaxations in the restrictions on the transfer of sensitive information from the United States to the United Kingdom. One immediate result of this was that, in November 1958, Sandys could report that the Americans would be willing to supply full details of the design of the *Thor* warhead, which was also "entirely suitable for *Blue Streak*." In addition the Americans had let it be known that they regarded *Thor* as a technologically inferior first-generation system, which would ultimately have to be replaced in any case. Thus by the end of 1958 the Minister of Defence was satisfied that *Thor* was no substitute for *Blue Streak* and he began "rallying his Cabinet colleagues in support of a reprieve" for the missile.[23] However, the government had to admit that Polaris might be a better system, simply by virtue of its advantages in terms of mobility, flexibility and invulnerability, as well as its reducing the risk to the British Isles, which a weapon hidden in silos dotted over the country would entail. Prime Minister Macmillan thus decided to go ahead with the development of the IRBM on a provisional basis for another year and he set up a working party chaired by Sir Richard Powell to review the future of the British deterrent.

It was in this context of local and American pressure, along with the reprieve of the faltering *Blue Streak* project, that the Royal Society and the Ministry of Supply decided to look again at the question of a British satellite program. The result was two virtually identical papers produced in October 1958, and essentially drafted by Massey.[24] They embodied a comprehensive statement of what a British space program should involve, and as such imbued space with new meanings which it had not previously had in the United Kingdom.

In his paper Massey made a strong plea on scientific grounds for U.K. participation in space research with artificial satellites. Satellites, he insisted, had two advantages over sounding rockets. Firstly, whereas the former took useful data for "only a few minutes," the latter could carry out as many observations in a month as "several thousand vertical sounding rockets." Secondly, satellites could rise beyond the atmosphere, and so study radiation which was not accessible to sounding rockets because absorbed by the air. Starting from these advantages he then went on to list eleven foreseeable scientific uses of satellites, selected specifically to bring out the possibilities for systematic study, continuous observation, and world-wide coverage opened up by this new technology. To blunt the criticism that everything would have done by the time the satellites were built, Massey insisted that the variety and the extent of the data were so great, and the likelihood of making unforeseen discoveries so high, that "there is no risk at all that the usefulness of the satellites will be confined to a few years only."

Massey was emphatic that this program would only produce the desired benefits if it was an independent, purely British effort, a "thoroughgoing British enterprise" as he put it. Could one not simply build instruments and include them in the payload of an American satellite? No said Sir Harrie. There was in the U.K. "a long tradition of scientific research and the British approach to scientific problems contains something unique born of this long tradition and experience." If British instruments were flown on American satellites, they would "have to conform very largely with American designs in order to be acceptable. This would destroy the main advantage of a different approach to the experimental problems." Granted that Britain should build her own satellites, could she not then ask another country to launch them? Massey felt that only one line was needed to dismiss this possibility. Such an option, he wrote, "seems to offer no advantage at all, quite apart from the obvious loss of prestige." He also considered the possibility of western European collaboration in the development of satellites. This idea was enjoying some popularity in the Foreign Office at the time. Concerned about Britain's lukewarm attitude to the development of the new European Economic Community, the Foreign Office suggested that Britain might take the lead in proposing the joint development of a western European satellite as a sign of its "solidarity" with countries across the Channel and to show that she was not only interested in collaborative schemes with the United States.[25] Massey ruled this option out immediately on the grounds that Britain's lead was so great that "there would seem to be big disadvantages in collaboration at this stage."

What of the use of space for applications satellites, notably those of military interest? Massey's paper briefly mentioned the possible uses of meteorological, telecommunications and navigational satellites only to dismiss such technologies as being far in the future, their "practical benefits" being still too "difficult to predict." As for the use of space by the military, Massey explicitly excluded a discussion of such aspects from his review. Whyte and Gummett have however, recently analyzed the debate on this issue in Britain at the time in some depth.[26] Their overall conclusion is that, at least for the period covered by this article, the British government was never convinced of the military importance of space, systematically refused to use the military argument to justify a space program and, which amounted to the same thing, was not willing to spend significant amounts of defense money on satellites or launchers. The reactions of the Chief Scientists of both the Ministry of Supply (Wansborough-Jones) and of the Ministry of Defence (Brundrett) to Massey's paper of October 1958 are typical in this respect. The former had serious doubts about the military interest of space, was convinced that a military space program would be far beyond Britain's means, and suspected that the military information she might need could be acquired from the U.S. anyway. Brundrett, for his part, recognized that there might be military benefits in the future, especially in the reconnaissance, meteorology and telecommunications areas, but he felt that these applications were "unpredictable" and that the Americans had made exaggerated claims for them. Early in 1959 the Pentagon called a joint meeting of American, British and Canadian military R & D officials to facilitate collaboration in defense research projects in the space area. U.S. representatives stressed the "value and importance" of a military space program. Brundrett briefed the British delegate to the talks to "try to keep space out of the defence field ... ."[27] Although of course this view was contested, it did summarize the general attitude that prevailed at this time: for Britain at the end of the 1950s space (unlike the upper atmosphere) was not a domain of military importance.

This is not to say, of course that there was not a military component to the program. On the contrary, Massey's and the Ministry of Supply's pleas for an all-British space effort, and the short shrift they gave to collaborative ventures, were premised on the assumption that one could "adapt ... military vehicles to launch useful satellites ...," as Massey's paper put it. What he had in mind, although his paper did not mention this explicitly, was the possibility of developing a British satellite launcher by combining the "reprieved" *Blue Streak* with *Black Knight*, a rocket initially developed to test re-entry of warheads into the atmosphere, plus a small third stage.[28] This was technically possible. It also seemed to be relatively cheap. Indeed Massey assumed that the development costs of these weapons would not be carried over into the civil space program. Thus British science could have a British launcher capable of putting satellites weighing one ton into orbit "for the cost of adaptation" only, which he estimated at £9 million for the first five satellites.

Let us now take stock. At the end of 1958 Britain had a sounding rocket program which was just getting into stride. Funded 50–50 by the civilian science and the defense budgets, its aim was to study the properties of the upper atmosphere over the south Australian desert. A debate was also under way on the nature and scope of a British space program. This program was perceived primarily as a nonmilitary scientific research effort which would extend a unique British tradition in upper atmosphere research and which would contribute to British scientific prestige.[29] The proposed program had one other important feature: it was to be a "thoroughgoing British enterprise." This independence was only possible because, hovering in the wings of the space program was the IRBM *Blue Streak*, a component of Britain's independent nuclear deterrent which had been given new life by the U.S. offer to fit it with *Thor* nuclear warheads. Massey's, and the Ministry's, idea was that *Blue Streak* could be "adapted" to a space program, so ensuring British autonomy—and its status as a significant world power. By reinforcing his argument for a British space program with an appeal to the independence that would come from using a nationally built launcher, Massey was at once confirming the import-ance of *Blue Streak* for non-military purposes, adding an additional argument for its reprieve, and appealing to the sense of pride that would ensue from an all-British effort. In the minds of scientists and policy makers alike, then, at the end of 1958 space was a place where Britain would harvest scientific data and gain scientific prestige, but also a place where she would confirm, to herself and to others, that she was still a major power to be reckoned with.

1959: FROM A "THOROUGHGOING BRITISH ENTERPRISE" TO A OLLABORATIVE EFFORT

These dreams of British grandeur soon proved to be illusory. Indeed by April 1959 what was said rejected only a few months before—collaboration with the United States—now looked distinctly attractive. The European option was also given a new lease of life as Britain painfully adjusted to the realities of being a medium-sized world power.

The first setback to the idea of developing an all-British program was the reaction to Massey's proposals in about February and March 1959 by the Advisory Council on Science Policy (ACSP). This body, which was chaired by Sir Alexander Todd, reported directly to the "science minister." It supported Massey's suggestion for a scientific space program. But it backed off from using a British launcher for this purpose on the grounds of cost and the complications arising from the military connection. Better, said the committee, to make "an immediate approach" to the authorities in the United States "to ascertain the terms under which suitable rockets could be supplied."[30] Only if this avenue proved unsatisfactory should one consider adapting *Blue Streak* to a satellite launcher. In sum the ACSP felt that the civil science budget could not afford the £2 million per annum needed to modify Britain's IRBM. The Ministry of Defence, for its part, had made it quite clear that it would not finance the changes. That left collaboration as the only alternative.

These recommendations coincided with an offer from NASA to encourage space research in countries other than the United States. At a meeting of COSPAR[31] in The Hague on March 14, 1959, the American delegate announced that the U.S. would be prepared to launch "suitable and worthy experiments proposed by scientists of other countries." NASA was prepared to consider single experiments to be inserted along with others into larger payloads or groups of experiments comprising complete payloads weighing from 100 to 300 lb., which could be placed in an orbit ranging from 200 to 2,000 miles in altitude. In the former case the collaborator would be invited to work in a U.S. laboratory on the "construction, calibration, and installation" of the equipment in the spacecraft. In the latter case NASA was "prepared to advise on the feasibility of proposed experiments, the design and construction of the payload package, and the necessary pre-flight environmental testing." And although the official letter did not explicitly say so, during the public announcement it was said that NASA intended to use its newly developed Scout rocket, and that the launching would be free of charge.[32] Within six weeks the British National Committee on Space Research, established at the end of 1958 with Massey as chairman, had established working groups to define the experiments which could be proposed to NASA.[33]

In addition to the NASA offer there were also moves afoot inside NATO to exploit the military potential and the prestige of space to protect the interests of the western alliance. In December 1957, in the wake of Sputnik, the NATO heads of government decided to set up a Science Committee to "speak authoritatively on science policy" and to ensure that science and technology flourished, this being essential "to the culture, to the economy and to the political and military strength of the Atlantic community." Within a year this committee had proposed to launch "a satellite for peaceful outer space research, bearing the emblem of the Atlantic community and circling the earth by 1960." NASA's proposals in March 1959 rendered this option redundant. However, Fred Seitz, who took over as NATO Science Committee Chairman from Norman Ramsey in June 1959 was not prepared to letter matters rest. He was strongly against the development of a European space organization then being widely discussed (see immediately). He thought that the possibility of such a body coming into being was "improbable and, in fact, impracticable." Instead he wanted NATO to establish a space agency in western Europe resembling NASA and which would collaborate with it in planning the utilization "for scientific purposes of the best missiles available for space research in the NATO family."[34] One potential member of that family was a new, second-generation solid-fuelled "European" IRBM produced under NATO auspices which the United States was promoting at the time, and for which it was prepared to offer technical and financial support.

The British government was not particularly keen on this scheme. They feared that it would provide the Federal Republic of Germany with a missile capability.[35] There was another alternative to total reliance on the Americans though, and one more befitting of a major power: taking the lead in a collaborative European space effort with one's partners across the Channel.

1959–1960: THE EMERGENCE AND CONSOLIDATION OF THE JOINT EUROPEAN EFFORT

The possibility of having a collaborative European civilian space science program was actively canvassed during 1959 by Edoardi Amaldi and Pierre Auger, two cosmic ray physicists and scientific statesmen who had played a key role in the foundation of CERN (the

European Organisation for Nuclear Research) earlier in the decade.[36] Their idea was that Europe should develop its own space agency modeled on CERN, i.e. a collaborative scientific effort, cofinanced by up to a dozen European governments, independent of the military, and in which scientists would have the freedom to pursue the research they chose. Auger's and Amaldi's efforts were given an added impetus by the enthusiastic support shown by Massey, not simply for the scientific program, but for providing the envisaged agency with its own launcher. Indeed at a meeting of about two dozen space scientists from ten western European countries in the rooms of the Royal Society on April 29, 1960, the British suggested that a new European organization could make use of *Blue Streak* as the first stage of a civilian satellite launcher, with a modified version of *Black Knight* atop of it. And they asked those present to give some idea of the kind of financial support their governments might be willing to make to such an organization if it were built around the British missile adapted for civilian use.

The British proposal was inspired by the government's decision, announced two weeks before the meeting at the Royal Society, to cancel the development of the IRBM program.[37] Following on the report of the Powell committee it seemed that the missile was now technologically obsolete and strategically unacceptable. It had been overtaken by developments in solid-fuel propellants and in inertial guidance. And being land-based and liquid-fuelled *Blue Streak* would have to be a "fire first" weapon. This, as the Chiefs of Staff pointed out, was incompatible with the government's deterrent policy, which was based on retaliation, not first strike. In October 1958 the proposal for a "thoroughgoing British" space effort had been premised on the "reprieve" of Blue Streak as a missile. In April 1960 the proposal for a "joint European satellite program" was premised on the recycling of *Blue Streak* as the first stage of a collaboratively built satellite launcher.

"The decision to cancel Blue Streak was one of a small handful of momentous defence decisions taken by Britain in the post-war period and has remained justifiably controversial ever since."[38] To sweeten the pill, the government decided to try to persuade its neighbors across the channel, and particularly the French, to collaborate with the U.K. in the development of a civilian satellite launcher using *Blue*

Streak stripped of its military characteristics as the first stage. This would serve to maintain the teams of engineers and industrialists involved in the program in the event that Britain might re-enter missile development, would be a way of "saving" the £50 million already spent on the scheme, and would be a gesture of political goodwill towards the new European Economic Community which that self-same government had previously tried to sabotage by setting up a rival European Free Trade Association.[39]

The French were not uninterested in the idea of jointly developing a rocket with Britain. De Gaulle was putting in place his *force de frappe*, which included a major missile program. In parallel there were steps under way, in which Auger himself was heavily involved, to establish a French national space agency CNES (Centre national d'études spatiales), dedicated to civilian space research.[40] However, the French space scientists insisted that the development of the rocket was not to be at the expense of their programs and should preferably be paid for from the defense budget. The Minister of Defence, for his part, insisted that he could only justify developing a civilian rocket if the British, at the same time, passed on technological information of military importance to France. In particular the French would want to know about inertial guidance and re-entry systems. However, these were just the technologies which London had promised Washington not to give away when British officials first had the idea of recycling *Blue Streak* in the European theatre. By the end of 1960 then it seemed increasingly unlikely that Britain would be able to save her rocket program.[41]

The deadlock was broken in January 1961. Towards the end of the month Macmillan met with de Gaulle in the Château de Rambouillet just outside Paris. The "unity of the West" was uppermost on the General's mind; "at important moments of history," he told the British Prime Minister, "solidarity had to take precedence over immediate political concerns." The communist menace was even greater than before. The new administration in the White House (Kennedy had just taken over as President) would have its hands full with domestic issues and in Indochina, and could not be counted on to make European affairs a priority. Europeans would have to join forces as never before. "In an economically united and perhaps even politically confederated Europe," said Macmillan, "France and Britain together, in their relations with the United States and the rest of the world, would have the possibility of playing their rightful role." De Gaulle saw a notable change of tone in the British premier's attitude.[42] And walking with Macmillan in the gardens of the Château that afternoon, he said that he was attracted by the idea of Europe becoming "the third space power." The French government would take a positive attitude towards the U.K.'s suggestion that they develop a satellite launcher together. No mention was made of the military aspect.[43] With that the "launcher problem" was solved, and within three years Europe was endowed with two organizations dedicated to the exploitation of space, one for civilian scientific research and the other for the collaborative development of a three-stage satellite launcher.

CONCLUDING REMARK

A striking feature of the development of the space effort in Europe is the leading role played by scientists in launching the program and the particular place of the military in it. Whether we are speaking of Britain or France (Germany was disqualified from taking the lead), or the joint European initiative, scientific statesmen like Massey, Auger and Amaldi are among the key actors. One of their main concerns was to keep NATO out of European space affairs and, for practical reasons, to restrict the place of national military establishments in their space science research programs. This is not to say that military developments were not crucial to the civilian space programs in Europe. Indeed first Britain's and then France's wish to develop an independent nuclear deterrent using rockets as delivery systems provided the essential industrial infrastructure and engineering expertise required for the "civilian" European space program. That program would have been impossible without "a massive transfer of technologies, and without having recourse to the skills and the installations developed by the military," to quote André Lebeau.[44]

There was no "Sputnik shock" in Europe. European governments did not treat space as a domain of military importance in the late 1950s and early 1960s, and it was only in the mid-1960s that some of them became aware of its commercial potential, particularly for telecommunications. The conquest of space was essentially a scientific challenge, though one imbued with the Cold War imperative of maintaining the technological leadership and prestige of the western alliance. Unlike the situation in the U.S., that a civilian program was not a cover for, nor a complement to, a military satellite program. But it was embedded into an expanding aeronautical and electronic military-industrial complex. It was that complex which supplied the satellites, but also the missiles and communications systems needed to ensure autonomy and to build Western Europe as a "third space power" between the United States and the Soviet Union.

1. The classic account is of course Walter A. McDougall, ... *The Heavens and the Earth: A Political History of the Space Age* (New York: Basic Books, 1985). For a fine summary of American reactions to Sputnik see Roger D. Launius, *NASA: A History of the U.S. Civil Space Program* (Malabar, FL: Krieger, 1994), chapter 2.

2. For general political developments in the United Kingdom see A. Sked and C. Cook, *Post-War Britain: A Political History* (London: Penguin Books, 1980), chapters 5 and 6. On the deterrent see A. J. Pierre, *Nuclear Politics: The British Experience with an Independent Strategic Force, 1939–1970* (London: Oxford University Press, 1972).

3. For the French program see M. Vaïsse, ed., *L'Essor de la politique spatiale franc[,]aise dans le contexte international, 1958–1964* (Paris: Editions des archives contemporaines, 1997); Lorenza Sebesta, "La science, instrument politique de securité nationale? L'espace, la France et l'Europe," *Revue d'histoire diplomatique*, n° 4, 1992, pp. 313–41; Walter A. McDougall, "Space-Age Europe: Gaullism, Euro-Gaullism, and the American Dilemma," *Technology and Culture* (1985): 180–203.

4. Negotiations were under way with the United States early in 1957 for the installation of *Thor* IRBMs on British soil and cooperation had already begun on nuclear submarines. For details see I. Clark, *Nuclear Diplomacy and the Special Relationship: Britain's Deterrent and America, 1957–1962* (Oxford, England: Clarendon Press, 1994). On NATO see, typically, Lawrence S. Kaplan, *NATO and the United States: The Enduring Alliance* (Boston: Twayne Publishers, 1992).

5. Quoted by Neil Whyte and Phillip Gummett, "Far Beyond the Bounds of Science: The Making of the United Kingdom's First Space Policy," *Minerva* (1997): 139–69, where the original source can be found.

6. On the Anglo-American "special relationship" see, e.g., J. Baylis, *Anglo-American Defence Relations 1939–1984* (London: Macmillan, 1984); M. Gowing, "Nuclear Weapons and the 'Special Relationship'," in W.M. Roger Louis and H. Bull, eds., *The "Special Relationship". Anglo-American Relations Since 1945* (Oxford, England: Clarendon Press, 1986), chapter 7; Clark, *Nuclear Diplomacy and the Special Relationship*; Pierre, *Nuclear Politics*.

7. Quoted in Whyte and Gummett, "Far Beyond the Bounds of Science."

8. See Gowing, "Nuclear Weapons and the 'Special Relationship'," p. 124.

9. On Jodrell Bank see Jon Agar, *Science & Spectacle: The Work of Jodrell Bank in Post-War British Culture* (Chur: Harwood Academic Publishers, 1998). Sputnik's orbital inclination near 65° carried it over the British Isles.

10. *Le Monde*, Bulletin de l'tranger, 6–7 octobre 1957.

11. See H. Massey and M.O. Robins, *History of British Space Science* (Cambridge, England: Cambridge University Press, 1986); and Whyte and Gummett, "Far Beyond the Bounds of Science." Some of the key papers cited in this article are easily accessible as published appendices in Massey and Robins. A fully researched history of the U.K.'s sounding rocket program remains to be written.

12. For Europe see D. Pestre, "Studies of the Ionosphere and Forecasts for Radio communications. Physicists and Engineers, the Military and National Laboratories in France (and Germany) after 1945," *History and Technology*, 13(3) (1997): 183–205, and O. Wicken, "Space Science and Technology in the Cold War: The Ionosphere, the Military and Politics in Norway," *History and Technology*, 13(3) (1997): 206–29. See also David H. DeVorkin, *Science with a Vegeance: How the American Military Created the Space Sciences in the V-2 Era* (New York: Springer-Verlag, 1992); David H. DeVorkin, "The Military Origins of the Space Sciences in the American V-2 Era," in Paul Forman and J.M. Sanchez-Ron, eds., *National Military Establishments and the Advancement of Science and Technology* (Dordrecht: Kluwer Academic Publishers, 1990), pp. 233–60; Bruce Hevly, "The Tools of Science: Radio, Rockets, and the Science of Naval Warfare," in Forman and Sanchez-Ron, "Nuclear Weapons and the 'Special Relationship'," pp. 215–32.

13. For a full account of that program see Stephen Twigge, *The Early Development of Guided Missiles in the United Kingdom, 1940–1960* (Chur: Harwood Academic Publishers, 1993).

14. Massey and Robins, *History of British Space Science*, p. 16.

15. Ibid., pp. 16 et seq.

16. P. Morton, *Fire across the Desert: Woomera and the Anglo-Australian Joint Project, 1946–1980* (Canberra: Australian Government Publishing Service, 1989).

17. Quoted in Whyte and Gummett, "Far Beyond the Bounds of Science," p. 145.

18. Ibid., p. 152.

19. Quoted in Sked and Cook, *Post-War Britain*, p. 153.

20. From Whyte and Gummett, "Far Beyond the Bounds of Science," p. 152.

21. For the birth of *Blue Streak* see John Krige, "What is 'Military' Technology? Two Cases of U.S.-European Scientific and Technological Collaboration in the 1950s," in F.H. Heller and J.R. Gillingham, *The United States and the Integration of Europe: Legacies of the Postwar Era* (New York: St. Martin' Press, 1996), pp. 307–38. See also Twigge, *Early Development of Guided Missiles*.

22. The quotation is from Clark, *Nuclear Diplomacy and the Special Relationship*, p. 165. This section relies heavily on his detailed analysis of *Blue Streak*'s political fortunes in chapter 5.

23. Clark, *Nuclear Diplomacy and the Special Relationship*, p. 172 for both quotations.

24. One version of Massey's paper is reproduced as Annex 3 in Massey and Robins, *History of British Space Science*.

25. See Whyte and Gummett, "Far Beyond the Bounds of Science," p. 152–53.

26. Neil Whyte and Phillip Gummett, "The Military and Early United Kingdom Space Policy," *Contemporary Record* 8(2) (Autumn 1994): 343–69.

27. Ibid., pp. 345–47.

28. For a description of *Black Knight*, a highly successful rocket first launched in 1958, which later served as an upper atmosphere research tool for both civil and military science, see Morton, *Fire across the Desert*, chapter 21. That Massey had in mind this combination of missiles is from Whyte and Gummett, "Far Beyond the Realms of Science."

29. Massey specifically excluded the launch of small satellites weighing a few pounds on the grounds that they would be outmoded scientifically once they were built, and would have no prestige value.

30. Quotes from Massey and Robins, *History of British Space Science*, p. 68. See also Whyte and Gummett, "Far Beyond the Realms of Science," pp. 155–60.

31. COSPAR (an international Committee on Space Research) had been set up in 1958 to take over space research with rockets and satellites after the IGY.

32. From Massey and Robins, *History of British Space Science*, Annex 4, which reproduces the U.S. offer, and pp. 67–69.

33. Massey and Robins, *History of British Space Science*, p. 69.

34. This paragraph is based on John Krige and Lorenza Sebesta, "US-European Cooperation in Space in the Decade after Sputnik," in G. Gemelli, ed., *Big Culture: Intellectual*

Cooperation in Large Scale Cultural and Technical System, An Historical Approach (Bologna: Clueb, 1994), pp. 263–97, where detailed references may be found.

35. Clark, *Nuclear Diplomacy and the Special Relationship*, pp. 175, 178–79.

36. For this paragraph see John Krige, *The Prehistory of ESRO, 1959/60* (Noordwijk: ESA HSR-1, July 1992). See also John Krige and Arturo Russo, *Europe in Space, 1960–1973* (Noordwijk: ESA SP-1172, 1994), chapter 2.

37. Macmillan secured at the same time an agreement in principle by Eisenhower that the U.S. would replace it with an alternative delivery system, perhaps (what turned out to be the ill-fated) Skybolt. For the cancellation see Clark, *Nuclear Diplomacy and the Special Relationship*, pp. 176–84.

38. Clark, *Nuclear Diplomacy and the Special Relationship*, p. 184.

39. For a general survey see Jacqueline Tratt, *The Macmillan Government and Europe: A Study in the Process of Policy Development* (London: Macmillan Press, 1996).

40. The law establishing CNES was voted by the French parliament on December 19, 1961. Pierre Auger was nominated its first Chairman and General Robert Aubinière its first Director General. For more on the French programme see C. Carlier and M. Gilli, *Les trente premières années du CNES, L'Agence franc[,]aise de l'éspace* (Paris: La Documentation franc[,]aise, 1994); G. Ramunni, "La mise en place d'une politique scientifique," in *De gaulle et son siècle. Actes des Journées internationales tenues à l'Unesco, Paris, 19–24 novembre 1990. Tome III. Moderniser la France* (Paris: La Documentation Franc[,]aise, 1992), notably pp. 677–81. See also the references cited in Note 3.

41. For a detailed account of the Anglo-French negotiations see John Krige, "The launch of ELDO" (Noordwijk: ESA HSR-7, March 1993). See also Krige and Russo, *Europe in Space*, chapter 3. The Anglo-American discussions about the transfer of sensitive technology to France is explored in Krige, in Heller and Gillingham, *United States and the Integration of Europe.*

42. The summary minute of the meeting held on 28/1/61 is available in the collection Affaires Etrangers, Secretariat Generale, 1965–65, Entretiens et messages, 1961, Box 62, Quai d'Orsay, Paris. See also letter de Gaulle to Adenauer, 9/2/61, in the same box.

43. See document headed "Rambouillet 3," January 28, 1961, File PREM11/3513, Public Record Office, London.

44. In his contribution after Ramunni's paper published in *De Gaulle et son siècle*, p. 743.

CHAPTER 11

Organizing the United States Government for Outer Space, 1957–1958

Eilene Galloway

DEFINING THE PROBLEM

The dramatic orbiting of *Sputnik 1* by the Soviet Union on October 4, 1957, was like a spark that ignited and speeded the process of developing the exploration and peaceful uses of outer space on a continuing and larger scale. People throughout the world were astonished by this phenomenal opening of outer space as a new environment, and surprised that the Soviet Union was first to accomplish this feat. But the news struck Capitol Hill like a thunderbolt because thrusting the 184-pound satellite into outer space was evidence of the capability of launching intercontinental ballistic missiles, and therefore instantly perceived as a crisis for U.S. national defense. This perception was reinforced on November 3, 1957, when a second Sputnik weighing 1,120 pounds began circling the Earth. The United States was still working on the civilian scientific Vanguard satellite of 3.25 pounds.

There was instantaneous reaction by the Senate Armed Services Committee. Senator Lyndon B. Johnson, Chairman of the Preparedness Investigating Subcommittee, brought forceful leadership to this challenge,[1] and on November 25, 1957, began the basic investigation of the nation's resources for achieving a superior space program. The "Inquiry into Satellite and Missile Programs" called upon experts representing government and industry, science and technology, education, military, and civilian fields of knowledge.[2] These hearings continued through November and December and into

January 1958, recording 2,476 pages of the facts essential for understanding the total situation as a basis for planning the future. The sense of urgency was a driving force during these hearings which began at 9 a.m. and often lasted until 9 p.m.

An understanding of the comprehensive nature of the task was revealed by the testimony of scientists and engineers. Instead of fearing rockets as weapons, they hailed them as advanced technology for producing many benefits for mankind. Thus, the problem that required solution was two-dimensional: to preserve outer space for peaceful exploration and uses, and to prevent its becoming a new arena for warfare. This situation was a classic case of presenting a choice between good and evil.

It became apparent that organization of the resources required for a space program would need to take the following factors into account: (1) satellites are inextricably international as they orbit over national boundary lines in 90 minutes or less; (2) all spacecraft require communications to send and receive information from the Earth; (3) a network of tracking stations is essential and requires international agreements; (4) the geostationary orbit used for global communications has been declared a "limited natural resource," already organized by the International Telecommunication Union; (5) nations must guard against pollution, contamination of the Earth and outer space, including celestial bodies; (6) provision must be made for protecting the health and safety of astronauts; (7) measures must be taken to cope with interference from space debris; (8) space vehicles can operate and produce benefits only when they comply with the specific rules of space technology and science; and (9) space exploration is expensive but large projects can promote patterns of international cooperation.

As the configuration of the total problem took shape during the Preparedness Hearings, major issues arose. What should be the roles of military and civilian organizations? How should national and international aspects be administered? How should the Congress be organized to handle space legislation?

On November 21, 1957, the Rocket and Research Panel submitted a proposal for creating a National Space Establishment, an independent civilian agency separated from the military and funded on a long-term basis. A similar proposal had been sent to President

Eisenhower by the American Rocket Society on October 14, 1957, only ten days after the Sputnik launching. The two proposals were combined and on January 4, 1958, Congress was urged to establish a new civilian space agency authorized to conduct manned and unmanned space missions, consider a permanent base on the Moon, flights to Mars and Venus, and develop a variety of peaceful uses. Scientists and engineers promised international leadership on this "endless frontier" and identified benefits which could be expected from space activities:

> There will be a rich and continuing harvest of important practical applications as the work proceeds. Some of these can already be foreseen—reliable short-term and long-term meteorological forecasts, with all the agricultural and commercial advantages that these imply; rapid, long-range radio communications of great capacity and reliability; aids to navigation and to long-range surveying; television relays; new medical and biological knowledge, and so forth. And these will be only the beginning. Many of these applications will be of military value; but their greater value will be to the civilian community at large.[3]

The Subcommittee concluded that the satellite was not yet a weapon, the Soviet Union had led the world into outer space, and it was now essential for the United States to make a tremendous effort to organize the resources for achieving preeminence in space. A finding was that the "same forces, the same knowledge, and the same technology which are producing ballistic missiles can also produce instruments of peace and universal cooperation": communications, meteorology, navigation, and remote sensing. The responsibilities of the Subcommittee were limited to defense and it was necessary to create a special committee to handle the civilian aspects of space legislation.[4]

INVESTIGATION BY HOUSE AND SENATE COMMITTEES

On February 6, 1958, the Senate established the Special Committee on Space and Astronautics with Senator Lyndon B. Johnson as chairman. Membership on the Committee was determined by the concern of existing committees with the variety of subjects involved in space applications that could cut across their jurisdictions. To solve what could become a complicated legislative problem, action had to be taken to avoid dispersal of the legislative process within the Congress, fix responsibility, and give outer space matters the highest priority.

In the House of Representatives a similar "blue ribbon" Committee on Astronautics and Space Exploration was created on March 5, 1958, with the Majority Leader, Hon. John W. McCormack as Chairman.

While these committees were considering how to organize the Executive Branch, Congress moved swiftly to pass interim legislation to speed U.S. space development. The Secretary of Defense was authorized to engage in advanced research projects leading to the creation of the Advanced Research Projects Agency (ARPA), and for one year to be responsible for space projects as designated by the President.[5] The Supplemental Military Construction Authorization Act provided $10 million in transfer authority to the Secretary of Defense for advanced research.[6]

Among the organizations in a unique position to furnish basic information and advice on planning for the future development of space activities were the National Advisory Committee on Aeronautics (NACA), the International Geophysical Year (IGY) coordinated by the International Council of Scientific Unions (ICSU), and the Department of State on U.S. foreign policy for arms control.

Congress established the National Advisory Committee on Aeronautics (NACA) on March 3, 1915, as an independent agency reporting to the President. NACA was given authority to "supervise and direct the scientific study of the problems of flight, with a view to their practical solution ... and to direct and conduct research and experiments in aeronautics."[7] Research included missiles and manned aircraft; rocket research was concerned with engines for space and ballistic missions. Although a small agency with limited funds, NACA was successful with its expert staff in relations with the Department of Defense and as a link between the government and the aviation industry.[8]

The International Geophysical Year (IGY) was established by the international scientific community from July 1, 1957, to December 31, 1958, for scientific research of the world's total environment:

the Earth, oceans, atmosphere, and outer space. This 18-month "year" was a period of peak sunspot activity favorable for inter-disciplinary research. Two such studies had been made in what were termed Polar Years, 1882–1883 and 1932–1933. Since that time, advances in communications and aviation technology had produced new tools for in-depth research and 67 nations took part in the IGY which was organized by scientific disciplines, national committees, governmental sponsorship, and international coordination.[9]

In February 1953, the U.S. National Committee for the IGY was created by the National Academy of Sciences-National Research Council, the scientific U.S. organization that adheres to the International Council of Scientific Unions. The role of the government in administering U.S. contributions to the IGY was undertaken by the National Science Foundation which received an appropriation from Congress of $43,500,000 for IGY research. Additional financial and logistical support came from other government agencies involved with geophysical projects, e.g., the Department of Defense, Weather Bureau, Bureau of Standards, and Coast and Geodetic Survey. The linkage between the scientific community and the government was effectively managed, both rationally and internationally, by organizations with clear policy objectives, implementing programs, and adequate funding.[10]

The United States and the Soviet Union each made a commitment to launch a satellite during the IGY. The U.S. decision was announced by the White House on July 29, 1955:

... the President has approved plans by this country for going ahead with the launching of small, unmanned Earth-circling satellites as part of the United States participation in the International Geophysical Year ... This program will, for the first time in history, enable scientists throughout the world to make sustained observations in the regions beyond the earth's atmosphere.[11]

Three days later, on August 2, 1955, a member of the USSR Academy of Sciences, L. I. Sedov, stated, during a press conference at the International Congress of Astronautics, that the Soviet Union planned to launch a satellite. This announcement was confirmed by the Soviet National Committee for the IGY which described the scientific mission as measurement of atmospheric pressure and temperature, cosmic rays, micrometeorites, the geomagnetic field and solar radiation.[12]

U.S. ARMS CONTROL POLICY IN THE UNITED NATIONS

U.S. policy on preventing rockets from being used as weapons in outer space, and limiting the new environment to peaceful uses, was being pursued in the United Nations and in diplomatic channels with the Soviet Union. In a memorandum to the United Nations on January 12, 1957, the U.S. called attention to the development of rockets and that it was "clear that if this advance into the unknown was to be a blessing rather than a curse, the efforts of all nations in the field need to be brought within the purview of a reliable armaments control system." On July 22, 1957, Secretary of State Dulles emphasized the opportunity to formulate in advance a system for ensuring beneficial scientific uses rather than destructive weapons.[13]

On November 14, 1957, the United Nations General Assembly passed Resolution 1148 (XII) calling for study of "an inspection system designed to ensure that sending objects through outer space shall be exclusively for peaceful and beneficial purposes." On January 12, 1958, President Eisenhower wrote Soviet Council of Ministers Chairman Nikolay Bulganin proposing "that we agree that outer space be used for peaceful purposes ... Let us this time, and in time, make the right choice," the peaceful choice.[14]

By the time the congressional committees began deliberations on future U.S. organization for space activities, the high priority objective had become peaceful purposes for the benefit of all mankind.

NATIONAL AERONAUTICS AND SPACE ACT OF 1958

While the House and Senate committees had been investigating the impact of Sputnik on the United States and the future for space development, analytical studies were under way in the Executive Branch. The President's Special Assistant for Science and Technology and the Science Advisory Committee combined their assessments with those of the Department of Defense, Department of State, the Bureau of the Budget, and the National Advisory Committee on Aeronautics as a basis for recommendations on legislative proposals.[15] On April 2, 1958, President Eisenhower sent his message to the Congress with draft legislation which was referred to the House Select Committee on Astronautics and Space Exploration and the Senate Special Committee on Space and Astronautics.[16] There was agreement by the committees with three main issues: that a new civilian space agency be created; that NACA should be the nucleus for transfer to the new agency; and that the Department of Defense should have jurisdiction over projects primarily associated with military requirements. However, as legislative inquiries progressed, it became apparent that there were differences between the Congress and the Executive Branch on some of the major problems of organization that required solution.

The draft proposal assumed that all agencies would cooperate without overall coordination; NASA was to have an internal advisory board which met infrequently and had no authority over other agencies with space and space-related programs and budgets. If a question arose as to whether an activity was military or civilian, it was proposed that NASA "*may* act in cooperation with, or on behalf of, the Department of Defense." Although the draft preamble provided that NASA should cooperate "with other nations and groups of nations," there was no provision implementing this objective in the draft bill.[17]

The House committee favored solution of the problem of overall coordination by civilian/military liaison committees of cooperation at the staff level with disputes settled by the President. The House committee also recommended an internal Aeronautics and Space Advisory Committee of 17 distinguished members to advise the President. The Senate committee, however, realized that the total U.S. space program involved coordination of a number of agencies and recommended a National Aeronautics and Space Board. This was changed by the Conference Committee to the National Aeronautics and Space Council whose function was to advise the President on the following duties: to survey all significant aeronautical and space activities, develop a comprehensive program, allocate responsibility, ensure effective cooperation and resolve differences. The Council was provided with a small staff.[18] The objective of Congress was to ensure that all U.S. space activities would be handled at the highest Presidential level. The bill as finally enacted required the President to send Congress an annual report.

Both House and Senate committees emphasized the necessity for international space cooperation. On January 14, 1958, Senator Johnson called for world leadership by the United States in the new dimension offered by space exploration:

> We should, certainly, make provisions for inviting together the scientists of other nations to work in concert on projects to extend the frontiers of man and to find solutions to the troubles of this earth ... it would be appropriate and fitting for our Nation to demonstrate its initiative before the United Nations by inviting all member nations to join in this adventure into outer space together.[19]

On May 13, 1958, Congressman McCormack introduced a House Concurrent Resolution calling for the Peaceful Exploration of Outer Space and expressing the sense of the Congress—

> That the United States should seek through the United Nations or such means as may be most appropriate an international agreement providing for joint exploration of outer space and establishing a method by which disputes which arose from the future in relation to outer space will be solved by legal, peaceful methods, rather than by resort to violence.[20]

The Concurrent Resolution was passed by the House on June 2, 1958, and by the Senate on July 23, 1958.

Emphasis on international space cooperation was further provided in the NASA Act by the Declaration of Policy and Purpose:

> The Congress hereby declares that it is the policy of the United States that activities in space should be devoted to peaceful purposes for the benefit of all mankind. (Sec. 102 (a)).

Furthermore, the aeronautical and space activities shall be conducted so as to contribute materially to—

316   RECONSIDERING SPUTNIK

Cooperation by the United States with other nations and groups of nations in work done pursuant to this Act and in the peaceful application of the results thereof. (Sec. 2 (c) (7)).

Section 205 of the Act provides for international cooperation:

The Administration, under the foreign policy guidance of the President, may engage in a program of international cooperation in work done pursuant to this Act, and in the peaceful application of the result thereof, pursuant to agreements made by the President with the advice and consent of the Senate.

When President Eisenhower signed the NASA Act on July 29, 1958, he stated:

I regard this section merely as recognizing that international treaties may be made in this field, and as not precluding, in appropriate cases, less formal arrangements for cooperation. To construe the section otherwise would raise substantial constitutional questions.[21]

When the Senate Committee issued its Final Report, it recommended that the

Congress should be kept informed of progress being made in studies undertaken by the United Nations Ad hoc Committee on the Peaceful Uses of Outer Space ... Particular attention should be paid to preserving and extending the patterns of cooperation which were formed during the International Geophysical Year .... The special committee commends the National Aeronautics and Space Administration for establishing an Office of International Cooperation ... and recognizes the need of the Administration to provide for various types of cooperation as approved by the President.[22]

Thus the way was open for NASA to engage in many kinds of scientific bilateral and multilateral projects with other nations, leading to hundreds of projects with over one hundred nations and international organizations.

Senator Johnson was determined that outer space activities receive continuing priority at all governmental decision levels. On August 21, 1958, two appropriation bills, one on military construction, and the other on supplemental appropriations, were amended to provide for an annual authorization of funds for NASA. The requirement was at first temporary but became a permanent feature of the legislative process for NASA's activities:

> Notwithstanding the provision of any other law, no appropriation may be made to the National Aeronautics and Space Administration unless previously authorized by legislation hereafter enacted by the Congress.[23]

CONGRESSIONAL ORGANIZATION 1958

The work of the Senate and House Special and Select Committees ended with the passage of the NASA Act and it was necessary to consider how Congress should be organized for jurisdiction over future space legislation.[24] Four options were examined: (1) to create a new joint committee; (2) to divide jurisdiction among existing committees; (3) to refer space legislation to the Joint Committee on Atomic Energy; and (4) to create new standing committees in the house and Senate. The fourth option was chosen and the House Committee on Science and Astronautics was established on July 21, 1958. The Senate created the Committee on Aeronautical and Space Sciences on July 24, 1958. This action was unusual and underscored the intent of Congress that all space activities must receive specific unified attention according to the bicameral system.

THE UNITED STATES AND UNITED NATIONS

In little more than a month after creating NASA with the strong statement of U.S. policy in favor of international cooperation, President Eisenhower asked the United Nations General Assembly to include on its agenda a draft U.S. resolution calling for the establishment of an Ad hoc Committee on the Peaceful uses of Outer Space. This committee was to report on United Nations's resources and organizations which could advance cooperation among nations to avoid national rivalries and preserve outer space for peaceful uses, including identification of legal problems that might arise. The resolution advocated continuation on a permanent basis of the scientific research being carried on by the International Geophysical Year.

President Eisenhower invited Senator Lyndon B. Johnson to address the United Nations and lend his influence for adoption of the U.S. resolution. This occasion, when the President sent a plane to Texas to fly Senator Johnson to the United Nations, is a dramatic event in American history, demonstrating the unity of the Government when the Republican president joined with the Democratic leader of the Senate for international action on a critical U.S. foreign policy.

On November 17, 1958, Senator Johnson urged the United Nations to adopt the U.S. proposal for the ad hoc committee:

> ... if nations proceed unilaterally, then their penetrations into space become only extensions of their national policies or earth. What their policies on Earth inspire—whether trust or fear—so their accomplishments in outer space will inspire also ... Today outer space is free. It is unscarred by conflict. No nation holds a concession there. It must remain this way ... . We know the gains of cooperation. We know the losses of failure to cooperate. If we fail now to apply the lessons we have or even if we delay their application, we know that the advances into space may only mean adding a new dimension to warfare. If, however, we proceed along the orderly course of full cooperation, we shall by the very fact of cooperation make the most substantial contribution yet made toward perfecting peace ... .[25]

Nineteen other nations joined the United States in sponsoring the Resolution 1348 (XIII) which passed the General Assembly on December 13, 1958. Membership on the 18-nation ad hoc committee was chosen on the basis of those most advanced in space technology and other nations representing fair geographical distribution. The USSR (joined by Czechoslovakia, Poland, India and the United Arab Republic) would not participate because of opposition to majority voting. This hurdle was overcome, however, when agreement was reached on making decisions by consensus.

On December 12, 1959, the General Assembly adopted Resolution 1472 (XIV) creating the permanent United Nations Committee on the Peaceful Uses of Outer Space (COPUOS) whose membership has now grown to 61 nations. COPUOS established two subcommittees: the Scientific and Technical Subcommittee analyzes and reports on technical subjects which are then taken up by the Legal Subcommittee. This is the organization and process by which five space treaties have been formulated. The Conference on Security and Cooperation in Europe made a useful definition of the consensus process:

> Consensus shall be understood to mean the absence of any objection expressed by a representative and submitted by him as constituting an obstacle to the taking of the decision in question.[26]

After attending many sessions of COPUOS and the Legal Subcommittee, this author observed:

> It is evident that consensus is a highly desirable way of achieving international accord because (1) the process of seeking agreement continues with patience and is not cut off suddenly by a vote which may defeat what might have come to fruition had more time been taken with the give and take process of consensus; (2) the situation may be such that a majority vote could not result in the adoption of a course of action, particularly if implementation of the decision in terms of funding, personnel and technological expertise, depended upon nations which had voted against the measure; and (3) group solidarity in decisionmaking ensures maximum compliance in establishing and maintaining an activity of general benefit. There is also a positive psychological effect when members of a group feel together with sympathy for differing viewpoints, motivated by a desire to bring about harmony in their collective judgment. If a member has not objected, a proposal can be adopted but this unspoken consent should not be interpreted as negativism; there is a positive willingness to settle the issue in question.[27]

The creation of COPUOS as a permanent space organization within the United Nations advanced the movement toward international space cooperation for protecting outer space from conflicts. The method of establishing the facts of operational space technology as a basis for formulating legal principles, and the process by which consensus was obtained on decisions by political representatives strengthened compliance with essential guidelines. The United Nations became the forum for coalescing action at a critical time in history. Strong agreement on the objective of maintaining outer space for beneficial uses kept nations negotiating with patience until their disagreements were overcome. Within this United Nations context which the United States worked with dedicated leadership to establish, the United Nations COPUOS formulated the basic Treaty on Principles Governing the Activities of States in the Exploration and Use of Outer Space, including the Moon and Other Celestial Bodies,[28] now ratified by 93 nations which comply with its science/technology oriented guidelines. From the provisions of this treaty, and to keep abreast of changing conditions of space science and technology, four more treaties have been spun off: the Agreement on the Rescue of Astronauts, the Return of Astronauts and the Return of Objects Launched into Outer Space,[29] the Convention on International Liability for Damage Caused by Space Objects,[30] the Convention on Registration of Objects Launched into Outer Space,[31] and the Agreement Governing the Activities of States on the Moon and other Celestial Bodies.[32]

In addition, Principles were adopted on four subjects: The Declaration of Legal Principles Governing the Activities of States in the Exploration and Use of Outer Space with outstanding U.S. leadership, was adopted in 1963; the Principles Governing the Use by States of Artificial Earth Satellites for International Direct Television Broadcasting in 1982, Principles Relating to Remote Sensing of the Earth from Space in 1986, and Principles Relevant to the Use of Nuclear Power Sources in Outer Space in 1992.

All the treaties except the Moon Agreement have gained acceptance and compliance, so evidently the Moon situation requires more definition before there is movement toward acceptable solutions.

During the past 40 years we have been successful in extending international law into outer space, and to such an extent that we have created a special branch of space law. We do not need a police force to ensure compliance with these ratified legal provisions because they have been shaped to conform with the operational imperatives of space technology.

On December 31, 1958, the Senate Special Committee on Space and Astronautics published "Space Law: A Symposium" which was an expression of continuing concern with legal guidelines for orderly conditions in outer space.[33]

CONCLUSIONS

If the Vanguard had been orbited before *Sputnik 1*, we would not have been able to establish a coordinated national and international framework for developing the beneficial uses of outer space as quickly and successfully as we did; instead, we might have drifted into a period of instability without decisiveness on creating the conditions essential for maintaining outer space for peaceful uses. Vanguard first would have been an historical event hailed by engineers and scientists as tools for their projects, and it would have made front page news. But it would not have electrified the world, the U.S. public, in particular, would have taken for granted that we would be first in rocketry; and most of all, it would not have aroused the fear of orbiting Soviet weapons that galvanized political decisionmakers to take immediate action for achieving U.S. preeminence in outer space. Had the second Sputnik of 1,120 pounds been orbited after Vanguard, it would have aroused fear of those responsible for U.S. national defense, but information falling on the public in pieces over a period of time would not have had the same psychological impact on all groups simultaneously as happened with the timing of Sputnik.

Sputnik 1 was launched at exactly the right time, when space rocketry was in its earliest stages, to alert nations to the necessity for containing this global mechanism within the bounds of civilization. By striking such a complete lightning blow, Sputnik had a unifying effect, drawing together all the societal elements needed for finding ways to establish conditions to ensure peace and prevent war in this pristine environment. Groups that were working separately on space missions, national defense, arms control, and within national and international organizations, immediately fused on one objective: to maintain outer space as a dependable, orderly place for beneficial pursuits. They were strengthened by the nature of space science and technology which produces benefits only when operating under normal conditions. No one wanted any disruption of space communications, which became a multibillion dollar industry, weather predictions which saved lives and property, disaster relief, and the many practical uses of remote sensing.

The quickened pace of political action in 1957 and 1958 launched a regime that preempted outer space for the peaceful exploration and uses that we have enjoyed for 40 years.

In 1957 the United States had all the resources needed for developing a superior space program.

First, there was a thriving aviation industry ready to expand into aerospace.

Second, the NACA, a government agency already working on problems of flight and outer space, could be expanded into a new civilian space agency. NACA already had excellent relations with the Department of Defense and the aviation industry, and there was agreement that the Department of Defense could not develop all the civilian applications but needed jurisdiction over military space matters.

Third, the International Geophysical Year was organized by scientists and engineers working on outer space projects, and with a strong U.S. contingent.

Fourth, political leadership rocketed immediately on the day after *Sputnik 1* was launched. Senator Lyndon B. Johnson began planning for the full-scale investigation "Inquiry into Satellite and Missile Programs" which began in November 1957. He was a driving force for speeding U.S. space objectives. On January 31, 1958, the U.S. orbited *Explorer I*, the satellite whose data led to the significant discovery of the van Allen radiation belts.

Fifth, there was harmony between the Executive and Legislative Branches of the government; between the Republican president and the Democratic Congress on national and international space objectives.

Sixth, to implement space policies, organizations were quickly formed in the Executive Branch by dividing civilian and military uses and providing overall coordination, in the Congress with new standing committees, and in the United Nations by establishing the Committee on the Peaceful Uses of Outer Space. National and international organizations expanded to use space technology to improve functions they were already performing, i. e., the International Telecommunication Union (ITU), the World Meteorological Organization (WMO), and their counterparts in national governments.

During four decades of space development, some institutions made changes in their organization charts, but the general framework for achieving overall goals continued. Such changes are not unique to outer space and often occur because different managers like to alternate between centralization and decentralization practices.

As long as rules and regulations are effectively coordinated with scientific and technical facts, we can expect to maintain peaceful non-violent conditions. Difficulties could develop, however, if special groups were allowed to promote political and economic philosophies which ignore physical facts.

We have succeeded in expanding international law into outer space so that a new branch of international space law has been formed. Space law is remarkably self-enforcing because engineers and scientists have contributed to its formation by specifying the conditions they must have in order to operate and communicate between the Earth and outer space.

1. Glen P. Wilson, "Lyndon Johnson and the Legislative Origins of NASA," *Prologue: Quarterly of the National Archives* 25 (Winter 1993): 363–72.

2. *Inquiry into Satellite and Missile Programs. Hearings before the Preparedness Investigating Subcommittee of the Senate Armed Services Committee*, 85th Congress, First and Second Sessions. Part 1, November, December 1957; Part 2, January 1958; Part 3, February, April, July, 1958. 2,476 pp.

3. *Compilation of Materials on Space and Astronautics*, Committee print No. 1, Senate Committee on Space and Astronautics. 85th Congress, 2nd session. March 27, 1958: 14–19.

4. Inquiry into Satellite and Missile Programs, Part 3, pp. 2428–29.

5. Public Law 85–325 (H. R. 9739), February 12, 1958.

6. Public Law, 85–322 (H. R. 10146), February 11, 1958.

7. 50 U.S.C. 151.

8. *NACA Research into Space*, House Select Committee on Astronautics and Space Exploration, Hearings on H.R. 11881, 85th Congress, 2nd Session, April 1958: 404–410. See also, *Resume of NACA Space Research Contributing to the United States International Geophysical Year Program*, pp. 457–58.

9. *Historical Background and International Organization of the International Geophysical Year*, Hearings before the Senate Special Committee on Space and Astronautics on S. 3609, National Aeronautics and Space Act. 85th Congress, 2nd Session, Part 1, May 1958: 132–36.

10. Walter Sullivan, *Assault on the Unknown: the International Geophysical Year* (New York, McGraw-Hill, 1961), p. 27. See also, *Annals of the International Geophysical Year* (New York, Pergamon Press, 1959), 2:258–60.

11. *Department of State Bulletin*, August 8, 1955, p. 218.

12. *The U.S. Rocket-Satellite Program for the IGY*, a talk given on November 9, 1955, by H. E. Newell, Jr. Third meeting of the Comite Special de l'Annee Geophysique Internationale (CSAGI) Brussels. *Annals of the International Geophysical Year* (London: Pergamon Press, 1959); IIA:267–71. Also, "CSAGI Conference on Rocket and Satellite Observations," report by Dr. Joseph Kaplan, Chairman of the U.S. National Committee for the IGY, September 11, 1956, *Annals of the International Geophysical Year*, pp. 300–10.

13. *Documents on Disarmament, 1945-1949* (Washington, DC: Department of State Publication 7009, 1960), 2:733, 832.

14. *Ibid.*, pp. 914-15; 938-40.

15. James R. Killian, Jr., *Sputnik, Scientists and Eisenhower* (Cambridge, MA: MIT Press, 1977), p. 315.

16. *Hearings before the House Select Committee on Astronautics and Space Exploration*, 85th Congress, Second Session. on H.R. 11881. April-May 1958 1542 p. See also, *The Problems of Congress in Formulating Outer Space Legislation* on pp. 5-10; *Hearings before the Senate Special Committee on Space and Astronautics*, 85th Congress, Second Session on S. 3609. Parts 1 and 2, May 1958. 412 p.

17. *Final Report of the Senate Special Committee on Space and Astronautics*, Pursuant to S. Res. 256 of the 85th Congress. 86th Congress, First Session, Senate Report No. 100. March 11, 1959. 76 p.

18. *Conference Report on the National Aeronautics and Space Act of 1958*, House of Representatives, 85th Congress, Second Session, Report No. 2166. July 15, 1958. 25 p. All the changes are listed in this report.

19. Address by Senator Johnson before a meeting of the Columbia Broadcasting System Affiliates, Shoreham Hotel, Washington, DC, January 14, 1958.

20. *Final Report*, p. 17.

21. White House Press Release, July 29, 1958.

22. *Final Report*, pp. 28-31.

23. Public Law 86-45, Section 4, June 15, 1959 (73 Stat. 75, 422-460). See also, National Aeronautics and Space Act of 1958, as amended, and Related Legislation; Senate Committee on Commerce, Science, and Transportation, 95th Congress 2nd session. Committee print. December 1978. 185 p. See also Eilene Galloway, "The United States Congress and Space Law," *Annals of the Air and Space Law Institute* 3 (1978): 395-407.

24. *Final Report*, Congressional Committee Organization and Jurisdiction, pp. 21-48.

25. *Final Report*, pp. 58-62.

26. A history of the Ad hoc Committee on the Peaceful Uses of Outer Space and events leading to the creation of the permanent committee will be found in *International Cooperation and Organization for Outer Space*, staff report for the Senate Committee on Aeronautical and Space Sciences, Senate Doc. No. 46, 89th Congress, First Session; 183-193 (1965). See also *Conference on Security and Cooperation in Europe*, Final Act 6. Rules of Procedure (69) 4, August 1, 1975.

27. Eilene Galloway, "Consensus Decisionmaking by the United Nations Committee on the Peaceful Uses of Outer Space," *Journal of Space Law* 7 (Spring 1979): 3-13.

28. This Treaty entered into force on October 10, 1967. 18 UST 2410, TIAS 6347, 610 UNTS 205.

29. 19 UST 7570; TIAS 6599; 672 UNTS 119.

30. 24 UST 2389; TIAS 7762; 961 UNTS 187.

31. 28 UST 695; TIAS 8480; 1023 UNTS 15

32. 18 ILM 1434; 1363 UNTS 3.

33. Space Law: A symposium. Senate Committee on Space and Astronautics. Committee print. 85th Congress, Second Session. December 31, 1958. 573 p.

CHAPTER 12

A Certain Future: Sputnik, American Higher Education, and the Survival of a Nation

John A. Douglass

The nation's ascendance as the major military and economic power in the post-World War II period made the cult of science pervasive in American society. Science and technology, explained Daniel Yankelovich, "were almost universally credited with a decisive role in gaining victory in war, prosperity in peace, enhancing national security, improving our health, and enriching the quality of life."[1] In the late 1950s, Americans took comfort in the fact that they were the champions of this rational and mechanical art. Sputnik shattered that vision, not only creating the image of an enemy capable of launching missiles of massive destruction, but a widespread fear that America had failed to nurture the sciences and build advanced technologies, with potentially horrifying implications.

To a large degree, American popular opinion credited the Soviet educational system with Sputnik's success. Here was the source for its scientists and engineers—the labor pool required to pursue technology related research and for conceiving and constructing the rockets that propelled Sputnik. Conversely, the reason for America's apparent second place position in both the arms and space races was its faltering schools and universities. Among the American public, the correlation seemed obvious.

Supporters of a stronger federal role in education united with critics of America's schools systems, running roughshod over the long-standing reluctance to expand the influence of Washington in policy areas traditionally reserved to the states. "For several years independent observers have been warning us about what the Soviets were doing in education, especially in science education," explained Thomas N. Bonner in the *Journal of Higher Education*, "but they were crying in the wilderness until October 4, 1957, when the Russians punctured our magnificent conceit by making it clear that in a number of related areas of basic research and applied technology they have already outdistanced us ... Science and education have now become the main battleground of the Cold War."[2]

Bonner was not alone when he pronounced his belief that "It is upon education that the fate of our way of life depends."[3] The quick conclusion of many was that America's system of education was disorganized, it failed to provide sufficient training and research in the sciences, and it catered to mediocrity at the expense of the promising student. Higher education bore the brunt of a national failure. The elementary school and the high school nurtured the pool of talent necessary for technological advancement. But higher education, and the research university in particular, was the primary institution for creating the next generation of scientists and engineers, and for producing basic research—the basis for virtually all major techno-logical innovations. There was not only the perception of a potentially catastrophic missile gap, there also was a perceived "education and technology gap."

The translation of Sputnik from a scientific into a political event changed the dynamics of policymaking. A sense of urgency elevated to new heights the importance of the research university as a national defense tool. The utilitarian image of the university also offered a window of opportunity for the higher education community: a chance to gain significant new resources from the federal govern-ment. Advocates of a greater federal role in higher education gained significant political influence, empowered by the Soviet success. Sputnik did not determine the evolving role of the federal govern-ment in higher education. But it did alter the way that Americans viewed the importance of advanced training and research, generating a blitzkrieg of media attention, and compelling the President and lawmakers to quickly pass judgment. Post-Sputnik jitters resulted in unprecedented levels of federal funding to schools, colleges, and universities.

As the following essay describes, the outgrowths of the post-Sputnik frenzy of policymaking are significant. Sputnik prompted a significant expansion in the training of scientists and engineers, and acted as a catalyst for large scale federal funding for higher education. It also resulted in the federal government becoming the nation's primary source of R&D investment. The result was a greatly accelerated shift in scientific research toward a multidisciplinary approach. Each also formed the foundation for today's technological innovations, and for long-term economic growth that may well exceed in importance the trials and tribulations of the Cold War itself. What will be the future pattern of the nation's R&D investment? Sputnik provided a political catalyst that created a large infrastructure for basic research that remains a market advantage for the United States within the world economy. Yet as this narrative outlines, future investment in R&D, including basic research, may actually decline. Science policy in the U.S. is once again at a crossroads.

THE FEDERAL GOVERNMENT AND THE USES OF THE UNIVERSITY

The tidal wave of federal funds that followed the Soviet's success in 1957 not only elevated higher education as a key component in the ideological and military battle between capitalism and communism. It also threatened to alter the purpose of the academy. Many leaders within the higher education community had used Sputnik to advocate greater federal funding. Members of the academy debated the merits of a corresponding growth in the federal influence. The influx of funds helped to generate a virtual revolution in the sciences and engineering—a revolution inextricably tied to promoting a new age of military technology, and resulting in expanded influence by Washington. The post-Sputnik period inflamed a growing, if not entirely new, concern of the higher education community: How could the academy, whose espoused purpose is the quest for knowledge unfettered by political, religious, or economic influence, sustain itself in this new environment?

Federal involvement in higher education dates back to the colonial period in American history with the passage by Congress of a series of land-grants to state governments. The concept of a national "University" generated heated debate among the members of the first Continental Congress. Benjamin Rush, Thomas Jefferson, and others urged the creation of such an institution, announcing it as a requirement for national development. Within the vacuum of a small number of private and primarily ecclesiastical colleges, a federal university promised to educate new political and economic leaders—to establish an "aristocracy of talent" that would stand in stark contrast to the world of privilege in Europe. While it would have an egalitarian purpose, there was also a practical side: a federal university might stop or at least alter the flow of students from America's elite families to European institutions. Equally important, a federal institution would serve as a center for a new nation's cultural development.

In the battle over the proper scope of a national government, however, the notion of a federal university vanished as a salient political issue. Instead, direct control and responsibility for all forms of education fell to state governments and the initiative of primarily sectarian interests. The Northwest Ordinance of 1787 provided the precedent of granting federal lands to states for their subsequent management and sale as a financial source to promote both private and public education.

Disbursing federal land, the one resource that Congress had plenty of, was the model used in the 1862 Morrill Act. Allocating federal acreage largely in the expansive American West, the Morrill Act provided the financial basis and incentive for the creation of the nation's infrastructure of "Land-Grant" universities. State governments had a deadline to submit a proposal for the usage and management of the land-grants, or lose the potential grant. And unlike the earlier Northwest Ordinance, the Morrill Act focused entirely on higher education.

The Morrill Act added two significant ingredients to the federal role in promoting higher education. One, it required that colleges and the relatively new idea of the American research university use federal derived funds for the establishment and maintenance of agricultural and mechanical arts colleges—the two major growth areas of the nation's economy. And two, it promoted the adoption of research that would bolster local and regional economies.

To the general concept of expanding access to higher education, hence, the Morrill Act added the values of promoting research and public service. The federal role was to provide subsidies for this mission, leaving the organization of higher education to state governments, and the job of developing academic programs to the academic community and its growing ranks of supporters—primarily businessmen and a rising class of professionals. The blatant link of public higher education to economic interests promoted by the Morrill Act at first took on the banal appearance of helping the individual: promoting agriculture in the age of the yeoman farmer, and creating labor and innovation for small businesses. By the late 1890s, however, the increasing value of higher education in the industrial age raised relatively new and important questions regarding the purpose of the academy.

As a source of institutional support, federal land grants declined rapidly after 1880. Federal funding then came in bits and pieces, notably a second Morrill Act of 1890 providing grants for instruction in agriculture, and the Smith-Hughes Act of 1917 supporting the development of vocational education. While these and other acts of Congress were important, federal influence in higher education had long been eclipsed by the first major infusion of state funding, and the unprecedented gifts of capitalists such as Andrew Carnegie, John D. Rockefeller, and in California families such as the Hearsts and the Stanfords. As a result, the lay boards governing institutions of higher learning and research had, in the view of Charles and Mary Beard, come to increasingly read like a corporate directory.[4]

Within the seemingly apolitical world of the sciences and applied fields, the relatively new marriage with corporate philanthropy and the business interest in serving on governing boards appeared healthy enough. It was within the emerging world of the social sciences, and its desire to explain and critique the problems of a changing capitalistic society, that first posed serious questions regarding the nature of academic freedom—questions that would be reiterated in the era of the Cold War primarily in the sciences. Thorsten Veblen thought the corporate influence a scourge on higher education. It made college and university presidents merely hirelings of the industrialist, it created litmus tests for professors and made them essentially hired hands, and ultimately corrupted a profession dependent on free inquiry.

Veblen's vehemence and conspiratorial tone made him a largely ignored critic. An academic environment secluded from the influences of the outside world was viewed for what it was: an impossibility. America's emerging network of research universities were engaged in a process of expanding their public service functions in areas such as agricultural extension, and in assisting in the development of local and state social and economic policy.[5] But the growing number of instances in which philanthropists, governing board members, and university presidents attempted to fire faculty engaged in non-mainstream research provided the spark for the formation of the American Association of University Professors (AAUP). By 1917, the AAUP articulated the parameters of academic freedom, and argued for institutional processes of peer review for hindering the overt dismissals of faculty and for promoting free scholarly inquiry.

The corporate influence on the academy would remain a salient issue throughout this century. Yet the age of big science, the rise of the Cold War, and the labor and training needs of the Post-World War II era would revive and change the federal role in American higher education. The fear of corporate interests would fade, dwarfed by the rise of a relatively new federal dependence and influence on the academy. The realities of post-war politics would bend the purpose of higher education to a degree unfathomable to Veblen and his cohorts forty years earlier. While the pressure to expand access to higher education would prove a profound influence on the structure and activities of the academy, the federally subsidized push for science would prove an equally powerful force—particularly within the university. "Scientific investigation," explains Peter David in a critique of universities, "has become the key function of the modern research university, the wellspring of its self-confidence—almost, in the minds of many of its employees, its *raison d'être*."[6] The invention of the Cold War altered the production of knowledge in the academy, and launched the initial stages of what Clark Kerr would term the "federal grant university." [7]

THE BEGINNING OF THE COLD WAR

In the post-World War II era, federal support of science and engineering became a central means to maintain technological superiority in domestic and international markets, and to support the nation's relatively new military dominance. America's Cold War strategy of containment depended on technological superiority. Within the context of a general rejection of New Deal politics and a celebrated return to a free-market economy, the federal government would enlarge its role in basic and applied research assumed vital to the economy and national defense.

As the director of the wartime Office of Scientific Research and Development, Vannevar Bush provided a blueprint for post-war science policy. He argued for continuing the flow of federal funds to universities, and to the university managed federal laboratories.[8] In 1945, the federal government was already funding 83 percent of all research in the natural sciences. Should this dominant role continue? Bush recognized the predilection of the private sector to invest in research that promised quick returns as commercial products, and argued that research universities should remain the primary engine for basic research. The generation of new knowledge vital to techno-logical change could be nurtured best in an environment that supported the free-market of ideas, with or without recognized com-mercial application. The seemingly impossible invention of the atomic bomb, the development of jet engines, and other innovations reinforced the notion that basic research would drive new and unforeseen revolutions in products and manufacturing processes. The emergence of the Cold War and the subsequent race for technological superiority made any other model obsolete. America simply could not afford to leave basic research to the private sector.

Two years later, the President's Scientific Research Board reiterated many of Bush's recommendations. Bush had advocated a single federal agency to set science policy and distribute funding. While this model never came about, the pragmatic argument regard-ing the primary role of the federal government in promoting basic research took hold. Both Bush's report and that of the Scientific Research Board led to the organization of the Office of Naval Research in the Department of the Navy, and later the establishment in 1950 of the National Science Foundation. The resulting prolifera-tion of federal agencies created to fund and manage the nation's scientific advancement provided a new source of influence on American higher education. In 1950, over a dozen federal agencies funneled over $150 million to a select group of universities for con-tract research. Some 13 institutions garnered over 85 percent of the federal research contracts, and created the semblance of a network of national research universities that has remained dominant in securing federal research funds.[9]

The assumed importance of higher education to national security and to economic growth would translate into relatively new forms of federal involvement. In 1942, the Roosevelt administration was already making plans for what would become known as the G.I. Bill of Rights. Legislation to benefit veterans had long been a political tradition. Yet added to the mix of pensions, medical benefits, and subsidies for housing, was a new commit-ment to support and encourage veterans to attend college. The motivation was not only the welfare of veterans. The surge of veterans return-ing to a peace-time economy, it was feared, would drive up unemployment rates and exacerbate the likelihood of a severe post-war recession—and possibly induce a depression. Training in post-secondary institutions, from vocational schools to research universities, promised to reduce unemployment rolls, and train a new army of laborers suited to an increasingly technologically driven economy.

Reflecting the traditional role of the federal government regarding education, the G.I. Bill purposely provided no direct funds for supporting postsecondary institutions. Instead, financial support went to the individual for the payment of tuition and fees at an accredited school, college or university. The contemporary system of self-accreditation of higher education institutions was created in the post-war period specifically in reaction to the G.I. Bill. The result of this relatively new intervention in higher education was a huge surge in enrollment in both public and private colleges and universities that peaked in 1947. In California, for example, just over half of all students enrolled at the University of California, 70 percent of students at the state colleges and 36 percent of junior college enrollment were veterans.

In the aftermath of the war, the federal government also provided the first infusion of property and funds for the capital needs of higher education. Previous federal legislation, including the Morrill Act, stipulated that federal funding be reserved for operating costs. The Surplus Act of 1944 donated federal lands and buildings, primarily former military bases, for expanding the infrastructure of colleges and universities. By 1950, Congress also authorized the Housing and Home Finance Agency to provide $300 million in loans to public and private colleges for the building of dormitories.

A post-war structure of support for higher education had emerged, linked to a growing faith in America's colleges and universities as a determiner of socio-economic mobility, economic prosperity, and as a key component in winning the Cold War. A major federal study commissioned by President Truman and issued in December 1947, articulated the national needs in higher education, and argued for the continued expansion of federal support. *Higher Education for American Democracy* proposed a doubling of higher education enrollment in ten years. Completed by a 26 member commission, the study boldly stated that at least 49 percent of our population has the mental ability to complete 14 years of schooling. At the time, only 20 percent of the nation's high school graduates went on to a post-secondary institution. Further, the study insisted that "at least 32 percent of our population has the mental ability to complete an advanced liberal arts or specialized professional education." Estimating the mental ability of an entire nation is, to say the least, problematic. Yet the faith in scientism and desire to expand access to higher education overrode such methodological questions.

How could the nation encourage these levels of access? For one, the commission urged state and local governments to expand the number of public institutions, and primarily the number of junior colleges. The major role of the federal government, it was explained, should be a significant expansion of its scholarship programs beyond veterans. Both state and federal government, it was argued, needed to help remove not only economic barriers to access, but geographic, racial, and religious barriers primarily by providing financial support to needy students.

Yet the push for a larger federal role in higher education essentially stalled by the mid-1950s. The G.I. Bill and a second bill to benefit veterans of the conflict in Korea kept the U.S. government in the business of providing scholarships. The net effect was an indirect subsidy to both public and private higher education through the collection of tuition and fees. Despite the lobbying by higher education leaders, Congress refused any further expansion of the federal government's scholarship program. In 1956, some 50 bills where proposed that would either grant non-veteran loans or scholarships to students. All were defeated or killed in subcommittees as inappropriate forays into state control and responsibility for higher education. The American

Council on Education proposed the creation of federal tax credits, one of numerous schemes that failed to gain significant support among lawmakers.[10] The enrollment bulge of veterans, and the flow of federal dollars for tuition, dissipated.

Federal loans for dormitories had proven of great benefit to public and private colleges and universities; but advocates for higher education had failed to convince Congress to expand this program to include loans, and perhaps even direct support, for the construction of buildings for academic programs. Funding was also needed for - equipment. Expanding enrollment and research activity required a significant expansion in infrastructure—a burden that federal officials and politicians claimed was not the responsibility of the federal government.

The one area of federal funding that continued to grow was related to supporting basic research. In 1955, federally funded organized research at American universities and a select number of colleges had climbed to $169 million. Another $180 million went to university managed laboratories, such as Los Alamos. By early 1957, federal contracts for research reached $229 million, with university managed laboratories consuming an additional $240 million. The result was the first phase of a significant transformation in the activities and perceived purpose of higher education, and research universities specifically. In 1939, organized research consumed only 4.8 percent of all expenditures in American higher education, both public and private. By 1945, that number increased to 9.4 percent, and by 1955 to 15 percent.[11]

The search for agricultural advancements had dominated the pre-war research activities of universities.[12] The increase in post-war research was primarily in state of the art technologies like microelectronics, pharmaceuticals, and engineering fields with strong relationships to national defense needs. Despite these significant changes, the greatest period of federal investment in R&D was yet to come.

SPUTNIK'S ARRIVAL

The seventeen year period between the end of World War II and the launching of Sputnik represents the first phase of a growing federal relationship with higher education. A general framework of policy had been created, funneling federal resources to institutions. The majority of these funds, some 45 percent (not including the university managed federal laboratories) went to faculty directed projects in research universities, reflecting the federal government's continued post-war investment in basic research.

As noted previously, the arrival of Sputnik did not revolutionize the post-war pattern of federal engagement with higher education. Clearly, the dependence on higher education for training scientists and engineers, for creating new technologies and investigating their applications, had already been formed, and precedents set, such as the GI Bill. But the event of Sputnik did provide a tremendous spark for enlarging federal investment on an unprecedented scale, and with tremendous implications for hastening the development of new modes of scientific research. No other Cold War event, including the Soviet attainment of the atomic bomb, so shocked and galvanized American lawmakers and the public in their joint resolve to invest in and reposition higher education. Sputnik created an urgency for further investment and introspection, heightening the sense among the public that education, and specifically the academy, provided the key ingredient for beating the Soviet's space age war machine. Promoting scientific knowledge now become a mainstream issue for lawmakers, not just the personal interest of academics and a select number of government officials.

In September 1957, a national recession and a court order to desegregate schools in Little Rock, Arkansas, dominated the national news. The pressures of the Cold War, while ever-present and ingrained in American political culture, seemed sufficiently distant. Soviet ideology had been the largest concern of Americans. A vast ocean and a continent separated the nation from the Soviet empire and battlegrounds such as Korea. In the public eye, the thought of communist subversives at home appeared the most salient component of the Soviet threat. Political careers were built on the fear of communist infiltrators.

The announcement on October 4, 1957, by TASS of the launch of *Sputnik 1* jolted the public and caught Congress and the President by surprise. President Eisenhower knew of the Soviet bid to beat the U.S. in launching a satellite. In the midst of racial strife, a recession, and plans for a summit with Khrushchev, however, the political ramifications of such an event were not fully understood. The president was ill-prepared for the onslaught that followed. Democrats and Republicans embraced the "devastating blow" of Sputnik to blame each other for what appeared to be a colossal failure to beat the Russians into space and, ultimately, to invest in technology. "The big significance for the U.S. is not the object itself, but the power of the rockets used," explained one national magazine: "The Soviet Union today possesses rockets capable of launching missiles with hydrogen-bomb warheads that—if guided accurately—can strike the U.S. sixteen minutes after being launched from Soviet soil."[13] Although the communist enemy had the bomb, Americans were confident of their technological superiority. Sputnik changed that. Suddenly, there was the threat of Soviet technology and possibly military superiority.

President Eisenhower had long doubted the validity of the so-called missile gap. Even with the news of Sputnik, he remained skeptical about the scientific accomplishment of the Soviets. Eisenhower at first attempted to deny that America was now in second place in missile technology. He also defended his administration's policies that had separated the development of missiles for military use from that of scientific experiments, including the plan to launch America's first satellite as part of the International Geophysical Year. But the lack of a seemingly coherent attempt to beat the Soviets drowned his pronouncements. Eisenhower's sparse defense budgets and lame duck status provided fodder for both the right and left of the political spectrum. Democratic Senator Henry Jackson characterized Sputnik as a symbol of a national failure of leadership, a "devastating blow" and a "week of shame and danger" that, implicitly, could be traced to the White House. Senator Styles Bridges, a Republican from New Hampshire, admonished Americans to "be prepared to shed blood, sweat and tears if this country and the free world are to survive."[14]

The political stakes required a strong response from the Eisenhower administration, and made federal resolve and legislation inevitable. But what form would it take? Eisenhower resisted calls for a massive infusion of funds for the military: "The American people," he would state before Congress, "could make no more tragic mistake than merely to concentrate on military strength." On October 15, Eisenhower met with his Science Advisory Committee on Defense

Mobilization. Two major recommendations came out of that meeting that inextricably linked winning the Cold War with the educational establishment.

First, Eisenhower agreed to a full-time science adviser in the executive branch. The President would later appoint MIT president James R. Killian to the post, who then chaired the newly formed President's Science Advisory Committee (PSAC). From this group would come proposals for reorganizing the federal governments funding of R&D, for disentangling the military service rivalries that, it appeared, were one major explanation for coming in second in the space race, and for the creation of new agencies, notably NASA in 1958. And second, the federal government would need to invest immediately in education, from the elementary school to the research university, to expand the number of scientists and engineers, and to substantially increase America's research prowess.

The Soviet success brought an unprecedented desire in American society to analyze the purpose and functions of education, and to seek its reform—a desire magnified by the launch of *Sputnik 2* just one month later. Both "shocked our citizens and our government out of their complacent faith," noted one observer, "in our ability to maintain a scientific and military lead over the Russians and in the superiority of our educational program."[15] The impulse was to compare the competing Soviet educational system to our own. Sputnik has "imparted a sense of urgency and, indeed, at times almost an atmosphere of panic to a searching examination of the techniques, methods, and philosophy which have enabled the Soviet Union to achieve so dramatic a sequence of achievements and, at the same time, have aroused a widespread demand for an equally comprehensive reevaluation of American education."[16]

Nicholas DeWitt had warned of the Soviet Union's massive investment in education for Cold War purposes, and his study of their education system became the source of popular sentiment: reduced to its fundamentals, he explained, the Soviets had realized that the "Advancement of science and technology is best promoted through central planning of education and research ... that scientific and educational efforts are primarily a means for the advancement of the social, economic, political and military interests of the nation."[17] "What the Russians have done is nothing very mysterious," claimed educator Thomas Bonner, "they simply prized brains ... opened [the educational system] to all who could profit from it, and provided mammoth incentives to excel. America had simply produced a curriculum made for the norm, and our society lacked respect for learning and the teaching profession."[18] The Russians made their "tremendous leap forward in science and technology," explained Clarence Hilsberry, president of Wayne State University, by making almost everything subservient to this end within the education system: "they have given student-in-training and teachers of science and technology the prestige of both position and salary."[19]

While there were indeed lessons to be learned from the Soviets, noted one congressional advisory committee under its "Proposed Program of Federal Action to Strengthen Higher Education in the Service of the Nation," an "unwise imitation of Russian education could be as disastrous for the United States as imitation of our educational system would certainly be for the ruling clique in the Kremlin." Composed of professional educators and administrators, this special committee openly noted its fear of wholesale reform and federal intervention in higher education. The committee contended that federal and state government needed to bolster the existing system and encourage curriculum development in the sciences. "The fundamental problem facing the American people is how to improve and strengthen our educational structure, not how to remake it in blind admiration of the Russian model."[20]

Others warned against the demand for vocational and professional instruction at the cost of a liberal education, particularly at the post-secondary level. The political reaction to Sputnik, it appeared, promised to convert much of higher education "toward training people to make bombs and carriers and television sets and automobiles," according to Alex Bedrosian and Bruce Jackson. Such a reorientation, already begun in the post-World War II era, would not give students "the maturity to determine the place and relative importance of these devices in society ... Too little accent is being put on the position of the sciences in relation to other disciplines."[21] C. P. Snow wrote about the growing cultural and resource divide between humanists and scientists, and argued for mitigation. Yet the saliency of this concern seemed rather remote to lawmakers caught in the frenzy of bolstering the science and technology abilities of the nation. The two general nonscience areas that did seem to deserve greater financial support again related directly to national defense needs: political scientists and others studying the nature and predilections of the Soviet bloc and vulnerable countries, and the study of foreign languages necessary for such understanding.

The Soviets' success was not the only event that set the stage for lawmakers returning to Washington in early 1958. The Naval Research Laboratory's attempt to launch a *Vanguard* rocket carrying America's IGY satellite in December 1957 lifted off only to come crashing down in a plume of smoke and fire. The successful launch of America's first satellite by the Army a little over a month later made little difference in the political environment. For the public and lawmakers, *Vanguard's* failure reinforced the resolve for a bill that would alter the landscape of American science and research. "Congress has repeatedly turned down educational assistance," remarked Congresswoman Martha W. Griffith at a meeting of the American Association of Land Grant Colleges and State Universities. "Now, for the first time, it will undoubtedly be one of the first matters on the agenda."[22]

Wholesale reform of the nation's education system was virtually impossible in light of decentralized state and local control of educational policy. Though some advocated it, centralized authority and planning, one of the apparent secrets of the Soviet system, was not a politically or operationally viable option in the United States.[23] State governments, not Washington, retained control over the chartering and organization of higher education in the United States. Centralized control threatened the sense and reality of institutional autonomy crucial to the very idea of the academy. Most importantly, a blatant embracement of Soviet style central planning was anathema to the public and lawmakers. Yet the Soviet example provided a new focus for the federal government, creating a crucial impulse to use federal largess toward a national purpose and in ways that promoted the nation's scientific and technological prowess. The future of America's decentralized education system engulfed the discussion on domestic policy to such an extent that the nationwide interest in education was perhaps at its twentieth century peak.

At the federal level, a consensus formed, shaped in large part by the science and education community empowered by the spectacle of

Sputnik. Funding was needed to increase support for our most promising students. Curricula in the sciences and in applied fields needed to be promoted, and data needed to be collected and analyzed on the activities and performance of America's education system. Federal policy, it was decided, would promote state level planning of higher education and create the semblance of a national policy. And finally, Congress agreed to substantially increase the flow of federal research dollars primarily to higher education. The federal government was going to shift substantial resources and influence toward education in the name of defense.

The result was the most important federal bill related to education since the 1862 Morrill Act: the National Defense Education Act of 1958.[24] The general provisions of the act stated the Cold War motives of Congress: "an educational emergency exists and requires action by the federal government. Assistance will come from Washington to help develop as rapidly as possible those skills essential to the national security." Federal expenditures for education more than doubled. For higher education, this included funding for federal student loan programs, graduate fellowships in the sciences and engineering, institutional aid for teacher education, funding for capital construction, and a surge of funds for curriculum development in the sciences, math and foreign languages.

"A careful reading of the Act," wrote Philip Coombs, the director of the Fund for the Advancement of Education in 1960, "makes it obvious that Congress knowingly took positions on issues which the educators themselves are still debating. Congress," he explained, "took a stand in favor of differential programming for abler students in the schools and colleges; it took a stand on debatable curriculum questions by giving special attention to foreign languages, science, and mathematics ... . And it also took the stand, with which many educators do not yet agree, that modern communications such as films and television should be given a much larger role in the learning process."[25]

R&D AND THE UNIVERSITY

The National Defense Education Act was only one part of the reaction to Sputnik. While new funds for science curricula, for graduate fellowships and the like flowed through the Department of Health, Education and Welfare, and the Office of Education, other agencies received large infusions of funds specifically directed toward expanding the nation's R&D effort. In the name of defense, congressional allocations provided a flood of taxpayer money to the Department of Defense (DOD), the Atomic Energy Commission, NASA, the National Science Foundation (NSF) and other agencies to fund basic and applied research.

A dramatic and quick transformation occurred in both the amount and in the source of the nation's R&D investment. Federal expenditures for R&D swelled from $2.7 billion to more than $15 billion between 1955 and 1965. And while the vast majority of funding flowed under the rubric of developing the nation's defense and space capabilities, two important forces spread the wealth and created a much broader and complex national research engine. Each involved and benefited America's research universities and expanded the role of basic research.

First, the very nature of technological innovation, both in military hardware and space travel, required an ever expanding base of scientific knowledge. The serendipitous nature of scientific research and its growing relevance to technological innovation required a diffusion of funding for basic research that was substantially different from earlier decades. It also required a new model for managing federal R&D efforts that would reallocate power to new civilian led agencies such as NASA and the NSF, and direct the nation's resources in a way that would promote new fields of science. "The technological revolution that is now fully upon us involves all areas and disciplines," stated James Webb, NASA's aggressive second administrator appointed in 1961 by President Kennedy. "No nation that aspires to greatness, or to use its power for good, can continue to rely on the methods of the past," he pronounced. "Unless a nation purposefully and systematically stimulates and regulates its technological advances and builds the fruits of those advances into the sinews of the system, it will surely drop behind."[26]

And second, the political realities of American government required spreading the surge of funds among competing constituents that were not directly part of the expanding military-industrial complex. Program research such as the Manhattan Project had dominated the federal allocations thus far. A philosophical shift placed greater emphasis on providing funding for research by individuals that might have direct or indirect relationships to agencies such as NASA or specific defense programs. NASA were literally awash in funds and sought mechanisms to not only spend money, but to broaden the purpose of the agency and expand political support.[27] "Unless the United States [was] to transmogrify into a high-tech military economy while civilian industries went begging," explains Walter McDougall in his political history of the space age, "the beneficiaries of the boom [needed] to demonstrate that R&D for space and defense also energized the larger economy."[28]

The new, post-Sputnik structure for R&D investment wrested, in effect, policymaking from the traditional military services, and articulated a larger vision of how science and technology might be promoted and sustained. The military and space objectives remained dominant; but they also provided a platform for the creation of a relatively new scientific community. And as a result, the immediate post-World War II tendency toward large, monolithic organizations funded by a single federal agency, and focused on the research and production aspects of a project, such as the national laboratories, gave way to a relatively new framework: multiple federal agencies funding individual scientists within their existing institutions. The federal government focused less on investing in research that might lead to a product useful for military purposes—such as the computer—to a more holistic approach of supporting a broad range of scientific endeavors.

The size of expenditures in federally funded R&D rose spectacularly, and the wealth was spread to a degree unimaginable before October 1957. With its primary and celebrated objective to transport a human to the moon, NASA employed this diffusion of workload as an essential component in its operations. NASA, explained Webb, was spreading "all of the problems, scientific and technical, over a very large number of able minds in educational institutions ... and throughout industry," and sought to "do all [we] can to build up the university research, teaching, and graduate and postgraduate quality and quantity of education."[29]

James Killian and the PSAC had pushed hard to have the federal government increase funding for basic research, and more generally to expand federal support for research universities. A PSAC appointed panel chaired by University of California at Berkeley chancellor and Nobel laureate, Glenn Seaborg, also argued that federal funds support a greater share of direct costs for contract research projects.[30]

The vested interests of public and private universities had long been relatively powerless to influence federal research policy. But the sense of crisis now transformed the politics of federal aid to higher education. Prior to Sputnik, university presidents found that the increase in federal allocations for basic research projects failed to cover the full and indirect costs of research. Each time a university accepted a new federally funded project, institutional funds from other sources needed to be found to support the development of appropriate facilities, the hiring of administrative staff, and other costs. The result was a growing drain on university resources. Yet opposition in Congress and the Eisenhower Administration's persistent efforts to limit the growth of the federal budget made any change in federal policy unlikely.

The centrality of the research university in the Sputnik era offered a unique opportunity for the scientific community to lobby for an expanded notion of federal obligations. The Seaborg Report, in particular, argued that federal funds support the entire research-related infrastructure of universities, from scientific buildings, laboratories and equipment, to graduate fellowships. Without an increase in federal support for indirect costs, research universities would discourage federal grant projects proposed by faculty.

Seaborg's panel also linked graduate education to any legitimate research enterprise in the sciences and engineering. The National Defense Act's support of approximately 4,000 graduate students per year was not enough. "Whether the quantity and quality of basic research and graduate education in the United States will be adequate," stated the report, "depends primarily upon the government of the United States. From this responsibility the Federal Government has no escape."[31]

Sputnik also helped to create a consensus within industry regarding the role of basic research. "We might liken our pool of basic scientific knowledge to a savings account," explained Crawford H. Greenewalt, the president of DuPont in 1959, "from which we make withdrawals as we convert that knowledge through applied research to new products and processes ... Applied research and development are concerned primarily with the present. For the future, we must place our reliance on basic research."[32] While Greenewalt and others in business advocated greater basic research investment by industry, the dominant role of the federal government and research universities was both rationalized as critical to national security, and welcomed by the private sector. It was work and investment that the private sector would never fully fund.

DuPont had created one of the first industry based research laboratories and pioneered cooperative relationships with university researchers. Yet some twenty-years later, Greenewalt's successor reiterated the value of university based research. "While we at DuPont consider research collaboration to be important," he explained, "we consider as much more important the university as the principal source of basic research and of future leaders."[33]

The immediate post-Sputnik years mark a major transition for the expanding business of research that is best demonstrated by following the flow of dollars. The majority of new federal funds for R&D continued to go to programmatic research and development for defense purposes and the space program, feeding America's growing and massive defense and aeronautics industry. Adjusting for inflation, total R&D funding grew by 200 percent between 1955 to 1965—the largest single period of increase in this century. Outlays for research had grown from 5 percent to 15 percent of the federal budget. Some 70 percent of the nation's entire research effort came from federal coffers. By 1970, an estimated 75 percent of all engineers and scientists who entered the field of scientific research had gone into federally subsidized undertakings in both public and private sectors. In 1976, *Fortune* magazine stated the obvious, "science and technology have become the wards of the federal government."[34]

As shown in Figure 12.1 post-Vietnam budgets and a national recession caused a reduction in federal investment in the early and mid-1970s in real dollars, a recovery in the 1980s buttressed by President Reagan's "Star Wars" program, and since then a significant decline. Federal R&D policy was profoundly shaped by Cold War fears; in the opening years of the immediate post-Cold War era, demilitarization and fiscal constraints appeared the primary influence on federal policymaking. The net effect has been a substantial erosion in the federally funded military-industrial

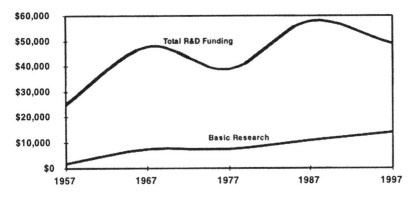

FIGURE 12.1   TOTAL FEDERAL FUNDING FOR R&D AND FOR BASIC RESEARCH: 1957–1997
(IN MILLIONS 1987 DOLLARS)

complex, and a difficult period of transformation for defense dependent industries and federal research laboratories.

Federal investment in basic research has largely avoided the
vacillations found in overall federal R&D funding. Reflecting the
centrality of basic research to technological innovation, and Vannevar
Bush's proclaimed government role in promoting and sustaining basic
research, this sector of the federal investment in R&D has steadily
grown. Funding for basic research quickly increased by some 320
percent during the same ten year period beginning in 1955, and grew as
a percentage of federal R&D expenditures from 9 to 14 percent. NASA
provided the single largest investor in basic research in 1965, spending
$790 million. The second largest source, the Atomic Energy
Commission, spent $268 million, HEW $237 million, the DOD
$220 million, and NSF $143 million.[35]

At the same time, and influenced by the Seaborg Report, federal
agencies substantially increased their funding for indirect costs. Prior
to 1957, the overhead rate that federal agencies would fund was
around 12 percent above the proposal budget for a research project.
By the mid-1960s, more elaborate formulas for overhead rates had
been developed, with each university campus negotiating the final rate
with federal officials. At major research universities such as Harvard
and Columbia, overhead rates had reached 36 percent—a rate that
higher education leaders still argued was insufficient.[36]

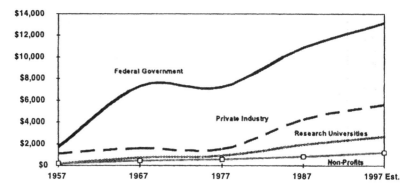

FIGURE 12.2   SOURCE OF U.S. EXPENDITURES FOR BASIC RESEARCH: 1957–1997 (EST.)
(IN MILLIONS 1987 DOLLARS)

By 1995 basic research accounted for 28 percent of all federal R&D funding. Overhead agreements at some research universities had climbed to 60 to 70 percent, reflecting not only increased federal support for indirect costs of institutions, but also the increased costs of laboratory science. In the decades after Sputnik the federal government emerged as the major source of basic research investment on a scale few thought imaginable in the immediate post-war years. As shown in Figure 12.2, while federal funding surged, industry investment in basic research remained at approximately the same level for twenty years following the launch of Sputnik. By the 1980s, however, private sector investment would grow with the arrival of relatively new industries focused on high-tech areas such as computers, pharmaceuticals, biotechnology, and medical diagnostic devices—commercial applications of technologies fostered by the federal basic science investment.

The majority of this large and growing federal investment in basic research, not surprisingly, went to research universities. Prior to Sputnik, federal and private sector investment in basic research was approximately the same. In addition the amount of basic research being conducted by research universities and non-profit agencies versus private industry was also about the same (as measured in actual expenditures). However, by 1967, expenditures for basic research in research universities outpaced those in the private sector by a ratio of 3 to 1. As shown in Figure 12.3, this ratio would be maintained throughout the 1980s and into the 1990s.

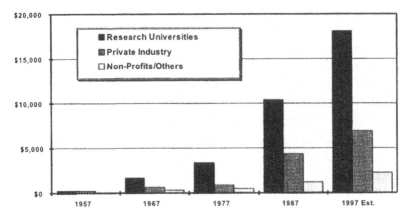

FIGURE 12.3   U.S. BASIC RESEARCH EXPENDITURES: 1957–1997
(IN MILLIONS CURRENT DOLLARS)

A FEDERAL AND STATE GOVERNMENT SYMBIOSIS

In the aftermath of Sputnik the federal interest in higher education was driven largely by national security interests. Here was the rationale for intruding into the affairs of state governments and public and private institutions. State and local support for higher education tended to focus on socio-economic mobility and economic development. The marriage of these interests created a powerful force that would not only expand access to higher education, but also buttress the growth of sciences and engineering within the academy.

Supported by federal grants and loans, states began a massive expansion of public colleges and universities. With the rapid growth of the Cold War economy, access to higher education increasingly became a matter of economic competitiveness. In states such as California, approximately 50 percent of the state's economic growth from 1945 to 1960 was tied directly and indirectly to the defense industry and federal outlays for research. The labor needs for engineers, technicians, and others in rapidly growing and federal subsidized industries, such as electronics and aeronautics, alone provided a tremendous catalyst for state investment in higher education.

Post-Sputnik federal policy encouraged larger state expenditures on higher education, particularly in areas that related to science education and basic research. Portions of the National Defense Education Act of 1958 required matching funding from states, and

stipulated that state governments establish agencies for coordinating
the dispersal of federal funds, for reporting back to Congress, and for
creating a new national data base on educational expenditures and
programs. In California the single largest benefactor of federal aid to
higher education, the Bureau of National Defense Education, was
established for this purpose, doling out NDEA funds. Similar agencies where created in all other states. And because the allocation of
federal funds was correlated to the number of students enrolled, colleges and universities gained an added incentive to increase student
numbers.[37]

The post-Sputnik surge to finance the nation's expanding network of
college and universities culminated in the passage of the 1965 Higher
Education Act by Congress. Passed at the height of President Johnson's
Great Society program, the Act expanded the federal government's
subsidization of college and university operating costs. Grants were
provided for university community services programs, college library
assistance, instructional equipment, and to enlarge the student loan
program even further. The Act, however, added a new aspect to
federal support. Growing concern that the new wave of federal assistance and contract research benefited largely a few select research universities provided the basis for new grants to "developing institutions."
Studies on the dispersal of federal funds among states and institutions
also led to changes in the allocation of contract research. In 1963, 20
universities received nearly 80 percent of all federal research funding
allocated to higher education—including university managed national
laboratories.

Despite the attempt to democratize the dispersal of federal funds, the
institutions that benefited the most from post-Sputnik legislation continued to be a select group of public and private universities. Within the
relatively new world of federal funding to higher education, private
research universities gained the most from the post-Sputnik policy
changes. On average, and benefiting from both the surge in federal
support for research and student loan programs, by 1965 the
national government provided 35 percent of the operating budgets of
private research universities—up from a total of around 20 percent
in 1957.

At public institutions such as the University of California, the surge
of federal funds proved decisive in both expanding access and increas-

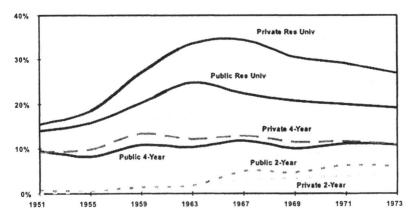

FIGURE 12.4   PERCENTAGE OF FEDERAL FUNDS IN HIGHER EDUCATION OPERATING BUDGETS:
1951–1973

ing the quantity and quality of scientific research (see Figures 12.5). In the 1960s, the largest period of enrollment and programmatic growth for California's land-grant university, federal funds for research, student scholarships, and capital construction, helped to build six new campuses. By 1965, $95 million, or some 22 percent of the university's budget (excluding the federal laboratories at Los Alamos, Livermore and Berkeley) came from federal coffers. At campuses such as Berkeley and UCLA, the percentage of their operating budget coming from Washington was closer to 35 percent.

The decline in the percentage of funds coming from the federal government to the University of California in the 1970s, and its subsequent growth in the 1980s, reflects national trends. By 1972, Congress had not renewed those aspects of the 1965 Higher Education Act that provided funds for capital construction, and began a shift away from scholarships administered by higher education institutions to a program of loans to students. Direct aid to higher education declined. Yet the eventual growth in the student aid program, and more importantly the continuing increase in federally funded research at the University would once again bring federal support to approximately 25 percent of the institution's operating budget. Within the contemporary context of a decline in state funding for higher education—a national phenomenon—federal funds have provided a key source of support.

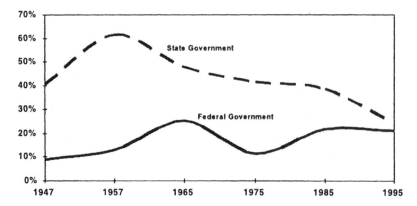

FIGURE 12.5   PERCENTAGE OF UC CAMPUS INCOME FROM STATE AND FEDERAL SOURCES: 1947–1995

THE PRODUCTION OF KNOWLEDGE

Born out of the fear of a rival foreign power, federal funds flowed to American research universities in the early 1960s. What was the impact on the academy, and more generally on the production of knowledge?

The explosion of federal funding had been largely embraced, indeed encouraged, by the academic community. At perhaps no other time in the nation's history had higher education, and the research university specifically, been elevated to such a pivotal place in the society. The studious avoidance of centralized planning and federal control over higher education institutions provided the semblance of independence for the academy. Indeed, the National Education Act explicitly stated that "Nothing in this act shall be construed to authorize any agency or employee of the United States to exercise any direction, supervision, or control over the curriculum, program of instruction, administration, or personnel of any educational institution ... ." Federal funds were dispersed to encourage science education, to promote and expand scientific research.

Yet the surreptitious influence of federal grants particularly for research once again raised important questions regarding the purpose of the academy. With the emergence of the Cold War research university, the president of the University of California, Clark Kerr, remarked it was perhaps ironic that America's universities, "which pride themselves on their autonomy ... which identify themselves either as 'private' or as 'state' should have found their greatest stimulus in federal initiative ... that institutions which had their historical origins in the training of 'gentlemen' should have committed themselves so fully to the service of brute technology."[38] A number of academicians lamented that the transition was not a conscious decision of the educational community. "Powerful precedents have been set," remarked Ross L. Mooney in 1963, "the universities have been by-passed as agencies adequate in their own right to determine what research to do."[39]

With the federal presence at its height in the mid-1960s—funding not only research and a growing federal scholarship and loan program for students, but a significant portion of capital construction in private and public institutions—there was some conjecture that the post-Sputnik era would eventually lead to greater federal control and authority over higher education, creating a national system analogous to those found in Britain and France.

The concept of academic freedom, of free inquiry, appeared on the wane, particularly in the select major research universities which garnered the vast bulk of federal grants. While substantial sums were now "available to support scientific investigation in higher education," remarked Richard J. Barber in an analysis of the politics of research in 1966, "it has been bought at a high price." For one, noted Barber, with a project or field of inquiry set by a federal agency, the choice of the investigator is limited, and the development of new areas of research restricted. "The result has been to favor certain fields of science, particularly the physical sciences, at the expense of others, especially the psychological and social sciences. The medical schools, benefiting from the largess of the NIH, have become much more attractive to students than, say, biology, so that those who might have embarked on graduate work in biology troop off to medical school instead."[40]

The servitude of the sciences and engineering to the national security agenda of Washington, however, did not lead to any serious opposition to federal funding of higher education. The windfall of funds helped major research university's expand in the sciences and increase the quality of programs at a time of large scale enrollment growth— expansion that could have significantly eroded academic quality. The subsequent growth of the sciences and engineering at university campuses also created a constituency not only dependent on the grant

JOHN A. DOUGLASS

structure in Washington, but deeply involved in influencing the flow of funds. Sputnik launched a movement that had partially disenfranchised the military leadership in allocating R&D funds. The creation of NASA was the first major break. The plans of Robert McNamara and other members of the Kennedy administration to improve the management and shape the direction of America's R&D effort further increased civilian control over DoD allocations.[41]

The decentralized and multiagency approach to funding R&D not only fed the scientific community, it allowed scientists to infiltrate and control much of the research agenda. The concern by academics, often nonscientists, regarding the influence of the federal government and compromises over academic freedom was accompanied by another worry that came from outside of the academy. Eisenhower's reflection on the rise of the military-industrial complex also included a fear that federal R&D increases would lead to the "domination of the nation's scholars." Shortly after leaving the presidency, Eisenhower noted not only the great importance of scientific research, but also urged awareness of the "equal and opposite danger that public policy could itself become the captive of a scientific-technology elite."[42] Barber, a persistent critic of federal R&D policy, noted that the post-Sputnik rush of Congress to more than triple federal outlays for research had neglected to establish a structure for analyzing and shaping national research policy. "We continue to leave science to the scientists and engineers. To me this makes no more sense than leaving war to the generals."[43]

The specter of a national system of federally controlled and funded research universities would never arrive.[44]

A crowded, federally subsidized, and, ultimately, more productive research engine emerged, which in turn hastened important changes in how scientific research is conducted. Buttressed by a series of post-Sputnik bills, including the 1963 Higher Education Facilities Act, scientists and engineers gained access to new laboratories with state of the art equipment. Scholarship funds provided support for graduate students and a growing cadre of professional researchers and support technicians. Team research began to replace the single scientist working in a laboratory. The number and variety of research consortiums at university campuses grew, enhancing communication between scholars and forming one basis for the expansion of multidisciplinary research. The speed in which new discoveries led to new areas of research also increased considerably.

While shaped by the international science community, America's rapid emergence as the leading scientific state fast-tracked the production of knowledge toward a multidisciplinary, trans-diciplinary, and collaborative model that, today, is increasingly obscuring the distinction between basic and applied research in fields such as biotechnology and communications. "While knowledge production within traditional disciplinary structures remains valid," explains Michael Gibbons, Martin Trow and other co-authors of *The New Production of Knowledge*, a relatively new mode of research "is growing out of these structures and now exists alongside them." The burst of new knowledge subsidized and encouraged by federal outlays to research universities and industry has, in turn, created new technologies such as the internet which, in turn, have fed this new mode. "In transdisciplinary contexts, disciplinary boundaries, distinctions between pure and applied research, and institutional differences between, say, university and industry, seem to be less and less relevant," state the authors. "Instead, attention is focused primarily on the problem area, or the hot topic, preference given to collaborative rather than individual performance and excellence judged by the ability of individuals to make a sustained contribution in open, flexible types of organisations ... ."[45]

Both reflecting and influencing this trend, the market for scientific research has diversified considerably. Private sector investment in both industry and university basic research has increased dramatic-ally in the last two decades. The melding of basic and applied research has also been encouraged by changes in federal policy, providing funding and tax incentives for the establishment of Industry-University Cooperative Research Centers in areas such as engineering, and in 1984 giving universities patent rights for discoveries connected to federal subsidized research. In 1986, the NSF funded the first Engineering Research Centers (ERCs) with private industry focusing on team research in areas that promised the next generation of technological advances. There are now twenty-five such centers all located on major research university campuses.

The 1986 Federal Technology Transfer Act permitted, for the first time, federally funded researchers to collaborate with industry to develop commercial patents. "In effect, the federal government was limping toward a sort of industrial policy," claims Norman E. Bowie. "Since American industry was failing to invest in sufficient research and development to bring new products to market that could compete internationally, especially with the Japanese, the government provided public funds to universities to help move the fruits of basic research into the marketplace."[46] The federal government had created a tremendous national infrastructure for science and innovation in the post-World War II era. In the mid-1980s and reflecting international trends, it now created new mechanisms to tempt the academy and industry into a closer alliance.[47] In attempting to gain a greater understanding of the role of basic and applied research in the economy, NSF director Richard Atkinson funded a series of studies on the link of R&D investment with economic growth. "As late as the mid-1970s, there was no substantial economic data, no reliable economic analysis of the relationship between investments in R&D and economic development," Atkinson later explained. He and others at the NSF realized that the lack of such information made it more difficult to gain support for research in Congress.[48] Atkinson's successor, Erich Bloch, proceeded to make economic competitiveness a part of the NSF mission.

Reflecting the evolving nature of technological innovation, and the development of university and industry based consortiums and partnerships, American research universities are now engaged in more applied and developmental research than any other time in their history (see Figure 12.6). Fields such as engineering and agriculture have long had applied and industry subsidized research components. The greatest expansion in applied research is in relatively new fields such as biotechnology that are, by their very nature, transdisciplinary and rooted in the hard sciences. As Roger Geiger has noted, the philosophical opposition to applied research in the academy, and the concerns of leaders such as Clark Kerr and later Derek Bok at Harvard, gave way to a new era dominated by new fields of scientific inquiry.[49] "The deadlock between pro- and anti-business forces was broken," writes Geiger. In no small part, the fast-paced development in biotechnology "overwhelmed the

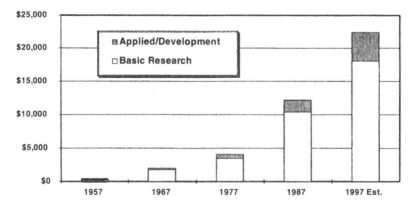

FIGURE 12.6 UNIVERSITY RESEARCH BY TYPE: 1957–1997 (EST.)
(IN MILLIONS CURRENT DOLLARS)

innate resistance of the universities to closer industrial ties," Geiger concludes.[50]

In one sense, the fact that research universities are increasingly relevant to the nation's economy is a reflection of an increasingly technology dependent private sector that needs both basic and applied research to remain competitive. In another sense, America's research universities are returning to the role they were appointed to by the Morrill Act of 1862: to be major sources of new knowledge and trained labor for dominant sectors of the economy. Agriculture and mechanical arts have been replaced by semiconductors and communications, by biotechnology and space exploration. It is a role in the economy for American higher education that never went away, but it has become more active, more complicated, and more essential to the private sector. In his study of technology transfer at four major research universities, Gary Matkin noted a number of trends that illustrate this shift: for one, every major research university, he predicts, will "eventually become a financial partner in start-up companies." As long as there is a balance of basic and applied activity, and a balance of sources for funding academic research from both public and private coffers, then the academy can be preserved. However, defining and regulating this balance of activity, and hence keeping at bay a dominating subservience to the commercial world, will continue to be a challenge.[51]

THE SPUTNIK LEGACY AND A NEW ERA

Sputnik provided the political currency for the federal government to inject money into the nation's education system, and to take on the burden of funding the nation's lagging applied and basic research effort. The race for technological superiority, essentially, made any other model obsolete.

Forty years later, a cadre of economists project that perhaps 50 percent of America's post-World War II economic growth is directly or indirectly related to private and federal R&D investment. A 1995 report of the Council of Economic Advisors to the President linked increasing productivity of America's workforce to future economic growth, and then stated that "Investments in research and development are the key to increasing productivity. The social rates of return," noted that report, "exceed the high private rates of return, of 20 and 30 percent, by a considerable amount because of 'spillovers'—benefits that accrue as other researchers make use of new findings, often in applications far beyond what the original researchers imagined."[52] Reflecting on some twenty-five years of studies, many through the NSF, Richard Atkinson, now the President at the University of California, notes that there is now real evidence that as investments in university research increase, a corresponding increase in private-sector investment results—but often with a time lag of five to fifteen years.[53]

The percentage of return is debatable, particularly since so much of the nation's R&D effort has gone toward military hardware with only indirect benefits to the economy and new industries. The bridge between new scientific discoveries to the actual reiterative process of design, and then commercial production and, finally, economic growth is also a topic of significant debate among economists: For example, the role of commercial development, argue scholars such as Terence Kealey, is the most important part of the story, not basic science.[54] Yet it is clear that there is a strong relationship between R&D investment, including the funding of basic research, and economic growth—a relationship that will only get stronger.

The Sputnik legacy remains, but the world has changed. The post-Cold War era has brought with it an overall decline in federal support for both higher education and the nation's R&D effort. At the same time, and as noted previously, the private sector has significantly expanded its R&D investment, essentially replacing declining federal dollars. Now it is the private sector that funds approximately 70 percent of all R&D activity in the U.S. It is exactly those industries that have benefited most from government funded basic research—high tech businesses focused on, for example, computers, communications, biotechnology and pharmaceuticals, even aeronautics—that are investing their own dollars in research. Biotechnology and digital communication industries owe their very existence to university based research and the scientists produced in the academy. In a sense, the federal government primed the pump, helping to create new market opportunities for the private sector, and which now encourages the evolution of university and business partnerships.

Is America investing enough in R&D? Between World War II and 1957, America's investment in R&D as a percentage of GDP averaged about 1 percent. In the post-Sputnik era it averaged about 2.6 percent. Total investment in R&D as a percentage of GDP, however, is now nearly a quarter less than the peak in the mid-1960s. While other industrial nations are increasing investment in R&D as a percentage of GDP, in the U.S. it may decline.[55] Despite the increase in privately funded R&D, most of the investment is in applied areas, with only a small percentage funneled to basic research. The vast majority of America's basic research activity, including the training of the nation's scientists and engineers, is still dependent on federal funding and carried out by the nation's research universities—confirming Bush's earlier contention that the private sector might never adequately fund or conduct basic research. The linear model outlined by Bush, of public sector basic research and private sector applied and development research, is frayed and perhaps metamorphosizing, but for now remains valid.

The path of new knowledge and the seeds for technological innovation still flow largely from the academy to industry, not the other way around—although collaborative modes between public and private sectors are expanding and new structures for promoting scientific research are developing. Most of the businesses engaged in both applied and basic research are still dependent on the findings and trained personnel that come from the academy. Indeed, many of the new high tech companies are closely linked with research universities, often started by academics, and purposely located at the perimeter of university campuses.

With nearly two-thirds of all new patents in the United States based on public sector basic research, the pace and mechanisms for encouraging the transfer of new knowledge from the academy to the private sector is growing in importance.[56] For the immediate future, the health and growth of a significant portion of the American economy relates directly to the activity level in basic research, and in turn the financial support of research in universities—what remains America's market advantage for technological innovation.

Recent federal belt tightening intended to balance the budget, however, and the rising cost of hard science research present serious problems for research universities that private-public consortiums and patent income cannot, thus far, compensate for.[57]

While the modes, mix, and quality of research is extremely important, one might conjecture that overall R&D investment in the United States will need to increase by a full 1 percent of GDP over the next decade, up from the current rate of 2.4 percent to approximately 3.5 percent. This would be an increase of about one-third over current investment, and is simply an extrapolation of previous investment trends. More of the R&D pie should also go to basic research, both in public and private sectors of the economy. If past history is any indicator, society and our economy will become more technologically bound, not less. New industries will rely even more on R&D, not less. The development of new modes of collaborative research will increasingly disregard national borders, and the globalization of the economy provides new parameters for developing national research policy. There may also be declining rates of research productivity and innovation as the level of R&D investment rises. However, some sizable increase in R&D is inevitable. The question is when and how.

Sputnik the political event made it relatively easy: a sense of urgency jolted Washington and the public into action. The results have been dramatic. The new era is less sanguine, less directed and, ultimately, more complacent. The pattern of investment today, however, may have an even larger impact on the quality of life, the environment, and the economy of tomorrow. While the federal government no longer plays the dominant role in applied R&D, there is a great need for a general

360   RECONSIDERING SPUTNIK

discussion of the future pattern of investment, and the appropriate role of the private and federal sectors—in applied and basic research. Science and technology are no panacea. Yet it is clear that the jobs and prosperity of the next century will depend in no small part on today's mix and level of R&D investment.

1. Daniel Yankelovich, "Science and the Public Process," *Issues in Science and Technology* 1:1 (Fall 1984): 6–12.
2. Thomas N. Bonner, "Sputniks and the Educational Crisis in America," *Journal of Higher Education* 29:4 (April 1958): 177–84.
3. Ibid.
4. Charles A. Beard and Mary Beard, *The Rise of American Civilization* (New York, 1927), 2: 470.
5. In the case of California, see John A. Douglass, "Californians and Public Higher Education: Political Culture, Educational Opportunity and State Policymaking," *History of Higher Education Annual,* Pennsylvania State University, Vol. 16, 1996.
6. Peter David, "The Tower of Science," *The Economist,* October 10, 1997.
7. Clark Kerr, *The Uses of the University* (Cambridge, MA: Harvard University Press, 1964) p. 49.
8. Vannevar Bush, *Science: The Endless Frontier* (Office of Scientific Research and Development, Washington D.C., 1945); see also Pascal G. Zachary, *Vannevar Bush: Engineer of the American Century* (New York: Free Press, 1997).
9. See Roger Geiger, *Research and Relevant Knowledge: American Research Universities Since World War II* (New York: Oxford University Press, 1993).
10. Symour E. Harris, *Higher Education: Resources and Finance* (New York: McGraw-Hill Book Company, Inc., 1962), p. 316.
11. June O'Neal, *Resource Use in Higher Education* (Berkeley: Carnegie Commission on Higher Education), p. 22.
12. See Roger Geiger, *To Advance Knowledge: The Growth of American Research Universities, 1900–1940* (New York: Oxford University Press, 1986).
13. "Around the World in 96 Minutes—What Does it Really Mean?" *U.S. News and World Report* 43 (October 18, 1957): 16.
14. Quoted in Barbara Barksdale Clowse, *Brainpower for the Cold War* (Westport, Conn.: Greenwood Press, 1981), p. 9.
15. "A Time for Greatness," *Journal of Higher Education* 29 (February 1958): 105.
16. H. E. Salisbury, "The Soviet Educational System," *Journal of Higher Education* 29 (November 1958): 462–64.
17. Nicholas DeWitt, *Soviet Professional Manpower* (Washington, D.C.: U.S. Government Printing Office, 1955), pp. 3–4; DeWitt's study was made before Sputnik but was highly influential. The dearth of other studies and reports included such books as George S. Counts, *The Challenge of Soviet Education* (New York: McGraw-Hill Book Company, Inc., 1957); Division of International Education, U.S. Department of Health, Education, and Welfare, *Education in the U.S.S.R.* (Washington, D.C.: U.S. Government Printing Office, 1957); Alexander G. Korol *Soviet Education for Science and Technology* (New York: Technology Press of MIT and John Wiley and Sons, Inc., 1957).
18. Bonner, "Sputniks and the Educational Crisis in America," pp. 177–84.
19. Clarence B. Hilsberry, "Sputnik and the Universities," *Journal of Higher Education* 29 (October 1958): 375–80.

20. Committee on Relationships of Higher Education to the Federal Government, "A Proposed Program of Federal Action to Strengthen Higher Education in the Service of the Nation," *Higher Education and National Affairs*, 11 (January 1958): 1–2; see also David D. Henry, *Challenges Past, Challenges Present* (San Francisco: Jossey-Bass Publishers, 1975), pp. 118–22, for a description of the concerns of university presidents regarding the growing federal role in higher education prior to Sputnik.

21. Alex Bedrosian and Bruce Jackson, "Intellectual Conformity: Not the Answer," *Journal of Higher Education* 29 (October 1958): 381–85.

22. Martha W. Griffith, "As I See it From Here," Proceedings of the Seventy-First Annual Convention of the American Association of Land-Grant Colleges and Universities, November, 1955; E.W. Kelley, *Policy and Politics in the United States: The Limits of Localism* (Philadelphia: Temple University Press, 1987) discusses the shift in federal policy toward education, and its impact regarding increased appropriations.

23. As Percival M. Symonds, "The Organization of Educational Research in the United States," *Harvard Educational Review* 27 (Summer 1957): 159–67 noted: "One outstanding characteristic of education in this country, presumably, is that it lacks organization, or at least, that whatever organization it has is determined at the local level and is not imposed from above by the central government."

24. National Defense Education Act, passed September 3, 1958. James Conant and Eliot Richardson helped to craft the bill which was carried in Congress by Democrats Carl Elliot and Senator Lister Hill, and gained the votes of twenty-four Republicans who had previously voted against similar legislation.

25. Philip H. Coombs, "Some Issues Raised by Recent Legislation," in Symour E. Harris, ed., *Higher Education in the United States: The Economic Problems* (Cambridge, Mass.: Harvard University Press, 1960), pp. 83–87.

26. James Webb, *Space Age Management: The Large Scale Approach* (New York, 1969), p. 117.

27. Roger D. Launius, *NASA: A History of the U.S. Civil Space* Program (Malabar, Florida: Krieger Publishing Company, 1994), p. 65.

28. Walter A. McDougall, *The Heavens and the Earth: A Political History of the Space Age* (New York: Basic Books, Inc., 1985), p. 385.

29. Cited in McDougall, *The Heavens and the Earth*, p. 385.

30. President's Science Advisory Committee, *Scientific Progress, the Universities, and the Federal Government*, (Washington, DC: U.S. Government Printing Office, 1960) pp. 10–11.

31. Ibid.

32. Crawford H. Greenewalt, "Basic Research: A Technological Savings Account," in Dael Wolfle, ed., *Symposium on Basic Research* (Washington, D.C.: American Association for the Advancement of Science, 1959), p. 130.

33. W.G. Simeral, "The Evolution of Research and Development Policy in a Corporation: A Case Study," in Langfit et. al., ed., *Partners in the Research Enterprise: University Corporate Relations in Science and Technology* (University of Pennsylvania Press, 1983).

34. "The Nationalization of U.S. Science," *Fortune*, September 1976, p. 158.

35. Budget of the United States, Fiscal 1965.

36. Harris, *Higher Education: Resources and Finance*, p. 649.

37. California Legislature, "Report of the Senate Fact Finding Commission on Education," 1961, CSA.

38. Clark Kerr, *The Uses of the University* (Cambridge, MA: Harvard University Press, 1964), p. 49.

39. Ross L. Mooney, "The Problem of Leadership in the University," *Harvard Education Review*, 33 (Winter 1963): 42–57.

40. Richard J. Barber, *The Politics of Research* (Washington, D.C.: Public Affairs Press, 1966), p. 60.

41. See McDougall, *The Heavens and the Earth*, p. 332.

42. Quoted in *The New York Times*, January 18, 1961.

43. Barber, *The Politics of Research*, p. v.

44. "Government-University-Industry Research Roundtable: Industrial Perspectives on Innovation and Interactions with Universities," National Academy of Sciences, 1991.

45. Michael Gibbons, et al., *The New Production of Knowledge: The Dynamics of Science and Research in Contemporary Societies* (London: SAGE Publications, 1994), p. 30.

46. Norman E. Bowie, *University-Business Partnerships: An Assessment* (Lanham, Maryland: Rowman & Littlefield Publishers, Inc., 1994), p. 19.

47. See Gary W. Matkin, *Technology Transfer and the University* (New York: Macmillan, 1990); and Roger Geiger, "The Ambigious Link: Private Industry and University Research," In William E. Becker and Darrell R. Lewis, eds., *The Economics of Higher Education* (Boston: Kluwer, 1992), pp. 265–97.

48. Richard C. Atkinson, "The Role of Research in the University of the Future," paper given at The United Nations University, Tokyo, Japan, November 4, 1997.

49. See Derek C. Bok, "Business and the Academy," *Harvard Magazine* 83 (1981): 23–35.

50. See Geiger, *Research and Relevant Knowledge*, and "What Happened After Sputnik? Shaping University Research in the United States."

51. Matkin, *Technology Transfer and the University*, p. 318.

52. "Supporting Research and Development to Promote Economic Growth: The Federal Government's Role," A Report Prepared by the Council of Economic Advisors, October, 1995.

53. Atkinson, "The Role of Research in the University of the Future."

54. Terence Kealey, *The Economic Laws of Scientific Research* (New York: St. Martin's Press, 1996); see also Stephen J. Kline and Nathan Rosenberg, "An Overview of Innovation," in Ralph Landau and Nathan Rosenberg, ed., *The Positive Sum Strategy: Harnessing Technology for Economic Growth* (Washington D.C.: National Academy Press, 1986).

55. See "Easing the Squeeze on R&D," *Science* Vol. 278 (November, 1997).

56. Francis Narin, Kimberly S. Hamilton and Dominic Olivastro, "The Increasing Linkage Between U.S. Technology and Public Service," CHI Research Inc., March 17, 1997.

57. See Albert H. Teich and Bonnie Bisol Cassidy, *The Future of Science and Technology in California: Trends and Indicators*, Center for Science, Technology, and Congress, American Association for the Advancement of Science, May 1996.

CHAPTER 13

Sputnik: A Political Symbol and Tool in 1960 Campaign Politics

Gretchen J. Van Dyke

It is common in the United States for presidential candidates and opposition parties to criticize intensely the policies of the incumbent president and party. This tactic is often used to garner support for the challenger and opposition party; moreover, it provides a basis on which the party can formulate its platform and a foundation on which the individual, if elected, can develop and implement new policies. This is precisely what the Democratic Party and its prospective presidential candidates did in the late 1950s in preparation for the 1960 presidential election. A major focus of the Democrats was U.S. national security and American power, prestige, and leadership in the international system. In fact, the 1957 Sputnik launching became the primary symbol of the Democrats' argument that Eisenhower administration defense policies had allowed a missile gap to develop that favored the Soviet Union and had precipitated a supposed decline in American power, prestige, and leadership internationally.

This paper will examine how and why the Democratic Party in the late 1950s used the Sputnik launching, and subsequent questions about the missile gap and American power and prestige, to raise national attention about American national security and international leadership for the 1960 presidential campaign. It will investigate the Party's efforts, including those of individual members within and outside of Congress, to articulate a new approach to national defense with the express purpose of regaining the White House in 1960. Specifically, it will study the role of the Democratic Advisory Council (which published the Party's pamphlet on national defense in 1959), the Congressional Democrats' use of budgeting and oversight authority (which served as a stage to promote their version of national security and defense), the conflicting interpretation of intelligence data by Eisenhower administration personnel and the Democrats (which fueled the Democrats' allegations about national defense), and John Kennedy's merging of the missile gap and American power and prestige questions during the 1960 campaign (which allowed him to establish his legitimacy as a potential national leader). As a result, this paper will clearly demonstrate how Sputnik "launched" a political battle that ultimately caused a shift in national defense policy—from Eisenhower's massive retaliation strategy to Kennedy's flexible response strategy.

SPUTNIK AND THE MISSILE GAP MYTH

As the Cold War intensified in the post-World War II era, American nuclear superiority became the primary means for deterring aggression by the United States' chief ideological and military adversary—the Soviet Union. Containment of Soviet expansionism had emerged as the paramount goal of the United States in the late 1940s and truly had become synonymous with the national interest. In turn, the Eisenhower administration had adapted its particular defense policies to this overriding national security goal. The Eisenhower strategy—sometimes called the 'New Look' or the 'Long Pull'—was a form of containment that combined massive retaliation (any Soviet aggression would be met with a swift, massive nuclear retaliatory strike on Soviet territory) with fiscal conservatism (the emphasis of the limited military budget would be placed primarily on nuclear weaponry needed for a credible nuclear deterrent).[1] This particular strategy gradually lost credibility in the 1950s as nuclear force was not used in limited non-nuclear conflicts, such as, in Korea, Indochina, and Hungary. Furthermore, the viability of the massive retaliatory threat was lessened further as Soviet strategic nuclear capabilities continued to progress during this period, symbolized by the Sputnik launching in October 1957.[2]

Sputnik was crucial to initiating and promulgating the missile gap issue for several reasons. First and foremost, it precipitated an hysterical public reaction in the United States because Sputnik was interpreted as evidence that the United States, for the first time, had fallen behind the Soviets in the nuclear missile and space races. In his memoirs, Dwight Eisenhower acknowledged the surprise and grave concern that enveloped the country and the world because of Sputnik. "Newspaper, magazine, radio, and television commentators joined the man in the street in expressions of dismay over this proof that the Russians could no longer be regarded as 'backward,' and had even 'beaten' the United States in a spectacular scientific competition ... . There was no point in trying to minimize the accomplishment or warning it gave that we must take added efforts to ensure maximum progress."[3] Kennedy and the Democratic Party were able to manipulate this public fear and promote their perception of national security and defense.

Second, the Sputnik launching confirmed the fears and reports of some scientists (Edward Teller, for example) and private analysts (particularly those at the RAND Corporation) who had been warning the administration that it had not been doing enough for the American strategic defense program. In Spring 1957, Eisenhower had appointed an ad hoc commission, the Gaither Committee, to investigate a Federal Civil Defense Administration proposal for a $30–$40 billion civil defense program. The Gaither Committee Report, submitted to Eisenhower on November 7, 1957—barely a month after Sputnik, recommended that "a massive civil defense program ... should take a back seat to what they saw as the much more pressing need of building up a much larger offensive missile force and protecting it from an attack through dispersal and hardened shelters, so that SAC [Strategic Air Command] might survive an attack."[4] It maintained that the Soviets had "probably surpassed the U.S. in ICBM [intercontinental ballistic missile] development" and it clearly underlined the existence and danger of the missile gap.[5] The Gaither Committee Report was followed by a January 1958 publication of a private study that had been commissioned by the Rockefeller Brothers Fund. The Rockefeller Brothers Report, titled *International Security: The Military Aspect*, provided a similar assessment of American strategic defenses, and it, too, recommended an increase in allocations for strategic forces and SAC's protection.[6] Both the Gaither Committee Report and the Rockefeller Brothers Report validated public and private fears about the state of American defense. Yet, while Eisenhower participated in several briefings, particularly on the Gaither Report, he nonetheless refused to increase his budget substantially to implement the report's conclusions because of his commitment to fiscal conservatism.[7] Thus, the

Democrats seized an opening and began formal discussions about American national security and the implications of Sputnik.

On Capitol Hill in November 1957, Senate Majority Leader Lyndon Johnson's military preparedness subcommittee initiated an eight-month series of hearings in direct response to Sputnik, investigating every facet of American defense. Moderate increases in allocations for defense programs resulted from these hearings as well as intense competition for those funds. During the 86th Congress (1959–60), Johnson remained at the forefront of the Democratic criticism of the Republican administration, serving as the chairman of joint hearings on the defense budget of the Senate military preparedness subcommittee and the Senate Aeronautical and Space Science Committee. Moreover, the Sputnik launching precipitated an intense debate within Eisenhower's Defense Department, on Capitol Hill, in the press, and in the private sector about whether Eisenhower's finite (minimum) deterrent strategy was receiving adequate funding within the administration's budgeting strategy.[8] Administration critics (some for very different reasons) argued that inadequacies in current defense policies, particularly for ICBMs, were allowing the Soviets to develop a credible ICBM counter-force capability. Furthermore, they argued that by the early 1960s the United States would face a "gap" "... in which the balance between Soviet offensive and defensive forces, on the one hand, and the American strategic forces, on the other, would be such that the Soviets might conclude that a surprise attack would reduce their losses to acceptable limits."[9] Subsequently, questions about a missile gap as well as American power and prestige were common sources of debate both within and outside of governmental circles. In essence, the question was whether the Eisenhower administration was effectively protecting American national security.

The Sputnik launching also confirmed for the public the fierce inter-service rivalry that was plaguing the Pentagon and the competition over possible new appropriations that the missile gap allegations might produce. The Army, in particular, had faced drastic budget cuts under Eisenhower's "New Look," and its Chiefs of Staff, Matthew Ridgway and Maxwell Taylor, had been critical of the administration's defense budgeting strategy. Lt. Gen. James M. Gavin, who had served under Ridgway on the Army staff, exposed Army concerns about the administration's missile and satellite programs in his 1958 book, *War and*

Peace in the Space Age. After his retirement in 1959, Taylor published his much heralded work, *The Uncertain Trumpet*, in which he, too, suggested the imminent danger of a missile gap; more importantly, he introduced a new flexible response strategy that promoted a gradual and primarily conventional response to enemy aggression.[10]

The Army's criticism, however, was not new. Both the Army and the Navy had consistently opposed the attention and funding that were given to the Air Force in the 1950s. Eisenhower's massive retaliation strategy, which emphasized the deterrent value of nuclear weapons, had depended on the Air Force's manned bomber as the sole means for carrying out an actual strategic nuclear strike; further, the Air Force's budgeting needs were always met at the expense of the Army's and Navy's. A flexible response would spread appropriations out across the services and resolve the missile gap as well. Moreover, the Air Force, which had usually supported Eisenhower budgets during the early and mid-1950s precisely because of generous Air Force appropriations, also became an Administration opponent as the missile gap issue emerged in the late 1950s. It believed it could exploit the issue on Capitol Hill and, in turn, gain greater allocations for its missile and long-range bomber programs.[11] Again, the alleged missile gap was used by various sectors to advance parochial interests, to undermine rhetorically the Administration's defense strategy, and to offer a different direction for American national security. Of course, Sputnik was at the root of these claims. Unfortunately for the Eisenhower administration and the Republican Party, the intensity of the missile gap debate and their inability to counter their critics convincingly helped to turn national defense strategy into an electoral liability by the end of the decade.

THE DEMOCRATIC PARTY'S RESPONSE

The National Party Committee
A year before Sputnik, the Democratic National Party began working on an effective counterforce to the Republican White House. Smarting from the 1956 presidential loss, national party activists proposed the Democratic Advisory Council (DAC) to act as the formal opposition— or the "presidential wing," as James MacGregor Burns once called it.[12] DAC proponents argued that the Democrats had lost the 1956 presidential race because they "had not undermined the Eisenhower

prestige by forcefully pointing out to the country the mistakes and folly of his policies, and they had not developed and presented to the country a distinct and liberal party program that would have given the voters a clear and attractive alternative to Eisenhower." Moreover, they maintained that "attacks on the Republicans had to be made, and the program had to be assembled and presented, day by day and week by week in the long years between presidential elections," not just during the short campaign every four years. Furthermore, these "presidential Democrats" blamed the party leadership on Capitol Hill (House Speaker Sam Rayburn and Senate Majority Leader Lyndon Johnson) for "blurring the image of the Democratic Party" through legislative compromises, which had often forced the party "to try to share the middle of the road in a bipartisan embrace with the conservative Republican President." History had clearly illustrated to them that bipartisan compromise was a no-win situation for the Democratic Party, particularly if it intended to regain control of the White House in 1960. It was paramount, therefore, that the Democrats find a way for presenting their interpretation of pressing national issues well before the 1960 election season. Thus, the DAC's primary role would be that of policy developer, on questions in both the domestic and foreign policy realms.[13]

Originally, the DAC was to be composed of the Congressional leadership, recognized party leaders (former presidential and vice presidential candidates, for example), and other party officials and activists; yet its formation was stymied by Sam Rayburn, who perceived national party agenda-setting as an infringement on his leadership territory in the House. Thus, Party Chairman Paul Butler assembled a council with northern-western, liberal members that worked with the liberal activist blocs in Congress to formulate a national party agenda and promote liberal legislative measures— in direct contrast to both the Republican White House *and* Democratic Party leaders in Congress.[14] The DAC also came to represent those party members who were not in Congress but who had interests in national party and legislative issues; further, it produced a series of reports and pamphlets on a wide variety of national problems and questions that emphasized the party's policy perspectives. By June 1960, those various publications numbered well over sixty.[15]

While much of the DAC's work focused on the domestic agenda (controversial legislative issues such as civil rights, unemployment, and education), it tackled national defense and American international power and prestige, in light of the Sputnik launching and ensuing national debate in the late 1950s about its implications. Arguing that Eisenhower's national security policy had not protected the national interest, as Sputnik supposedly symbolized, the DAC enlisted former Secretary of State Dean Acheson and Paul Nitze, whose role in NSC-68 and the Gaither Committee was well known, to write a party pamphlet on national defense. That publication, "The Military Forces We Need and How to Get Them," was released on the anniversary of Pearl Harbor in 1959 and would serve as a blueprint for the 1960 party platform position on national security. It accused the Eisenhower administration of "first, failing to take adequate precaution to ensure the invulnerability of American strategic weapons; second, for treating the tactical atomic weapons as a cheap substitute for strong conventional forces; and third, for failing to build strong mobile forces for brush fire wars." Moreover, the Acheson-Nitze pamphlet estimated that $7.3 billion was needed to correct U.S. national security policy so that all American vital interests could be protected.[16] As Richard Aliano has noted, the pamphlet clearly stated that Eisenhower defense policies had put American national security in grave danger; the alleged missile gap was its prime target. "Charging that the administration was pursuing a "second-best" defense policy which would give the Soviets a 3 to 1 ICBM advantage until well into 1963 and preclude the possibility of the United States fighting limited wars, the DAC called for the repudiation of a party which believe[d] money to be more important than the military security of our country."[17] The Democratic Party had articulated clearly its position on national security and set a definitive tone for the 1960 presidential election. Moreover, Sputnik remained as the symbol for Party arguments.

One might argue that the DAC pamphlet came rather late in the ongoing national debate about the missile gap and the most effective national security policy. After all, Lyndon Johnson's hearings on military preparedness in the 85th and 86th Congresses (1957–59) had already addressed the issue. Further, Democratic Senators, including John Kennedy and Stuart Symington (both of whom would become presidential candidates in 1960 along with Johnson) had begun to criticize openly administration defense policies in Senate debates, particularly on defense budgeting matters. Yet, the DAC's role was to present the *national party* position for those non-Congressional Democrats who were attracted to the council's agenda precisely because it was not controlled by the party's Congressional leadership. The ongoing mission of the DAC had emerged from "a conviction that the national party need[ed] to articulate a policy and a program that [could] form some basis for future campaigns and for uniting the party around it."[18] This was the exact purpose of the DAC's defense pamphlet. It was imperative that the Democrats articulate a strong stance on national defense in Sputnik's aftermath if they wanted to challenge the Republicans' leadership on national security. As far as the timing of the statement was concerned, the chosen day (Pearl Harbor Day 1959) and its proximity to the 1960 election brought the party prominent media attention and had the ultimate unifying effect.[19] In fact, the national party's pronouncement became the centripetal force for both its broad national membership and the individual candidacies that emerged in early 1960. In many respects, Sputnik clearly could not have been launched at a more opportune time in terms of Democratic Party objectives.

The Congressional Democrats
The Democratic Advisory Council was not the only vehicle in the late 1950s for Democratic attacks on Eisenhower defense policies, debate about the missile gap myth, and the protection of American interests. As already suggested, Democratic members of Congress continuously vocalized dismay with the Republican approach to national security, particularly after the Sputnik launching. These attacks, which occurred within the confines of Congress' appropriations and oversight authority as well as in the Senate military preparedness subcommittee hearings from late 1957 to mid-1958, also precipitated an intense debate in the national media about national defense, the alleged missile gap, and American power and prestige in the international system.[20] Not surprisingly, the defense budgeting process in the 86th Congress (1959–1960) also involved these themes, particularly the possibility of a missile gap; in fact, the Democrats' questioning would become quite fervent.

The January 16, 1959 *Congressional Quarterly Weekly* reported that Stuart Symington had criticized openly and harshly a Richard Nixon assertion that the United States was ahead of the Soviet Union in ballistic missile development and was catching up rapidly in other phases of the space program (Nixon had reiterated the Administration's position concerning the alleged missile gap in an interview with some newsmen). Symington "told the Senate that if Nixon had made such a statement, it is not correct, and I do not know a single impartial expert in the missile field who could support it." Symington further argued that "there seems to be a continuing effort on the part of high officials in this Administration to lull the people into a state of complacency not justified by the facts."[21] In fact, Symington's assessment was just the beginning of the seemingly relentless criticism to be inflicted by the Democrats.

It continued during a January 29, 1959 Joint Senate Preparedness Subcommittee and Senate Aeronautical and Space Committee hearing, in which Lyndon Johnson argued for a clearer understanding of the United States' capability in relation to Soviet ICBM development. Johnson pointed out that in a secret background briefing earlier that month, Defense Secretary Neil McElroy had allegedly reported a three to one Soviet advantage by 1961–1962. Yet, in his defense budget testimony before the Johnson committee, McElroy maintained that the Soviets were not ahead but could conceivably catch up and pass the United States if their ICBM production program was implemented at full capacity. Johnson was naturally incensed with McElroy's, and thus the administration's, apparent contradiction and lack of clarity.[22] It is not surprising, therefore, that the tension between the Republican executive and the Democratic legislature continued throughout these hearings and in the FY 1960 budget debate in 1959. Partisanship emerged early in the process because of the controversial nature of the issues, and the public outcry after Sputnik compelled Congress to attack the White House.

The FY 1961 budget process in 1960 ignited even more profound criticism about Republican defense policies than in 1959, precisely because of confusing and often conflicting testimony on the part of Administration witnesses. Moreover, the hysteria that followed the Sputnik launching had further hampered the extraordinarily difficult task of determining accurately the Soviets' defense capabilities. Yet considering the nature of election year politics, controversy could hardly have been unexpected. The February 5, 1960 *Congressional Quarterly Weekly* reported that "Democrats, digging diligently for soft spots in the Eisenhower Administration's defense and space programs, maintained a drumfire of criticism as a series of Congressional inquiries continued for a third week."[23] The missile gap allegations, the deterrent power of existing U.S. forces, and the pace of the American space program were vehicles for the Congressional Democrats' attacks. Further, Sputnik was a perfect symbol for their arguments.

The testimony of Secretary of Defense Thomas Gates before the House Appropriations Subcommittee on January 13, in which he maintained that the missile gap would be smaller than expected over the succeeding two to three years if one based that estimate on Soviet intentions and capabilities, precipitated a terse reply from Johnson. "[To] rely upon hunches concerning the thoughts that skip through the Kremlin minds is incredibly dangerous," argued Johnson on January 23. In an attempt to defend himself, Gates told the House Science and Astronautic Committee on January 25 that "our intelligence information has improved" and that "we have never been relying on what their intentions will be in reference to specific actions." He continued: "If our best estimates prove wrong and the Soviet Union builds far more (missiles) than we expect, there will still be no 'deterrent gap'. Our total defense will still give a margin of safety." As *Congressional Quarterly Weekly* suggested, Gates received further support from the President, who maintained at a January 26 press conference that Gates had been totally misinterpreted on Capitol Hill.[24]

Gates' retort and Eisenhower's defense did little, however, to quell the debate in Congress, particularly because its members continued to receive contradictory information—from other executive branch officials, for that matter. A January 26 Senate Armed Services Committee meeting, at which Air Force Secretary Dudley C. Sharp and Air Force Chief of Staff Thomas D. White testified, prompted Chairman Richard Russell to comment how "woefully behind [we are] in this missile program." The following day Russell's fellow committee member, Stuart Symington, released a 2000 word statement accusing the Administration of balancing the budget at the expense of American defense. The February 5 edition of *Congressional Quarterly Weekly* also reported Senator Henry Jackson as saying that CIA Director Allen Dulles had testified before the Joint Senate Preparedness Subcommittee and Aeronautical and Space Sciences Committee that the Soviets had both a qualitative and quantitative lead in missile development. Furthermore, Air Force General Thomas Powers, Commander of SAC, apparently had painted an even grimmer picture by testifying that the Soviets would soon have such a lead that SAC would be destroyed with one single blow. Once again the White House tried to control the damage, realizing the *political* potency of the alleged missile gap: on February 1, Secretary Gates told a Senate Defense Appropriations Subcommittee that Powers was wrong; and, during a February 3 press conference, the President accused Powers of being "parochial." Lyndon Johnson did call Powers back to the Senate on February 2, but Powers continued to stand behind his earlier assertion.[25] The administration's efforts were to no avail; they were continuously undermined by its own membership.

The second week of February was an even more volatile one on Capitol Hill, and the partisan rhetoric on this issue was exceedingly dramatic. During a February 5 House Space Committee meeting, Pennsylvania Republican James G. Fulton charged that "the missile issue had been made a political football by Democratic presidential aspirants," a charge that *Congressional Quarterly Weekly* mentioned was aimed at both Johnson and Symington. Fulton's accusation evoked an ardent rejoinder from House Majority Leader John McCormack: "Anytime we Democrats don't agree with the Administration on defense we're accused of talking politics. I think you're getting on dangerous ground when you impugn the motives of anyone who questions defense policies." McCormack was implying that national security was as much the responsibility of Congress as it was of the White House. Symington also continued to push the Administration in a February 8 statement. He reported "that the [CIA] had estimated the Soviets would have a greater, not smaller, edge in long-range missiles over the next two years than had been estimated in 1959," and "threatened to reveal the true percentage figures unless the Administration admitted the outlook had worsened."[26]

Symington's statement sparked a counterattack by the Republican National Committee on February 9. In its publication, "Battle Lines," the RNC claimed that Symington would be committing "an act of total reckless irresponsibility" by releasing any such data, and that "it [was] the responsibility of the Democratic leadership to see that no such information reache[d] his hands." Yet, as *Congressional Quarterly Weekly* reported, Symington apparently had gained his inside information independently and directly from the Air Force, not from Senate hearings. As a result, there was little the Democratic leadership, or anyone else, could do to control what Symington actually did with that data.[27] There is no evidence, moreover, that the leadership was ever truly upset with Symington's tactics anyway, no matter what his political motives were at that point. Congressional Democrats were arguing consistently that it was their responsibility to address the public's national security and defense concerns, which had been ignited by the launching of Sputnik.

Both sides of this partisan debate were frustrated further when the Republicans' party leader, President Eisenhower, declared in a February 11 press conference that he refused to participate in this partisan battle; he cited his unrivaled military service and impending retirement from public office as justifications for staying above the fray. In fact, he argued that the Administration's was position clear, and maintained that if the Democrats did not respect his record and current position as ample qualifications for following his leadership on this issue (that there was no missile gap), then that essentially was their problem. He, however, was not going to be swayed from his stance. Although Eisenhower's inability and unwillingness to substantiate the administration's position more specifically essentially cost the Republicans their biggest, most qualified, and most reputable advocate, he had couched his refusal to participate in terms that were nearly impossible for the Democrats to surmount.[28] Nonetheless, the debate had lost the one person who could have definitively laid to rest the missile gap allegations and broader national security concerns, and because of that loss, the partisan rhetoric and accusations continued for yet another week.

Stuart Symington declared publicly on February 14 that the Administration had deliberately misled the American public on national defense issues and how best to protect American interests; this, in turn, raised the ire of the supposedly apolitical President, who charged in a February 17 press conference that Symington's accusation was "despicable." Yet Symington repeated the same allegation on the Senate floor on February 19. On February 21, newly announced Democratic presidential candidate John Kennedy entered the rhetorical debate, but tried to find a balance between Symington's harshness and stridency and Eisenhower's ostensible non-involvement. According to the *Congressional Quarterly Weekly*, Kennedy maintained "that the President's reports were made in good faith but that it was difficult to make accurate estimates of Soviet strength." He also contended that "in these dangerous times we should err on the side of safety" and mentioned that he personally advocated a "greater effort than this Administration seems willing to undertake."[29] Kennedy was yet another Democrat who was publicly pressing a Republican administration on an issue that, because of national security concerns, it could not adequately defend. Those same national security considerations—the symbol of which always was Sputnik— necessarily caused it to be a powerful political issue for the American electorate and, in turn, a useful rhetorical tool for the Democratic party in its attempt to challenge successfully the incumbent Republican White House.

What is particularly interesting about the missile gap controversy is that it was used by the Democrats when it could garner the most public attention. When Congress was heavily involved in investigating military preparedness, particularly for budgeting purposes, the issue was at the forefront of the debate in the media. Cold War tensions naturally made it a marketable issue for the media as well. Once the Congressional hearings were completed in March 1960, however, the missile gap issue, as part of the larger debate about American power and prestige in the international system, died down until the presidential campaign heated up in the summer and fall of 1960, another period when the American public and the national media would indeed focus on national issues.

What is even more surprising is that despite the volatile rhetoric on Capitol Hill (not only in 1960, but from Sputnik forward) Congress never added drastically to the defense budgets during this period, although there were moderate funding increases in the Administration's requests.[30] The lack of major increases leaves one questioning how committed the Democrats actually were to correcting the alleged problem that had them in such a frenzy, although correcting the problem in the budgeting process certainly would have been much more complicated than simply articulating the problem, especially considering Eisenhower's profound resistance to major budget increases. It is hard to deny that the Democrats used Sputnik and the alleged missile gap for the greatest political gains, at the most appropriate times, in a far bigger political game. That game would undoubtedly involve other equally important domestic questions, such as education, civil rights, unemployment, and health care. Yet defense issues could be mixed in as well so that a well-rounded national party agenda could be shaped and, ultimately, a national challenge be mounted to defeat the Republicans. The evidence provided here certainly suggests that pure politics played at least a partial role in the Congressional Democrats' approach to the missile gap controversy that occurred after Sputnik's launching, even if that issue was most often addressed within the confines of the appropriations and oversight authority that is vested in the institution by the Constitution. The end result was that the Congressional Democrats were in line with national party headquarters, an interesting standpoint when one remembers the Congressional leadership position in 1957 concerning the national party's role in defining the party's political agenda, particularly on American national security.

THE INTELLIGENCE DATA PUZZLE

It is eminently fair to question and criticize the Democratic Party, the various military sectors (particularly the Air Force), and Democratic members of Congress (especially Johnson, Symington, and Kennedy), for pursuing an issue for parochial, political gains. Yet, they had a powerful visual symbol—Sputnik—with which to contend, and public fears that demanded attention. Furthermore, their assertions concerning the missile gap and the broader question about American power and prestige were based on information that was, in many respects, very much beyond their control—that is, the confusing and contradictory intelligence data that dominated this period. Inaccurate and ill-defined intelligence on operational Soviet missile systems and Soviet production capability and progress, often articulated by various sectors of the executive branch itself, enabled the Democratic Party and Congressional Democrats to question legitimately the Eisenhower administration's national defense strategy. Even when the Administration began to downsize previous estimates, which in turn provide a more positive view of the United States' position vis-à-vis the Soviet Union, the multiplicity of data essentially allowed various actors to pick and choose among data depending upon what they believed, or wanted to believe, was true.

The seeds for this intelligence data puzzle were planted during Defense Secretary McElroy's testimony before Lyndon Johnson's post-Sputnik military preparedness subcommittee in late 1957 and early 1958. Despite the profound visual image that loomed over the hearings, McElroy presented a strong, sound overall picture of American defenses at that time and clear evidence of American superiority in long-range bombers, the chief strategic weapons system of the day.[31] Yet when McElroy was questioned in subsequent hearings about the specifics of the missile race, he was unwilling to concede American superiority. Instead, he maintained that he had no position concerning United States' missile development relative to the Soviet Union's, but that the United States "must accelerate our programs in order to stay ahead if we are ahead, and to get ahead if we are not ahead."[32] McElroy seemed to suggest that intelligence was not providing a clear answer on this particular issue, and it is hardly surprising that Lyndon Johnson and Stuart Symington took this opportunity to initiate a debate about American-Soviet comparative military strength, which also incorporated the missile gap question and a intense numbers game. Furthermore, critics always had the image of Sputnik to underpin their arguments. It is somewhat ironic that the Republican administration provided that opening itself and would, in turn, be forced to spend the succeeding two years trying to close the debate.

Newspaper columnist Joseph Alsop brought the comparative military strength question into the public realm with a series of articles in the summer after Sputnik, 1958. In an August article, Alsop predicted that the Soviets would hold a 2000 to 130 advantage in ICBMS by 1963, numbers which were believed to be quite close to the National Intelligence Estimate (NIE) at that time. Alsop's article precipitated intense rhetoric in the Senate, including a strong floor speech by Senator John Kennedy on August 14, 1958.[33] But the picture would become even more confusing in early 1959, and it was Administration actors whom precipitated the confusion.

Secretary McElroy's January 1959 secret background briefing for the Senate, in which he supposedly predicted a three to one Soviet ICBM advantage by the early 1960s, was apparently based on early 1959 CIA estimates. Those estimates, which were much lower than the 1958 NIE, suggested that the ICBM score by 1961 would still favor the Soviets: USSR-100 to 300, U.S.-80 to 100; by 1962: USSR-500, U.S.-100 to 300. What complicated intelligence matters further in 1959 was that two NIEs were prepared under two different premises: one based on an "orderly" Soviet ICBM production program (a low prediction); another based on the Soviets pursuing a "crash" program (a high prediction).[34] Later in 1964, then-Defense Secretary Robert McNamara confirmed that in 1959 the "orderly" prediction for Soviet ICBMs was 350 and the "crash" prediction was 640; those projections were for mid-1963.[35] The primary problem with two official sets of estimates was that it seemed to imply that both were equally legitimate and that a policy maker, in turn, could justifiably use either one of them. This twist naturally added to the existing confusion in the intelligence puzzle.

In his scholarly analysis of the Kennedy administration's strategic missile program, Desmond Ball has argued that despite the grave intelligence projections in 1959, most observers at that time still concluded that American national security was sound. The real concern for many analysts, according to Ball, was the security of the Strategic Air Command and the fact that even limited Soviet ICBM production, combined with limited American production, could quickly negate SAC's effectiveness in the case of a Soviet surprise attack:

> Neither side as yet [in early 1959] had an operational ICBM, but the United States had about 1,800 long-range and medium-range nuclear-armed bombers stationed within range of Russia, while Russia had only 150 long-range bombers capable of reaching the United States. These bombers would, however, be vulnerable to a surprise attack. At this time [SAC] had only about 44 major bases, with 29 overseas, and since it was assumed that it would take two to six Soviet ICBMs to destroy the effectiveness of an air base, it looked as though SAC could be negated by a surprise Soviet missile attack in the period of maximum danger, 1962–63.[36]

Even the most conservative estimates gave the Soviets a quantitative advantage, which reinforced the severe concern that the Gaither

Committee Report and the Rockefeller Brothers Report had previously expressed about SAC's vulnerability.[37] Again, the real concern for the Administration's critics—and especially the Air Force, which controlled SAC—was not the fact that the Soviets would actually have ICBMs in their arsenal; rather, it was the damage those ICBMs could do to the American counterforce capability if the comparative ratio was in the Soviet's favor and if action was not taken to protect that capability more effectively. One can, therefore, understand why intelligence data, which were the only real predictors of Soviet capability, were such a crucial and volatile aspect in this debate.

In mid- to late 1959, unconfirmed reports were apparently implying difficulties in the Soviet ICBM testing and development programs; moreover, the Soviet economy was supposedly struggling under the economic demands of its missile program.[38] In January 1960, new Secretary of Defense Thomas Gates subsequently testified before the House Appropriations Committee that the Soviets had undertaken an "orderly" production program, and he indicated that earlier projections of even this "low" estimate had been too high. He, in turn, reduced the low estimate for mid-1963 Soviet ICBMs, and the new NIE released in February 1960 (it was actually the annual NIE for 1959) apparently reflected this downward revision.[39]

What is particularly interesting about the February 1960 NIE was that for the first time intelligence concerning Soviet ICBMs was broken down into two categories: ICBMs for inventory, which would come to mean "capabilities"; and, ICBMs on launchers, which would come to mean "intentions." Gates' testimony was based on the "on launchers" category, as was Joint Chiefs of Staff Chairman Nathan Twining's January 13, 1960 testimony before the same committee. In a memorandum to President Kennedy in March 1963, Robert McNamara explained the reasoning behind this important modification. "Change to include 'on launcher' data was based on the belief of the intelligence community that by early 1960 the Soviets had acquired an initial operational capability [launchers] and that the development program was a useful estimative target," he said. Furthermore, it was an important recognition "that the construction of operational launchers, rather than the buildup of missile inventories, was the pacesetting factor in any deployment program, as well as the best measure of salvo capability."[40] While this alteration in the reporting of data was designed to clarify further the comparative missile strength question, it actually precipitated the exact opposite.

As discussed above, Lyndon Johnson's reaction to Gates' testimony was that it was dangerous and incredibly difficult to try to decipher the intentions of the Kremlin, and other members of Congress, as well as critics outside of the government, thought it was safer to think in terms of theoretical capabilities rather than intentions. Yet, even more significantly, the presentation of two data categories had a similar effect as the preparation of the high and low estimates in early 1959. It again seemed to imply that one could choose between the two, and the "renewed intensive controversy ... was described in the press as the 'missile gap' " and the "new method of working intelligence estimates—intentions VERSUS [emphasis added] capabilities."[41] Even though the Chief of Naval Operations, Admiral Arleigh Burke, tried to clarify the Administration's position before the joint hearings of the Senate Armed Services and Aeronautical Sciences Committees, saying that intelligence data "were based on Soviet missile production rather than on the maximum capacity of the Soviet Union to produce missiles," his efforts had little effect.[42]

Why, even after two well-respected, high-level administration officials presented a much less threatening view of Soviet missile strength that was seemingly evaluated on more legitimate intelligence data, did the Administration's critics remain unconvinced, and, as a result, continue to promote the alleged missile gap? Many scholars place most of the blame on the intelligence situation itself and on those who were responsible for collecting, interpreting, and disseminating that information.[43] There clearly was a plethora of data and little consensus on what it meant. Within the Eisenhower administration in 1960, the Air Force and SAC officials refused to accept the new intelligence figures and continuously stressed the strategic vulnerability of the American retaliatory forces. In a January 1960 speech in New York, SAC's commanding officer, General Thomas S. Power, "claimed that the 100 U.S. nuclear launching bases [in the United States and Europe] ... could be virtually destroyed by a force of only 300 ballistic missiles [IRBMs (intermediate range ballistic missile) and ICBMs] ... and that the Soviet Union could accumulate this number before the United States had developed an adequate warning system against missile attacks."[44]

In a May 31, 1963 memorandum to the then-Assistant Secretary of Defense for International Security Affairs, Paul Nitze, Defense Special Assistant Lawrence McQuade maintained that despite the fact that the NIE released in February 1960 reduced the estimates of actual Soviet missiles on launchers, there was still little room for comfort. Those estimates, said McQuade, "left open the possibility of an effective Soviet missile attack destroying our vulnerable SAC bases, particularly since we believed that improvements in the accuracy, reliability and CEP [Circular Error Probability] of Soviet ICBMs had sharply reduced the number required to attack our target system effectively."[45] McQuade argued further that another NIE released in August 1960 clearly indicated that "the judgements of the intelligence community on the Soviet ICBM capability were still based on insufficient direct evidence." It is therefore perfectly understandable why the intelligence picture remained ambiguous.

There continued to be, moreover, a range of estimates about when the Soviets would have the ICBM capability for destroying SAC, each of which garnered the advocacy of various sectors of the executive branch. The worst case scenario (called Program B), which "was adjudged to provide the Soviet Union with high assurance of being able to damage severely most of the SAC operational bases in an initial salvo by about mid-1961," was supported by the Air Force. Program A predicted that damage point to be late 1961, a judgement advanced by the CIA. Between Program A and B, but on the high side closer to Program B, stood the State Department, the Defense Department, and the Joint Chiefs of Staff. The best case scenario—Program C—estimated mid-1963 as the critical point, a position maintained by the Army and Navy. Even though the intelligence community had judged in both the "intentions" and "capabilities" estimates that the Soviets probably had not undertaken a crash program and that Soviet motives were probably based on the deterrent value of ICBMs, the fact that the community "did not have [precise] evidence of Soviet plans for production and operational deployment of ICBMs" helped to bolster the missile gap myth until the early 1960s.[46] The fact that the intelligence community and different governmental agencies provided ranges of estimates only helped to exacerbate an already murky picture.

According to McQuade, it was not until mid-1961, the middle of what had been thought of in 1957 as the "critical period," that more precise information concerning Soviet ICBM production and operational deployment, combined with important information on Soviet IRBM and MRBM (medium range ballistic missile) programs, became available. "Though we were still uncertain about the number of Soviet ICBMs [in the NIE released in June 1961], it was clear (a) that the Soviets had not made the choices and taken the actions since 1957 which would have produced for them the best possible strategic relationship vis-à-vis the United States for the critical period, and (b) ... the U.S. retaliatory forces had achieved a greater degree of survivability than it seemed to expect in 1957.[47] What made the problem even more difficult was that several administration officials—specifically Eisenhower and Nixon— seemed to base their statements on sources other than those outlined above.[48] In light of this incredibly confusing intelligence picture, it is hardly surprising that the missile gap persisted as a legitimate national security concern (legitimate, that is, in the eyes of the American public) until the early 1960s.

Probably the only means for effectively countering the missile gap proponents and alleviating the public fears, which had been initiated by Sputnik's launching, would have been to release the sensitive data that the Eisenhower administration had collected from the U-2 reconnaissance overflights in the late 1950s and 1960. In his memoirs of the White House years, President Eisenhower emphasized the value of the U-2 program, particularly with regard to the missile gap issue:

> During the four years of its operation, the U-2 program produced intelligence of critical importance to the United States. Perhaps as important as the positive information—what the Soviets DID—was the negative information it produced—what the Soviets DID NOT. Intelligence gained from this source provided proof that the horrors of the alleged "bomber gap" and the later "missile gap" were nothing more than imaginative creations of irresponsibility. U-2 information deprived Khrushchev of the most powerful weapon of the Communist conspiracy—international blackmail—usable only as long as the Soviets could exploit the ignorance and resulting fears of the free world.[49]

While Eisenhower's memoirs may be looked upon as a grand defense of his administration's actions, others have also confirmed the vital nature of U-2 intelligence in shaping Eisenhower's official position on the missile gap controversy.[50]

Yet the success of the U-2 program necessarily demanded secrecy; there was no way that U-2 data could be accurately reported to the public in any sort of specific detail because of the very nature of the program. In a rather ironic twist, however, it was in the very name of national security that the Republican administration essentially found itself incapable of convincing the rest of the American government, as well as the American public, that American national security was safely intact. In turn, the Administration's opponents were able to make effective use of the other varied intelligence data to uphold the alleged missile gap as a political issue and to question the Administration's handling of American national security. Even after the U-2 flights became public knowledge in May 1960, when the Soviet Union shot down Gary Powers' plane, the partisan politics of a general election season essentially allowed the missile gap controversy to remain in the forefront because, again, specificity was necessarily ruled out in the name of national security.[51]

AMERICAN NATIONAL SECURITY & LEADERSHIP, AND THE 1960 CAMPAIGN

The Democratic Party and Its 1960 Platform

No scholar would suggest that the Democrats and John Kennedy won the 1960 presidential election because of their position on national security and defense. Yet one can safely argue that by keeping the national security and defense issues in the public arena, particularly in terms of the decline of American power, prestige, and leadership internationally as symbolized by Sputnik and the missile gap, Kennedy and the Democrats also raised and underlined existing doubts about Richard Nixon's ability to protect American national interests. Actually, the Democratic Party (both the national committee and its leadership in Congress) was well positioned to address national security and defense during the 1960 presidential campaign. Because of the public's keen sensitivity after Sputnik to national security issues, the Congressional debates on national defense budgets, military preparedness, and the space program offered fertile ground for media attention, which fostered public concern in these issues. Furthermore, the Democratic Advisory Council's publication of position papers and pamphlets had, by early 1960, helped the party to "propose a preliminary draft of the party platform to the National Convention [and] interpret the platform in relation to current problems," and this was especially true in the case of national security and defense policy.[52] By February 1960, Platform Chairman Chester Bowles was finalizing the party's platform based on Advisory Council documents and testimony gathered at pre-convention hearings, which had been conducted around the country.[53] In fact, the first substantive item in the Democratic Party's 1960 platform was national security and defense policy.

The foundation for the Democrats' platform position on national security and defense was that the United States had lost its superiority in defense vis-à-vis its chief adversary—the Soviet Union; in turn, it had lost the respect of the international community as well. The alleged missile gap, about which debate intensified following Sputnik (the primary symbol of that fall from dominance), was blamed directly on the apathetic Republican policies of the mid- and late 1950s. The beginning of the Democrats' statement on national defense was, indeed, dramatic, urgent, and far-reaching in its tone:

> When the Democratic Administration left office in 1953, the United States was the pre-eminent power in the world. Most free nations had confidence in our will and our ability to carry out our commitments to the common defense. The Republican Administration has lost that position of pre-eminence. Over the past 7 1/2 years, our military power has steadily declined relative to that of the Russians and the Chinese and their satellites.
>
> This is not a partisan election-year charge [however]. It has been persistently made by high officials of the Republican Administration itself. Before Congressional committees they have testified that the Communists will have a dangerous lead in intercontinental missiles through 1963 [the missile gap]—and that the Republican Administration has no plans to catch up.[54]

Furthermore, the Democrats argued that the United States was losing the race in space research and in limited, conventional war tactics; the Republicans had even admitted their losses but, nonetheless, seemed unwilling to do anything about it.[55] Clearly, the lessons of Korea, Indochina, Hungary, the Suez, Berlin, and Laos, as well as the implications of Sputnik, had not gone unnoticed by the Democrats. By nominating John F. Kennedy as their presidential candidate in July 1960, the Democrats chose a candidate who not only was willing to endorse the Party's platform, but one who had continuously demonstrated his affinity with the Party's national security and defense position.

What Kennedy and the Democrats advocated instead was the maintenance of both strategic nuclear superiority—to ensure a credible retaliatory deterrent—and a strong conventional force capability. This new approach, which was in direct contrast to the Eisenhower-Republican massive retaliation strategy, was better know as "flexible response"; it involved increasing both the conventional and nuclear means needed to respond to any form of aggression at the appropriate level, in any location, and at any time, while raising the nuclear threshold as well. The flexible response strategy is most often attributed to General Maxwell D. Taylor, who would become President Kennedy's chief military advisor and Chairman of the Joint Chiefs of Staff. In his 1960 book, *The Uncertain Trumpet*, Taylor articulated the importance of the flexible response strategy as follows:

> The name suggests the need for a capability to react across the spectrum of possible challenges, for coping with anything from general atomic war to infiltrations and aggressions such as Laos and Berlin in 1959. The new strategy would recognize that it is just as necessary to deter or win quickly a limited war as to deter a general war. Otherwise, the limited war which we cannot win quickly may result in our piecemeal attrition or involvement in an expanding conflict which may grow into a general war we cannot avoid.[56]

By maintaining a large and strong conventional force, the United States would not have to rely on nuclear force to deter non-nuclear aggression. Yet a Soviet nuclear first-strike would also be deterred by more reliable and invulnerable American missiles; furthermore, the missile gap would be closed by proper, cost-effective defense budgeting and accelerated research and development of state-of-the-art nuclear weapons and warning systems.[57] The most fundamental value of the strategy was that it would finally free the United States from the potential embarrassment of massive retaliation's "all-or-nothing" approach, and this was, in and of itself, a good enough justification for a flexible response. In the 1960 campaign, Kennedy and the Democrats would promote this vision of national defense because of these theoretical benefits for national security and American power and prestige in the international system. It is hardly surprising that practical, and more difficult, questions of implementation would be avoided until after the election.

The Missile Gap, American Power, Prestige, and Leadership, and the 1960 Campaign

The alleged missile gap was the issue of the hour in the early part of 1960 as Congress prepared the FY 1961 defense budget. On the campaign trail, however, the issue tended to fall within the context of American power and prestige, which Candidate Kennedy consistently argued was on the decline because of eight years of Republican policies and leadership. The downing of the U-2 reconnaissance plane and the subsequent failure of the Eisenhower-Khrushchev Paris summit in May 1960 forced questions about American power, prestige, and leadership to the front pages. While Richard Nixon actually moved up in the polls immediately after these two events (49 percent of people polled chose Nixon to represent the United States in future summits compared with 37 percent who chose Kennedy),[58] Kennedy quickly went on the offensive. In mid-June, he gave a major foreign policy speech on the Senate floor. He charged the Eisenhower administration with woefully inadequate and weak foreign policy, and, in turn, presented a twelve-point program (which included a major defense appropriations increase) to address the Soviet challenge and to demonstrate his leadership ability in international affairs. It is hardly surprising that in his June 15 New York Times column James Reston suggested that Kennedy was preparing himself for his campaign against Nixon, who naturally would base his run for the presidency on his foreign policy expertise and leadership capability.[59] Clearly, the tenor of the impending presidential campaign was already being set.

The Democrats were assisted further in bringing attention to the power and prestige question by New York Governor Nelson Rockefeller, who, in his brief bid to unseat Richard Nixon for the Republican nomination, urged the Republicans to assert a more strident position on national security and defense in their platform. This would, in turn, foster a perception that the Republicans were strengthening the power and prestige of the nation as well, and, thus, make them less vulnerable to Democratic criticism in this sphere.[60] Rockefeller succeeded in pressuring Nixon and the Republicans to include platform language that emphasized the further development of an invulnerable second-strike capability, a commitment to modern, well-protected strategic missiles, and the production of "highly mobile and versatile forces ... to deter or check local aggression and brush-fire wars."[61] In this particular instance, some of the Republicans' platform language was very similar to that of the Democrats.

Yet, the same Republican platform continued to endorse Eisenhower's "Long Pull," which necessarily demanded fiscal conservatism and the emphasis on the deterrent value of nuclear weapons; thus, the platform was ultimately promoting the type of leadership that had been provided during the previous eight years. Moreover, the Republicans' section on national defense followed the preamble and a lengthy section on foreign policy, both of which stressed the power of the United States and the success of Republican leadership under the direction of Eisenhower and Nixon.[62] Nonetheless, the Republicans would face criticism on foreign policy because of perceived failures: Indochina, Cuba, the Middle East, Hungary, and especially national defense strategy because of the powerful symbolism of Sputnik and the alleged missile gap. Added to those perceptions were strong arguments against Republican leadership on the domestic front. In essence, by setting up American power, prestige, and leadership as the unifying themes of their 1960 campaign, the Republicans opened a Pandora's box that they were incapable of closing. Moreover, the missile gap and questions about American power, prestige, and leadership were all part of a larger, implied question: Who should be responsible for protecting American interests? It was this question that essentially underlay the entire 1960 presidential campaign.

Beginning with his acceptance speech at the Democratic Convention, John Kennedy emphasized the need for new leadership to restore American power and prestige; only then would the United States be able to meet the challenges of what he called the "New Frontier" of the 1960s. Once Congress adjourned in early September and the presidential campaign began in earnest, the press was constantly filled with language that stressed national security, the need to meet the Communist challenge, and the restoration of American strength and prestige to ensure world peace. Moreover, the Republican administration came under heavy Democratic criticism for allegedly increasing the possibility of war, and Nixon's leadership skills were consistently questioned. The attention that had been given to the power and prestige issue was so prominent that on September 21 Nixon asked that a voluntary moratorium on the issue be enacted so that Khrushchev would not perceive weakness and division on the part of the United States. Nixon suggested even further that it was the *duty* of all of the candidates to support Eisenhower's efforts for peace.[63]

Not only did Kennedy and the Democrats reject the moratorium, but Kennedy continued to attack Nixon's foreign policy experience and suggested that the United States had to do more than just *react* to Communist action.[64] Polls in late September gave Nixon a slight lead (47 percent to 46 percent) based on foreign policy experience; Khrushchev's presence in the United States had certainly elevated foreign affairs and national leadership as paramount issues at this point.[65] Yet Kennedy's performance in the first televised debate on September 26—one primarily dominated by domestic, not foreign, issues—seemed to reassure voters that he was capable of national leadership and, thus, protecting the national interest. Robert Divine, in his study on foreign policy and presidential elections, neatly summarized the impact of Kennedy's performance in that first debate:

> Still relatively unknown and unproven, [Kennedy] had displayed a remarkable degree of maturity, remaining calm and unruffled as he rattled off answer after answer with machine-like rapidity. "Kennedy was alert, aggressive and cool," summed up *Time*. Viewers realized that he was not the green immature challenger of the GOP stereotype, but rather a gifted man with remarkable poise and polish. "Kennedy did not show that he was Nixon's master," *Newsweek* acknowledged grudgingly, "but he did show that he was Nixon's match" ... . Above all the debate seemed to remove any doubt of Kennedy's ability to perform effectively under pressure.[66]

Because of Kennedy's seemingly skillful handling of Nixon, who had been touting his own ability to handle Khrushchev, Kennedy was also viewed as being capable of managing Khrushchev—or any other international leader or question of national security, for that matter.[67] While it is generally acknowledged that Nixon's performance improved dramatically during the three succeeding debates, specifically on foreign policy questions in the second and third debates, nothing he did seemed to damage the leadership image that Kennedy had attained because of the first episode.[68]

The power and prestige issue remained in the forefront in October 1960, not only because it was raised in the each of the remaining debates (on October 7, 13, and 21) but because various surveys and polls kept the issue in the media. An October 2 *New York Times* survey indicated that most voters were concerned about American prestige, and in the ensuing weeks information suggested that the Eisenhower administration was withholding reports that demonstrated a decline in American prestige. On October 20, Senator William Fulbright publicly accused the administration of suppressing unfavorable United States Information Agency (USIA) data—specifically, a August 29, 1960 USIA report—and argued that Nixon was deliberately misleading the American people by claiming the predominance of American power and prestige. During the final debate on October 21, Kennedy questioned Nixon on the report, and while Nixon maintained—albeit erroneously—that the report in question was for 1957 and that he would be willing to have it released, he also blamed Kennedy for contributing to that alleged decline by making it a prominent campaign theme.[69] This was only the beginning of a rather heated and intense discussion that would continue almost until Election Day.

In the days following the last television debate, Kennedy continued to press Nixon and the Administration publicly for the survey's release. His position was bolstered by an October 25 *New York Times* report that cited a summer 1960 USIA survey indicating an almost unanimous belief in nearly ten non-Communist countries that the Soviet Union was the world's leading military power and that the gap between the United States and the Soviet Union was widening. A CBS News poll reported similar data, which led Senator Fulbright to conclude publicly that the USIA data would be released if the CBS News report was *incorrect*. On October 27, the *New York Times* printed a secret June 1960 USIA report, which was based on data that had been collected in late May, that showed that both the

Americans and Soviets lost prestige in Great Britain and France after the Paris Summit collapse; but this still was not the report that Fulbright had requested and the White House continued to refuse to release. On October 29, *The New York Times* printed yet another confidential USIA report (dated October 10, 1960) on a global survey that demonstrated a world-wide belief that the Soviets were leading in the space race and that the American capacity for world leadership was on the decline.

Finally, on November 2—just six days prior to the election—*The Times* published a section of the August 29 secret USIA report that Fulbright and Kennedy had pressed for and, again, that global survey indicated a continuation and acceleration in the decline of American prestige. This series of reports could only help but add legitimacy to the arguments that the Democrats had been making almost on a daily basis. All the Republicans could do was to try to deflect the reports by questioning their accuracy and by suggesting that the Democrats were being unpatriotic and distorting the image of the United States; they hoped to paint the Democrats as being irresponsible and inexperienced, and, thus, incapable of handling responsibly American interests internationally.[70]

Twice in the final weeks of the campaign, Kennedy specifically raised the missile gap issue to demonstrate the loss of American power and prestige: once to the American Legion convention on October 18, and on November 4 during a major national defense policy speech in Chicago. There was little the Republicans could do, however, except to counterattack with their own rhetoric; yet it was to no avail at that point. Again, the primary problem for the Republicans was that national security constraints restricted their use of concrete data (specifically, the U-2 intelligence) that could have possibly shown that American power was far more stable than the Democrats were arguing. Because the Republicans necessarily were unable to do that, they were also incapable of countering effectively the outcry about the supposed decline in American prestige. After all, if there had been no legitimate questions in the public's mind concerning United States power vis-a-vis the Soviet Union—and unfortunately for the Republicans, the public's distress had never truly been relieved after Sputnik—the arguments concerning American prestige also could not have taken root.

Statistical Reflections on the 1960 Presidential Election
As mentioned above, few experts would argue that national security
and defense issues were the dominant themes of the 1960 presidential
campaign; in fact, it is generally accepted that the election was driven
by domestic concerns. Yet, one of John F. Kennedy's primary tasks
during the campaign was to convince the American public that he
could handle sufficiently the complexity of the international arena in
order to complement the voters' tendency to support the Democrats
on domestic affairs. By doing so, the Democratic ticket could maintain a hold on voters who might otherwise vote for Richard Nixon
because they misperceived Kennedy's youthfulness as necessarily
meaning he was inexperienced in foreign affairs, and thus incapable
of national leadership in either the domestic or international spheres.
A demonstration by Kennedy of at least an average level of competency in foreign affairs—that would then translate into national
leadership skills—was essential for ensuring that the Democrats did
not lose their traditional hold in the domestic sphere, and thus lose
the 1960 election as well.[71]

There are data that suggest that Kennedy and the Democrats made
just enough headway in foreign affairs, as well as on defense-related
issues, to affirm Kennedy's leadership ability and to secure the
Democrats' victory in 1960. The crosstabulation of two 1959
American Institute of Public Opinion (AIPO) polls, less than a year
before the 1960 election, indicated that while 31 percent of the
respondents had more confidence in the Republicans' handling of
major international issues (ie., keeping peace, dealing with the Soviet
Union, foreign policy, external communism), 28 percent sided with the
Democratic Party and 29 percent saw no difference between the two
parties. Even though only a small percentage of respondents in those
same polls identified "National defense preparedness" and "Space,
Sputnik, and missiles" as the most important issues of the day, the
Democrats garnered overwhelming confidence from the voters who did
recognize those particular problems (35 percent to 8 percent, and
50 percent to 21 percent respectively).[72] What is most interesting about
this data is that while the Republicans may have had a slight lead in
foreign affairs, they clearly were not dominating the Democrats in the
arena that is traditionally an asset for the incumbent party—and
should have been for the Republicans in 1960. The public seemingly

had begun to question the quality of Republican leadership in the foreign affairs realm, even prior to the commencement of the 1960 election season; the actual candidates had not even been formally announced at this point, yet the Republican Party was looking vulnerable.

In an October 18, 1960 AIPO poll, conducted just prior to the 1960 presidential election, switches in voters' preferences between 1956 and 1960 were evaluated. That poll indicated that among those voters who identified "national defense, defense preparedness, and the missile gap" as the most important problem (albeit a small group—about 5 percent), 59 percent who had voted for the Republicans in 1956 were switching to the Democrats in 1960. In that particular poll, six primary issues had been identified by voters, and on five of those issues the index of relative pulling power was positive for the Democrats. The issue with the second highest index of relative pulling power to the Democrats—second only to "unemployment"—was national defense policy. Additionally, the issue of "American prestige abroad" was raised by voters for the first time in AIPO polls (similar AIPO polls had been conducted several times throughout the 1950s), and this issue provided a positive pulling power index for the Democrats as well.[73] One cannot help but be reminded that Kennedy had emphasized the question of American power and prestige continuously during the campaign, and the polling data did seem to indicate that his effort was successful.

Even in the one area of the AIPO poll that registered a negative pulling power index for the Democrats—foreign policy issues (the largest category of the six)—25 percent of Republicans in 1956 were still switching to the Democratic ticket in 1960. Furthermore, 58 percent of voters in the foreign policy category who had not voted in 1956, who did not remember how they had voted, or who had voted for an "other" party candidate were also supporting Kennedy-Johnson in 1960. On the other hand, the 1960 Republican ticket was only gaining 8 percent of the Democrats in 1956 who were identified in the foreign policy category (meaning that 92 percent of the Democrats in 1956 were staying with the Democratic ticket in 1960).[74]

This evidence suggests that Nixon-Lodge had not garnered enough support from voters in the foreign affairs category to overcome the losses on other issues, even though the Republicans were still considered to be the predominant party in this particular area. Clearly, Kennedy-Johnson had been able to demonstrate at least an average level of competency in foreign affairs; the public, therefore, seemed willing to accept the Democrats' leadership in the office of the presidency in November 1960.

CONCLUSION

In January 1960, the importance of the public's awareness, as well as the adversary's awareness, of the reality of power was raised during the House subcommittee hearings on the defense appropriations for 1961. As previously discussed, Secretary of Defense Thomas Gates presented a positive overall picture of American military power based on an across-the-board analysis of operational and developmental weapons systems, force structures, and defense management in the United States. Yet, Daniel J. Flood (D-Pennsylvania) pressed Gates as well as his fellow committee members to remember the relationship between perception and reality, particularly in terms of deterrence; in fact, he chided both Gates and the Committee for seemingly neglecting this fact.[75] Flood continued by explaining why he thought the public had a negative perception of American power, prestige, and leadership, and his assessment was amazingly vivid:

I think the reason why there is no longer a public image of supremacy in these matters is because of an attitude here in Washington. There has been a preaching of "balanced forces." You say[:] "Do not get excited. Do not worry about this Soviet thing; we have balanced forces and catching up to the Russians missile for missile is not that important." Well, that may or may not all be true ... . You were completely satisfied that, because you understood [the argument] clearly, had stated it brilliantly, that you were presenting the public image, which you are not presenting. The public has no concept of that ... . This goes back to sputnik. When that sputnik flew around the globe and the desert tribes and the mountain tribes and the coastal tribes—black, white, red, and yellow—all over the world heard about it, all they knew was this was a public image, a public manifestation of the ascendancy and the primacy over America. And that has not been changed up until noon today, in the jungles, in the mountains and on the seacoasts of the world. They still think that ... the mere reality of power in your inventories and arsenals will not do. There must be a public image in the minds of the peoples of the world of that reality of power, and there is no image. Therefore, we are short one of the two legs that we must have.[76]

It was a leg that Richard Nixon and the Republicans desperately needed during the 1960 presidential campaign as well, but one that they were never able to attain.

Certainly, negative public perceptions that began with Sputnik's launching were fed by the seemingly constant Democratic rhetoric that emphasized the decline of American power and prestige and criticized the Republican leadership for allowing the nation to suffer that fate. That rhetoric took on an air of legitimacy because it was supported by statistical data, which could be questioned and countered—but then only by experts, who used equally questionable data to bolster their own cases. The only truly definitive data, which would have had the best chance for resolving the power and prestige debate (and the missile gap myth), was that being collected by the U-2 reconnaissance program. Yet, Nixon and the Republicans were restrained from using that specific evidence publicly because of the national security secrecy demanded by the very nature of the U-2 program. The Democrats put the Republicans on the defensive, and because the American public could not be given the definitive proof about American national security and power, the Democrats were able to question effectively the Eisenhower administration's defense posture; in turn, it became a liability for Richard Nixon and the Republicans during the 1960 presidential campaign. Ultimately, the American public had come to believe that the Republicans had mishandled American national security, and that perception helped cost Nixon the 1960 election.

1. John Lewis Gaddis, *Strategies of Containment: A Critical Appraisal of Postwar American National Security Policy* (New York: Oxford University Press, 1982), p. 147.
2. Richard D. Challener, "The National Security Policy from Truman to Eisenhower," in *The National Security: Its Theory and Practice, 1945–1960*, ed. Norman A. Graebner (New York: Oxford University Press, 1986), pp. 64–65.
3. Dwight D. Eisenhower, *Waging Peace: 1956–1961* (Garden City, NY: Doubleday & Company, Inc., 1965), p. 205.
4. Fred Kaplan, *The Wizards of Armageddon* (New York: Simon & Schuster, Inc., 1983), 134–5. Also see the Gaither Committee Report itself: U.S., Congress, Joint Committee on Defense Production, *Deterrence and Survival in the Nuclear Age (The "Gaither Report" of*

1957), Joint Committee Print, 94th Cong., 2nd sess., 1976, pp. 12–19. Hereafter cited as the "Gaither Committee Report."

5.   Gaither Committee Report, pp. 15 & 25.

6.   Rockefeller Brothers Fund, *International Security: The Military Aspect*, American at Mid-Century Series-Special Studies Report II (Garden City, NY: Doubleday & Company, Inc., 1958), 56. Hereafter cited as the "Rockefeller Brothers Report."

7.   Kaplan, 149–52; and, Gaddis, 185.

8.   Richard A. Aliano, *American Defense Policy from Eisenhower to Kennedy: The Politics of Changing Military Requirements, 1957–1961* (Athens: Ohio University Press, 1975), pp. 52–59 & 102–115; and, Gaddis, pp. 182–88.

9.   Samuel P. Huntington, *A Common Defense: Strategic Programs in National Politics* (New York: Columbia University Press, 1961), p. 104.

10.   Taylor's flexible response and the impact of *The Uncertain Trumpet* will be discussed in more detail later in this study. See Gaddis, pp. 198–236, for an excellent overview of the flexible response strategy.

11.   See Challener, pp. 39–75, for a good discussion of the Air Force's role in the massive retaliation strategy during the Eisenhower administration.

12.   James MacGregor Burns, *The Deadlock of Democracy: Four-Party Politics in America* (Englewood Cliffs, NJ: Prentice-Hall, Inc., 1963), pp. 195–203 & 253–54.

13.   James L. Sundquist, *Politics and Policy: The Eisenhower, Kennedy, and Johnson Years* (Washington. DC: The Brookings Institution, 1968), p. 406.

14.   Ibid., pp. 406–407.

15.   Cornelius P. Cotter and Bernard C. Hennessy, *Politics Without Power: The National Party Committees* (New York: Atherton Press, 1964), p. 220.

16.   Alastair Buchan, "Defense on the New Frontier," *Political Science Quarterly* 33 (April-June 1962): 130.

17.   Aliano, p. 220.

18.   Hugh A. Bone, *Party Committees and National Politics* (Seattle: University of Washington Press, 1968), p. 277.

19.   Cotter and Hennessy, 220, pointed out that the Acheson-Nitze pamphlet was not the first or only DAC statement on national defense; it was just the most prominent. The DAC also released several statements concerning national defense and the space race in the years in between Sputnik and the Acheson-Nitze pamphlet.

20.   In a June 17 1963 memorandum to McGeorge Bundy, Assistant Secretary of Defense Paul Nitze discussed the extensive public debate about the missile gap in the late 1950s. In turn, he attached a seven-page appendix that listed 76 articles, which had appeared in various newspapers around the country between 1958 and 1960, as a sample of the national debate. See: Memorandum, Paul Nitze to McGeorge Bundy, June 17 1963, National Security Files (NSF): Subjects: Missile Gap, 6/63–7/63, Box 298, John F. Kennedy Presidential Library (JFKL).

21.   *Congressional Quarterly Weekly*, January 19 1959, p. 73.

22.   Johnson's irritation with McElroy and the administration was deftly reported in *Congressional Quarterly Weekly*, February 5 1959, p. 215.

23.   *Congressional Quarterly Weekly*, February 5 1960, p. 212.

24.   Ibid.

25.   Ibid. One should also remember that the Gaither Committee Report, pp. 16–18, had also stressed the potential danger of Soviet ICBM capability for the survival of SAC, although it had did not provide precise figures about what it would take actually to destroy SAC. In a sense, therefore, Senators were not receiving totally new information but just that which was more alarming.

26.   *Congressional Quarterly Weekly*, February 12 1960, p. 240.

27.   Ibid.

28.   The public record of the Eisenhower press conference on February 11 1960 can be found in the *Public Papers of the Presidents of the United States: Dwight D. Eisenhower, 1960–1961* (Washington, DC: Office of the *Federal Register*, National Archives and Record Services, 1961), pp. 167–68 & 170–71. The February 12 1960 addition of the *Congressional Quarterly Weekly*, p. 240, provided a good summary and analysis of Eisenhower's remarks. As will be discussed later in this study, Eisenhower's position was based on data being collected under the heavily guarded U-2 reconnaissance program. Obviously, he could not reveal his source publicly because of national security considerations and the safety of the U-2 program. Also, McGeorge Bundy, in *Danger and Survival: Choices About the Bomb in the First Fifty Years* (New York: Random House, Inc., 1988), pp. 339–40, discussed Eisenhower's rejection of partisanship during the missile gap controversy. Bundy strongly suggested, moreover, that Eisenhower's relative silence actually lent itself to the almost self-perpetuating nature of the debate and, therefore, the issue as well.

29.   *Congressional Quarterly Weekly*, February 26 1960, p. 305.

30.   Aliano, pp. 59–60; and, Gaddis, pp. 184–86.

31.   Desmond Ball, *Politics and Force Levels: The Strategic Missile Program of the Kennedy Administration* (Berkeley: University of California Press, 1980), p. 6. Also see Edgar M. Bottome, *The Missile Gap: A Study of the Formulation of Military and Political Policy* (Rutherford, NJ: Fairleigh Dickinson University Press, 1971), pp. 51–61, for a good overview of the preparedness subcommittee hearings.

32.   As quoted in Bottome, p. 56.

33.   Ball, p. 7. It is interesting that Alsop's estimate was given so much weight or that it was believed that Alsop was approximating the NIE. In a memorandum to President Kennedy in early 1963, Robert McNamara suggested that the 1958 NIE projected Soviet ICBMs at 1000—that is, 1000 less than Alsop's prediction. See: Memorandum, Robert McNamara to the President, March 4 1963, NSF: Subjects: Missile Gap, 2/63–5/63, Box 298, JFKL. Kennedy's speech was devoted entirely to the missile gap issue and later reprinted in a book published by his 1960 presidential campaign committee: John F. Kennedy, *The Strategy of Peace*, ed. Allan Nevins, (New York: Harper & Brothers, 1960), pp. 33–45.

34.   Bottome, pp. 103, 127, 184.

35.   Ball, p. 9. McNamara's confirmation of these figures occurred in his February 1964 testimony before a Joint Session of the Senate Armed Services Committee and the Defense Appropriations Subcommittee on Defense.

36.   Ibid. Also see Bottome, pp. 103–105.

37.   Gaither Committee Report, pp. 16–18; and, Rockefeller Brothers Report, pp. 21–22.

38.   Edgar M. Bottome, *The Balance of Terror: Nuclear Weapons and the Illusion of Security, 1945–1985* (Boston: Beacon Press, 1986), p. 52.

39.   See Gates testimony before the Defense Appropriations Subcommittee: U.S., Congress, House, Committee on Appropriations, *Department of Defense Appropriations for 1961. Hearings before the Subcommittee on Defense Appropriations*, 86th Cong., 2nd ses., 1960, pp. 22–25. Also, Bottome, *Balance of Terror*, p. 52. Gates reconfirmed this position in March 1960 during the Joint hearings of the Senate Preparedness Subcommittee and the Senate Aeronautical and Space Sciences Committee. See: U.S., Congress, Senate, Committee on Armed Services and Aeronautical and Space Sciences, *Missiles, Space, and other Major Defense Matters. Hearings before the Preparedness Investigating Subcommittee in Conjunction with the Committee on Aeronautical and Space Sciences*, 86th Cong., 2nd ses., 1960, pp. 441–43.

40.   Memorandum, Robert McNamara to the President, March 4 1963. McNamara suggested in this memorandum that the NIE released in February 1960 estimated the number of Soviet ICBM operational launchers as 250–350.

41.   Ibid.

42. Ball, p. 10.
43. For instance, see Ball, p. 10; Bottome, *Balance of Terror*, pp. 52–53; and, Bundy, pp. 342–54.
44. Ball, p. 11.
45. Memorandum, Lawrence McQuade to Paul Nitze, May 31 1963, NSF: Missile Gap, 2/63–5/63, Box 298, JFKL.
46. Ibid. Also see the memoirs of Eisenhower's CIA Director, Allen Dulles, *The Craft of Intelligence* (New York: Harper & Row, 1963), p. 165.
47. Memorandum, Lawrence McQuade to Paul Nitze, May 31 1963.
48. Ball, p. 11.
49. Eisenhower, p. 547.
50. See: Bottome, *The Missile Gap*, pp. 135–36; Bundy, pp. 338–39; Gaddis, pp. 186–88; Theodore C. Sorensen, *Kennedy* (New York: Harper & Row, Inc., 1965), pp. 610–13; and, Ball, p. 15. Ball based his analysis of the importance of the U-2 data on a January 1973 interview with Dr. James R. Killian, who had been president of MIT and then appointed to be Eisenhower's Special Assistant for Science and Technology.
51. Desmond Ball has argued that the administration's intelligence estimates and defense policies should have gained considerable credibility because of the May 1960 revelation, but they just did not. See Ball, p. 15.
52. Committee on Political Parties, American Political Science Association, "Toward a More Responsible Two-Party System," *American Political Science Review* 44, Supplement (September 1950): 43; also, Aliano, p. 220. It is generally accepted that the APSA's report served as the modus operandi for the DAC. Cotter and Hennessy, pp. 214–15, noted that Charles Tyroler, the DAC's executive director, admitted in a letter to Hennessy that the report is "generally believed to have furnished the framework" for the DAC. Moreover, Paul Butler apparently referred to the report's recommendations on numerous occasions in his effort to enhance the DAC's effectiveness. James Sundquist, p. 391, also recognized the influence of the APSA's report on the emergence of the Democratic Party's activism in the 1950s. Leon D. Epstein, in *Political Parties in the American Mold* (Madison: University of Wisconsin Press, 1986), pp. 33–35, provides one of the best and most succinct discussions of the particulars of the APSA report. Also, Aliano, p. 220.
53. Chester Bowles, *Promises To Keep: My Years in Public Life 1941–1969* (New York: Harper & Row, 1971), 289–91.
54. Kirk H. Porter and Donald Bruce Johnson, ed., *National Party Platforms: 1840–1964* (Urbana: University of Illinois Press, 1966), pp. 574–75.
55. Ibid.
56. Maxwell D. Taylor, *The Uncertain Trumpet* (New York: Harper Brothers, 1960), pp. 6–7.
57. Ibid., p. 63; also, see pp. 130–64 for an extensive examination of the programs that Taylor thought should be implemented under flexible response auspices.
58. Robert A. Divine, *Foreign Policy and U.S. Presidential Elections: 1952–1960* (New York: New Viewpoint/Franklin Watts, Inc., 1974), p. 209.
59. Kennedy's speech was reported in *New York Times*, June 15 1961, p. 1; James Reston, "Kennedy Starts to Work on the Vice President," *New York Times*, June 15 1960, p. 40.
60. See Theodore H. White, *The Making of the President, 1960* (New York: Atheneum Publishers, 1962), pp. 180–88 & 191–98, for a good discussion of Rockefeller's debate with Nixon and the Republican Party. This conflict was also played out in the *New York Times* on an almost daily basis during June and July 1960.
61. Porter and Johnson, p. 608.
62. Ibid., pp. 604–607.

63. *New York Times*, September 21 1960, p. 1. Nixon's request coincided with a visit that Khrushchev was making to the United Nations during the two weeks of September and first days of October, 1960.

64. *New York Times*, September 21 1960, p. 1; 22 September 1960, p. 1; September 24 1960, p. 1; and, September 30 1960, p. 1. Senator William Fulbright openly scoffed at the moratorium idea and accused the Republicans of "a conspiracy of silence to mislead Americans." See *New York Times*, September 22 1960, p. 16.

65. Divine, pp. 250–51.

66. Ibid., pp. 254–55.

67. Ibid., p. 255; also, Bundy, p. 345.

68. Theodore White, pp. 279–95, provided an excellent analysis of the debates during the 1960 presidential campaign.

69. *New York Times*, October 2 1960, p. 21; October 21 1960, p. 1; and, October 22 1960, p. 1.

70. *New York Times*, October 25 1960, pp. 1 & 28; October 27 1960, p. 1; October 29 1960, p. 1; November 2 1960, p. 29; 5 November 1960, p. 1. It is interesting to note that the October 29th publication of the confidential USIA report followed the *Times'* previous day endorsement of the Democratic ticket based on foreign policy considerations. One of the front page stories in the November 2nd edition (the publication day of the August 29th USIA report) reported on a Kennedy speech in Los Angeles the previous day in which he linked national power and prestige to respect in the international arena.

71. Divine, pp. 252–57 & 270–72; and, Sundquist, pp. 467–68. Sundquist argued that Kennedy did not make huge inroads among voters who advocated Republican control of foreign affairs, but that Kennedy carried a strong enough mandate in domestic affairs to offset his losses on other issues.

72. Sundquist, p. 464. The polls asked two questions: "What do you think is the most important problem facing the country today?" and "Which political party do you think can do a better job handling the problem you have just mentioned—the Republican or Democratic Party?" Of the 7515 responses, over 3600 (about 48 percent) indicated foreign policy issues as the most important ones.

73. Ibid., p. 467. The poll was based on 2944 responses to the following questions: "What do you think is the most important problem facing the country today?" and "If the presidential election were being held today, which candidates would you vote for—the Democratic candidates, Kennedy and Johnson, or the Republican candidates, Nixon and Lodge?" If the voter were undecided, the question asked was: "As of today, do you lean more to Kennedy and Johnson or more to Nixon and Lodge?"

74. Ibid.

75. *Department of Defense Appropriations for 1961. Hearings before the Subcommittee on Defense Appropriations*, p. 120.

76. Ibid., pp. 120–21.

Sputnik and Technological Surprise

Glenn P. Hastedt

Sputnik 1 was launched on October 4, 1957. It was not a small satellite. Twenty-two inches in diameter and weighing 184 pounds, it dwarfed the six-inch, three and one-half pound satellite that was scheduled to be launched by the American *Vanguard* missile. *Sputnik 2* was even larger, weighing 1,121 pounds, and went into orbit on November 3 carrying a live dog. The Eisenhower administration put forward a low-keyed response. Eisenhower was at Gettysburg for a weekend of golf when the announcement of Sputnik's successful launching was made. He left it to White House Press Secretary James Hagerty and Secretary of State John Foster Dulles to put forward the Administration's response. They informed the press that the Administration had not been caught by surprise by Sputnik's success and that Eisenhower was being kept abreast of events because they were of "great scientific interest." Hagerty also offered the opinion that the Administration had "never thought of our program as one which was in a race with the Soviets." In the following weeks Secretary of Defense Charles Wilson would dismiss Sputnik as "a nice scientific trick" and trade advisor Clarence Randall called it a "silly bauble."[1]

The measured and condescending response of the Eisenhower administration was a stark contrast to the highly agitated response of the media and public. As Walter McDougall notes, the public's response to Sputnik poured forward in a series of chaotic waves whose crests and troughs overlapped to "reinforce alarm one week and confused inertia the next."[2] Frequent comparisons were made to Pearl Harbor. *Time* and *Newsweek* placed Sputnik squarely into a

Cold War context of competition with the Soviet Union. The *New Republic* likened Sputnik to Columbus' discovery of America and the *U.S. News and World Report* compared it to the splitting of the atom. Eisenhower's efforts to minimize the significance of Sputnik failed. By early November 1957 his standing in public opinion polls had fallen 22 points from a high of 79 percent in January 1957 to 57 percent.[3]

The loss of public confidence in Eisenhower was not due simply to the actions of circulation-hungry press or opportunistic political opponents who wished to make Eisenhower look bad. Sputnik touched a raw nerve that both excited and frightened the American public in a way that the Eisenhower administration had not anticipated. Robert Divine states that "at the heart of the problem was the popular belief in American supremacy in science and technology." Sputnik set off a highly public and visible debate on education, science, space exploration, national security, and fiscal policy that continued on into the 1960s. McDougall writes that with Sputnik, "a new political symbolism had arisen to discredit the old verities about limited government, local initiative, balanced budgets, and individualism."[4]

The Eisenhower administration's protests not withstanding, it was caught by surprise by Sputnik. The question examined here is why? Was it the product of forces that have produced other surprises or was the failure to predict Sputnik the product of a unique constellation of forces? This latter possibility exists because where existing studies of surprise have focused on military or diplomatic surprise, Sputnik—while it contains elements of each—is more accurately seen as a case of technological surprise.

Technological surprise is seldom studied. This in spite of the fact that one of the defining features of scientific and technological research is the search for new ways of doing things, new insights into how and why things work, and an unwillingness to accept the limits of the present as the boundaries of the future. Barry Hughes notes that one reason few studies of the future specifically address technological change is that "as difficult as forecasting population growth or energy demand over the next twenty years might be, such forecasts are trivial compared to the difficult task of anticipating technological developments."[5] Hughes then adds that "the rate of technological change is both largely unmeasurable and uncertain."[6]

The lack of attention to technological surprise is unfortunate given the increasingly important role that technology plays in world politics today. Not only does technology shape the traditional strategic agenda of states and shape the language in which strategy is discussed, it is also central to the dynamics of many of the issues that occupy positions of prominence on the post-Cold War agenda. By studying the surprise surrounding the launching of Sputnik not only can we gain additional insight into why it had such a significant effect on the American psyche, but we also can begin to lay a foundation for studying technological surprise as an integral and distinct part of world politics.

WHY SURPRISE

Studies of surprise approach the subject from a variety of perspectives. Some address the basic nature of surprise; some seek to uncover underlying forces that produce surprise or are at least conducive to it; some examine the actions of the "attacker" where others focus on the "defender"; and still others are concerned with particular types of surprise such as military surprise or diplomatic surprise. In order to better grasp the extent to which the surprise induced by Sputnik was unique, we need to place it within the context of these broader theoretical and practical concerns.

THE NATURE OF SURPRISE

On any given day foreign and national security policy makers have few reasons to expect surprise. Bureaucratic inertia, vested personal interests, domestic political pressures, and international system constraints conspire to prevent much more than incremental change from usually taking place. Yet surprise does happen. What is important about surprise is not that it happens, but that sometimes when it happens a surprise will fundamentally alter the strategic context within which future decisions are made. In these cases, surprise invalidates the assumptions on which diplomatic initiatives were premised and defense plans based. It exposes states to vulnerabilities they had not anticipated and are not prepared for. As Richard Betts notes, intelligence makes its impact through its "jolting originality" as policy makers are suddenly confronted with the realization that events are moving in an unexpected and dangerous direction.[7]

Several different frameworks exist for classifying surprise. At the most general level they can be grouped under two headings. One school of thought emphasizes causation. Surprises are defined in terms of the defender's unreadiness. Potential dimensions of unreadiness include whether the opponent would attack, when the opponent would attack, where it would attack, how it would attack, and why it would attack.[8] A second school groups instances of surprise in terms of their impact on world politics. Minor surprises are those which are unexpected moves that change the course of relations between states but do not alter the underlying balance of power in the international system. Major surprises are unanticipated moves that have a considerable impact on the real or expected division of power either at the regional or global level.[9]

WHY SURPRISE SUCCEEDS

Surprise is never total; bolts from the blue do not happen. There is always some warning. The repeated instances in which surprise takes place in spite of warning has caused researchers to look closely at the problems that the victim state encounters in trying to correctly anticipate the moves of its adversary. Michael Handel suggests that they might be grouped under three headings.[10]

First, are problems deliberately caused by the enemy. Foremost among these is deception. Those contemplating a surprise move often seek to cloak their actions behind a veil of secrecy. States contemplating surprise find deception to be an attractive strategy because it is relatively cheap and because it is virtually impossible to maintain prolonged secrecy for major diplomatic initiatives or military campaigns.[11] For example, in World War II the allies made no serious attempt to hide their intentions of invading Europe. Instead, they sought to deceive Hitler as to where the invasion would take place. Allied intelligence was so successful in focusing Hitler's attentions on Pas de Calais and away from Normandy that Germany still had most of its forces there even after the invasion of Normandy was well underway.

Deception is most effective when it seeks to reinforce a belief already held by the intended victim rather than when it tries to change a policy maker's mind. Such was the case when Stalin refused to believe the warnings of Hitler's pending invasion. A paradox also presents itself here. Findings suggest that the more alert a state is to the potential of deception, the easier it is to deceive it. The acknowledged possibility of deception provides a rationale for both dismissing as bogus all incoming pieces of information that do not fit with current expectations and for being skeptical of all information that does fit. The result is that policy makers are free to indulge their biases or engage in wishful thinking.

Quite apart from any planned program of deception, the actions of the attacker also complicate the task of anticipating a surprise move. The attacker always has the option of changing its plans. Warnings of an attack, therefore, may be correct even if no attack materializes. Japanese plans called for aborting the attack on Pearl Harbor if its attack fleet was discovered. Indecision on the part of the attacker similarly complicates the problem of predicting an attack or diplomatic breakthrough. On more than one occasion in the period leading up to the surprise announcement of President Nixon's trip to China, contradictory signals were sent by China to the United States because of conflicts within the Chinese Communist Party.

A second set of problems that stand in the way of anticipating surprise are inherent in the task of predicting the future. Major events simply do not come in nice neat packages. It is only with twenty-twenty hindsight that the correct interpretation of data is obvious. Most commonly cited as obstacles to the correct assessment of information are noise and the ambiguity of evidence. Noise is the opposite of deception. Where deception succeeds by increasing the adversary's certainty about the validity of false interpretations of data, noise confuses the adversary through the clutter of extraneous information that its intelligence services are picking up.[12] The problem at Pearl Harbor was not too little information but too many irrelevant pieces of information. Moreover, on the eve of the attack a great deal of evidence existed supporting all the wrong interpretations of the last minute signals being received. A chronology of over 100 events can be strung together pointing to the culminating announcement of Nixon's breakthrough trip to China. But the picture was far less clear as events unfolded. Diplomatic feelers often went unanswered because they were too subtle; sometimes more than one month separated the sending of a signal and the response to it.

And, building on points raised above, because neither side fully trusted the other, ambiguity was often purposefully inserted into messages in order to allow for an orderly diplomatic retreat.

Faced with these barriers, some intelligence analysts and observers argue that a distinction needs to be drawn between forecasting and fortune telling. Too often intelligence is asked to engage in fortune telling where they are asked to predict the occurrence of a particular event. Former Israeli intelligence officer Shlomo Gazit argues that intelligence organizations should only be tasked with producing estimates on three situations: 1) decisions already taken; 2) possible reactions to a certain situation combined with speculation on the most probable one; and 3) analyzing the outcome of a developing situation in terms of milestones, turning points, and possible outcomes.[13]

A final set of problems in anticipating surprise are self-generated. One obstacle to the ability of policy makers to appreciate and respond to warning intelligence stem from the blinding effect of current policy-maker preoccupations. Policy makers do not sit back and passively take in information. They interact with it, picking and choosing which pieces are relevant to their needs and which are not. One of the most important perceptual filters that determines what is seen and what is not are the immediate concerns that dominate the policy maker's thoughts (the evoked set).[14] Policy makers in Washington were doubly blinded to Japanese war plans. Not only were they absorbed with events in Europe but they were convinced that Japanese aggression would first take place in the Western Pacific. Secretary of the Navy Knox responded to news of the attack on Pearl Harbor by asserting "this can't be true...[it] must mean the Philippines." Communications between Washington and Pearl Harbor also show the impact of the evoked set on reading intelligence. Policy makers in Washington assumed their immediate concerns about a Japanese attack were shared by their counterparts at Pearl Harbor. This was not the case. Officials at Pearl Harbor were primarily concerned with the possibility of internal sabotage and they read the warnings from Washington in this light.

Contingency plans have the same blinding effect. Having spent considerable time and energy into putting together a contingency plan, the tendency is for it to color one's perceptions to the point where all future events are seen as being consistent with its assumptions. Committed to stopping the spread of communism, U.S. policy makers needed little information to "see" communism in Third World revolutionary movements. The response was highly predictable with the United States either coming to the aid of the threatened government or committing itself to overthrowing the newly installed regime. Given this pattern of response, many observers have concluded that the failed 1960 Bay of Pigs invasion of Cuba to overthrow Fidel Castro was in a sense inevitable because in carrying out covert operations against communist regimes the United States had come to rely too heavily on responses that had been successful in Iran (1953) and Guatemala (1954) and had stopped paying close attention to the unique features of each situation.

Also promoting surprise is a policy maker's commitment to a given course of action and the problems it creates for integrating intelligence and policy. Selecting a course of action and building support for it is an expensive undertaking and once adopted personal and institutional prestige become attached to its success. Into this setting walks the intelligence professional. To be effective, intelligence must know the policy maker's concerns and plans otherwise the intelligence they provide is likely to be found irrelevant. However, if intelligence is brought into close contact with policy it runs the risk of being corrupted. Intelligence will be expected to support policy and not allowed to perform its intended function of presenting policy makers with warning. This tension between intelligence and policy was evident in the period preceding the fall of the Shah. The complete confidence the United States had in the ability of the Shah to survive skewed U.S. intelligence efforts. Analysts complained that "you couldn't give away intelligence on the Shah."

At times the tunnel vision preventing policy makers from accurately reading the intelligence available to them grows out of biases deeper than the attachment to a line of policy or being pre-occupied with a problem. It may also stem from a false set of assumptions which are widely held throughout the policy-making process.[15] For Israelis in 1973 it was the assumption that Egypt would not go to war until it could control the air space over the battlefield; for the Arabs in 1967 it was that war would not begin until negotiations proved fruitless; for Stalin it was that Hitler would issue an ultimatum before attacking; and in 1950 the United States was surprised that North Korea attacked because it was assumed that this would only happen as part of a more general war.

In a similar vein, surprised states also mistakenly assume that other states are conducting their foreign policy on the basis of the same set of assumptions under which they are operating. They are especially likely to overestimate the value placed by others on maintaining the status quo. The United States failed to understand that Japanese leaders had posed their predicament in such a way that an attack on Pearl Harbor appeared reasonable. In 1973 Israeli leaders failed to appreciate the fact that Sadat found a continuation of the status quo to be so unacceptable that he was willing to start (and lose) a war in hopes of breaking the Middle East stalemate. Under these circumstances, the state carrying out the surprise is able to exploit "the logic of craziness." The more extreme the action, the more unbelievable it is, the less likely it is to be adequately defended against or prepared for. As such, the possibility of success is often better than if the state pursued a more "reasonable" course of action.

Given the forces of inertia that have been noted above, massive amounts of information are often needed to make policy makers reconsider their position. At the same time it is possible to be over-warned. Individuals and organizations cannot stay at high levels of alert indefinitely. The more often one is warned, the more accustomed one becomes to the situation. Familiarity causes one's response to warning to become routine as a "cry wolf" syndrome takes hold. Pearl Harbor was warned once in 1940 and twice in 1942. Israel counted on the numbing effect of routine to aid their surprise in 1967 and fell victim to the deadening impact of repeated mobilizations and demobilizations in 1973. The cry wolf problem is especially difficult to overcome in dealing with terrorism. In speaking to the problem of intelligence regarding bomb threats preceding the October 1983 attack on the Marine barracks in Beirut, one officer told a congressional committee that "since we have been here, we have, I think, counted over a hundred car bomb threats." Another noted that once a threat was received regarding a blue Mercedes and added "there are quite a few blue Mercedes over in Lebanon."

TYPES OF SURPRISE

In addition to laying out the fundamental tenets of why surprise happens, attention has also been given to developing a better understanding of the two major types of surprise: military surprise and diplomatic surprise. As Michael Handel observes, there are significant differences between them.[16] First, where military surprise is an inherent part of military planning, diplomatic surprise is not. In military affairs, surprise is treated as a force multiplier. In diplomacy continuity and predictability are valued, not the ability to surprise an ally.

Second, military surprise and diplomatic surprise present intelligence organizations with different types of problems. Due to their large size and need for long lead-time, military operations emit many signals and are difficult to "keep secret." At the same time, however, military actions present intelligence officials with multiple challenges in predicting the where, when, and how of possible activity. Because it involves fewer participants, and may even be unilateral in nature, diplomatic surprise is easier to keep secret. Anticipating it involves understanding a leader's perceptions more than it does comprehending organizational behavior. Moreover, in trying to anticipate the moves of an adversary, those concerned with military surprise must pay attention to both intentions and capabilities. Those concerned with preventing diplomatic surprise need only focus on the adversary's intentions since capabilities are not a significant limiting factor in the ability to carry out diplomatic surprise.

Third, the impact of military surprise is immediate because the national security challenge to the attacked state is direct and real. The impact of diplomatic surprise may be immediate or delayed because the consequences of the surprise may not place the state in immediate danger.

Fourth, military surprise is always a hostile and negative act. Diplomatic surprise may be positive and cooperative in nature.

Fifth, from the point of view of the initiating state, military surprise offers only advantages since it serves as a force multiplier increasing the power of its existing military forces. Diplomatic surprise involves a trade-off for the initiating state. Its action may alienate allies, provoke protests from domestic groups, or negatively impact on the achievement of other foreign policy goals.

SPUTNIK: THE SOVIET STORY

Soviet leaders made little attempt to keep the impending launch of Sputnik a secret. Public signs of growing official Soviet interest in putting an artificial satellite into orbit around the earth began to mount following Stalin's death. Responding to the call of organizers of the International Geophysical Year (IGY) for the international community to work together to launch a satellite in earth orbit in 1957, the USSR Academy of Sciences created a blue ribbon commission whose purpose was to "organize work concerned with building a laboratory for scientific research in space." This was followed by the Soviet Union's July 30, 1955 official announcement that, like the United States, it planned to launch a satellite during the IGY. Radio reports spoke of teams of scientists being formed to build such a satellite and Leonid Sedov, who chaired the blue ribbon panel, predicted a satellite launch in 1957 using a multistage rocket. At the First International Conference of Rockets and Guided Missiles in 1956, Soviet scientists spoke openly about high altitude experiments and launching dogs into space. And at an international meeting of scientists in Barcelona in October 1956, Soviet scientists briefed their American counterparts over dinner regarding many of the details of the upcoming Sputnik launch including its weight. This information was promptly passed on to American officials attached to the delegation.

McDougall notes that once the IGY began on July 1, 1957, "Soviet predictions of a satellite became a weekly occurrence."[17] The first test rocket exploded in failure in Spring 1957 and it was not until August 21 that a successful launch took place. Days later, on August 27, the Soviet Union announced to the world that it now possessed an intercontinental ballistic missile (ICBM). Sergey Korolev, one of the key figures in the design and testing of Soviet missiles, states that it was only at this point that Soviet leaders gave final approval for launching Sputnik. Moscow made its intentions public on September 17 and on October 1 it informed the world of Sputnik's radio frequency.

After *Sputnik 1*'s successful launch Soviet leaders were quick to exploit its propaganda value in the Cold War competition with the United States. However, it is worth noting that even prior to Sputnik high ranking party members and military officials had begun to emphasize the Soviet ICBM capability and its significance for world politics.[18] Between 1955 and 1957, Nikita Khrushchev, Nikolay Bulganin, and Anastas Mikoyan all made public comments which either extolled the virtues of long-range missiles or denigrated the military significance of long-range bombers. At the twentieth party congress in 1956, Defense Minister Marshall Georgiy Zhukov made reference to long range and "mighty" missiles. And a 1957 article cited the strategic virtues of ICBMs: it could take off from mobile launchers; it could operate under all types of weather conditions; and it would permit its possessor to launch surprise attacks.

Soviet interest in rocketry and space flight predated Stalin's death. The September 17 announcement coincided with the one hundredth anniversary of Konstantin Tsiolkovskiy's birth. Tsiolkovskiy was one of the founding fathers of Soviet rocketry having written a 1903 treatise on the mathematics of orbital mechanics and expounded on rockets powered from liquid oxygen and liquid hydrogen. Under the Tsars, Russian advances in rocketry were primarily in the areas of theory and design rather than testing and building; this changed with the Communist party's ascent to power. In 1924 it created a Central Bureau for the Study of the Problems of Rockets. By 1934, McDougall notes that the link between rocketry and revolution had become institutionalized as the drive for technological supremacy had become a major goal of the Soviet state.[19]

Step by step rocket research was "swallowed up in the belly of Stalin's leviathan." The Academy of Sciences was accused of counter-revolutionary activity and placed under the direction of the Council of People's Commissars. The first Five Year Plan established research and development priorities which replaced individual goals and visions as the driving force in rocketry. During the 1930s, the entire Reactive Scientific-Research Institute fell victim to Stalin's Great Purges. Still, by the end of the 1930s, Soviet scientists had managed to test air-to-surface missiles, surface-to-air missiles, surface-to-surface missiles, and launch the world's first two-stage rocket. Soviet advances in rocketry were such that for them, the German V-2 contained few technological breakthroughs.

Not only did the end of World War II produce a surge of Soviet activity directed toward the development of an atomic bomb, it also brought forward renewed interest in rocketry. The Chief of the

Soviet Air Forces observed that merely building more V-2 rockets would not be enough to ensure Soviet security in a future war. "They were good to frighten England, but should there be an American-Soviet war, they would be useless; what we really need are long range, reliable rockets capable of hitting target areas on the American continent." The head of the Aerodynamics Laboratory of the Moscow Military Air Academy echoed these thoughts: " ... we have no intention of making war on Poland. Our vital need is for machines which can fly across oceans!" By the end of 1947, almost two years prior to the detonation of the Soviet atomic bomb, "everyone wanted to design a trans-Atlantic rocket."[20] By 1949 results were beginning to appear. The first all-Soviet upgrade of the V-2 with a range of 550 miles was in production. An intermediate range ballistic missile (IRBM) was under construction by 1952 and the blueprints for the ICBM which the Soviets used to launch Sputnik were approved in 1954.

THE AMERICAN VIEW OF THE SOVIET STORY

Lawrence Freedman notes that early postwar estimates of the Soviet strategic threat to the United States were little more than guesses.[21] The focus of these estimates was the current and future size of the Soviet bomber force. Intelligence information was scarce and came primarily from two sources. The first of these were official Soviet government statements which were distrusted and often dismissed as propaganda. The second was personal observation. The main opportunity for viewing Soviet bomber strength were air parades such as the 1955 May Day parade in which the Soviets appeared to fly more *Bison* bombers overhead than intelligence estimates suggested they possessed. Where the 1954 estimate predicted that full production of the *Bison* bomber would not begin until 1956, by the end of 1955 intelligence estimates were now predicting that by the end of 1956 twenty-five bombers would be produced each month with the total Soviet inventory reaching between 600 and 800 by 1959 or 1960.

The "bomber gap" became the focus of a short-lived but highly charged political debate pitting Congress against the President, and the Air Force against the other members of the intelligence community. The issue was resolved largely through the development and deployment of a new intelligence collection system: the U-2 reconnaissance plane. The first flight began from Weisbaden on July 4, 1956, and overflew Moscow, Leningrad, and the Baltic Coast. Two additional missions were each flown on July 5 and July 9 before Soviet protests caused the flights to be called off for several months.[22] These five flights produced conclusive proof that Soviet bomber production was not proceeding at an alarming rate and a downward revision of the bomber threat began in December 1956.

What the U-2 flights did reveal was a growing Soviet commitment to ICBM testing and production with the construction of a second missile testing site at Tyuratam. The intelligence community had been aware of the Soviet's first missile testing facility at Kapustin Yar since 1947. Information on activities there was provided by defectors, returning German scientists, and aerial reconnaissance. In 1952 the CIA set up an Office of Scientific Intelligence and in 1954 predicted that the Soviet Union would be capable of launching an Earth satellite by the end of 1957 and that it would not possess an operational ICBM before 1960. As with the bomber gap, the Air Force disagreed and made use of its close ties with key congressional supporters to press the Eisenhower administration to do something about the growing "missile gap."

Information gathered from U-2 overflights of the Tyuratam missile testing site in 1957 led CIA analysts to conclude that the Soviets were preparing to launch a satellite into space using an ICBM. A U-2 overflight in the summer produced pictures of an ICBM sitting in its launching pad. The status of the Soviet ICBM program was brought to Eisenhower's attention at a May 10, 1957, National Security Council Meeting. Public attention was directed to the rapidly progressing Soviet ICBM program by a May article in *Aviation Week* and a July column in *The New York Times* by Steward Alsop, in which he stated that the Soviets had tested an experimental long-range missile. Between May and August 1957, eight long-range missile firings were observed. On September 12, 1957, the Office of the Army Chief of Research and Development estimated that the Soviet Union would launch a satellite within 30 days.[23]

THE AMERICAN STORY

Control over air power (both from a strategic and bureaucratic perspective) was a major focal point of U.S. military policy in the immediate post-war period. The weapon of choice was the manned bomber with rocketry, and its accompanying satellites running a distinct second. Not only was the manned bomber a known and trusted commodity to military officers and strategists, but real doubts existed over the ability of a rocket to deliver atomic bombs given their size. The case for the manned bomber was buttressed further by the desire of the Truman and Eisenhower administrations to limit funding for nonessential programs in pursuit of a balanced budget.

During the war the Army Air Force Services division tasked the Special Weapons Group at Wright Field to do work on guided missiles. As the military recognized the potential of the V-2, competition to control this program mushroomed and in 1944 the decision was made to give the Army Air Force control over all missiles dropped from planes, and the Army Services unit control over ground-launched ballistic missiles. Upon his return from Potsdam, President Truman authorized an increase in aviation research and development spending. Air Force spending on missiles went from $3.7 million in FY1945 to $38.8 million in FY1946. No sooner had this emphasis on missile research and development began that it ended. Instead of a projected FY1947 budget of $75.7 million on missiles, the Air Force's outlays were cutback to $22 million. Eleven programs were canceled including the MX-774, a 5,000 mile ballistic missile.

This reversal of field reflected four key assumptions that were widely shared among American policy makers.[24] First, funding military programs ranked third in priority behind domestic spending and overall fiscal conservatism. Second, it was assumed that the United States was ahead of the Soviet Union in aviation technology. Third, "blue sky" Air Force officers favored the manned bomber over missiles. And, fourth, there was a general air of scientific pessimism surrounding the development of an ICBM.

Research and development in the field of satellite technology fared little better. Just days after their surrender, German scientists had briefed some of their American counterparts about the possibilities of missiles and satellites. The Navy was first to seize the initiative establishing an Earth Satellite Vehicle Program in 1945. In 1946 it approached the Army Air Force about the possibility of collaborating on a joint project. Rebuffed by Vannevar Bush, Director of the Office of Scientific Research and Development, the Army Air Force commissioned the Rand Corporation for a study of the value of Earth satellites. The Rand Report, released on May 2, 1946, made the following three points. First, satellites would become one of the most potent scientific tools by the end of the century. Second, the launching of a satellite would have repercussions comparable to that of the atomic bomb. Third, "the nation which first makes significant achievements in space travel will be acknowledged as the world leader in both military and scientific techniques." Still, 1948 proposals for funding by the Navy and Air Force were rejected and the project canceled because of a failure to establish "either a military or a scientific utility commensurate with the presently expected cost of a satellite vehicle."

New life was breathed into the U.S. missile program by a series of events in the early 1950s. The (unexpected) Korean War led Truman appointed K.T. Keller to the newly created post of special advisor on missiles and in 1951 he authorized funds for work on a new Air Force ICBM that would become the *Atlas* missile. The successful detonations of hydrogen bombs by the Soviet Union and the United States put an end to the debate over whether such a weapon could be delivered by missiles. Evidence was now also mounting that the Soviet Union was making significant advances in missile research and development.

A 1950 Rand study released on October 4, 1950, also kept a spotlight on satellite research and development by directly addressing their military significance. It noted that while satellites were not weapons, they did possess great military utility because they could gather data not available from other sources. Rand predicted that their unconventional nature would also guarantee that satellites would become a factor in the global balance of power and thus have significant political-psychological effects as well as military ones. Rand also felt that the primary political problem that satellites would pose centered on the reaction of other states to the loss of sovereignty that resulted from their overflights.

A further boost to the U.S. missile and satellite programs was given by the findings of the Killian Report, commissioned by President Eisenhower in 1954 to study the problem of surprise attack on the U.S. The Killian Report identified the period from late 1954 until 1955 as one of American air-atomic advantage and one in which it was vulnerable to a surprise attack due to the lack of an early warning system. The following period, from 1956–57 until 1958–60, was anticipated to be one in which the United States would hold a great offensive advantage but one in which the Soviet Union would be testing new missiles and bombers. The Killian Report noted that the "single most important variable" in its scenarios was the Soviet development of an ICBM. It recommended that the U.S. give highest priorities to the development of an ICBM, and an early warning system. It was especially important, noted committee member Edwin Land, that the U.S. find ways of increasing the number of facts on which our intelligence estimates are based.

Three different satellite programs emerged almost simultaneously in the mid-1950s. One, put forward in secret, was the development of an advanced and technologically sophisticated Earth recon-naissance satellite, the WS-117L, which was a direct outgrowth of the Killian Report. The second was a joint Army-Navy proposal, Project Orbiter, which was to be a "no-cost" satellite using existing technology. The third was a proposal from the scientific community that the U.S. launch a scientific satellite to commemorate the IGY. Assistant Secretary of Defense Donald Quarles oversaw a study to evaluate these proposals from a military perspective. His staff recom-mended launching a small scientific satellite in the near future and acknowledged that the Soviet Union was now working on a satellite program of its own. The scientific satellite was recommended as a way of testing the principle of "freedom of space" and the legality of satellite overflight. The report also insisted that care needed to be taken so that nothing done would prejudice U.S. satellite efforts outside the scope of the IGY.

In July 1955, the Eisenhower administration publicly committed itself to a scientific satellite as part of the IGY. It was agreed that the IGY National Committee should take responsibility for work on the satellite and that the Defense Department would provide the missile that would be used in its launching. The choice was between the Army missile which was to be used in Project Orbiter, and a Naval Research Laboratory proposal, Project *Vanguard*, which involved designing a new four-stage *Viking* missile. Work on *Vanguard* got underway in fall 1956. In May 1957, the National Security Council approved a launch schedule that included flight testing of the three *Vanguard* stages in the remainder of 1957 with a fully instrumented test satellite to be launched in March 1958. Testing of *Vanguard* was on schedule when *Sputnik 1* went into orbit on October 4, 1957. The first launching of a U.S. satellite was set for December.

WHY SPUTNIK AS A CASE OF SURPRISE?

When placed within the context of the literature on surprise, a review of the events leading up to the launching of *Sputnik 1* points to the conclusion that the major reasons for surprise were self-generated within the United States. There is little evidence of Soviet deception. Soviet leaders made no effort to hide their commitment to launching a satellite. Vernon van Dyke goes so far as to argue that there were "relatively abundant and open reports concerning developing Soviet capabilities."[25] The principal problem lay in getting top level officials in the United States to believe these reports. The successful Soviet deception with regard to the bomber gap led many American observers to dismiss Soviet statements about their progress in missile and satellite technology as nothing more than propaganda-inspired boasting. Only in one sense did the actions of Soviet leaders directly contribute to surprise. This was the relatively late date at which a firm commitment to launch a satellite was made. It was only after the successful August 21 test firing of the ICBM was that decision made. Given the secrecy of the Soviet system, this did not permit the intelligence community much time to warn American leaders of the actual launching of Sputnik.

The second frequent cause of surprise noted earlier are problems inherent in the task of predicting the future. Here too, these problems were only of secondary importance in the case of Sputnik. According to Gazit, one of the three legitimate tasks that can be assigned to an intelligence agency is that of analyzing the outcome of a developing situation. Not only did the CIA manage to accurately see future developments in the Soviet missile and satellite program, so, too, did other elements of the national security bureaucracy. Furthermore, Rand correctly saw the political and psychological significance of launching the first satellite into space.

One can speculate that they were able to do so in part because, as specialists concerned primarily with developments in Soviet missile and satellite research, they were not as affected by the "noise" surrounding the many signals that Soviet leaders were sending regarding their intentions as were officials with more general areas of foreign policy responsibility. Soviet statements and actions regarding such early 1950s Cold War hot spots as Berlin, the Middle East, Central America, and Korea were far less likely to distract them or negatively color their evaluations of Soviet missile and satellite activity.

In identifying self-generated problems within the United States as the primary source of surprise, the case of Sputnik is not unique. It falls well within the confines of that body of literature which emphasizes that surprise occurs despite warning. The picture which emerges is one of knowledge of Soviet activity but an absence of any sense of urgency. At least four factors contributed to this false sense of security.

First, was the problem of mirror imaging—assuming that the Soviet Union was making decisions on the same set of assumptions that were guiding American policy. The most crucial assumption dealt with the continued superiority of the manned bomber as the vehicle for delivering long distance air strikes on the enemy. Freedman notes that in the 1940s American officials had come to dismiss the ICBM as a delivery vehicle, for reasons of cost and accuracy, and had assumed that Soviet officials would reach a similar conclusion.[26] They were not totally wrong. For a time Korolev agreed with the assessment that "winged rockets" and not missiles were destined to be the delivery vehicle of choice in the space age, and Soviet leaders expressed early interest in the Sänger project, a wartime German plan for constructing a piloted bomber with intercontinental capabilities.[27] What they failed to recognize was the shift in Soviet thinking away from the bomber to missiles, a shift signaled in Soviet publications and official statements.

Second, American officials were blinded to the significance of Sputnik by their own policy priorities. For the Eisenhower administration speed was not of the essence. If it had been the Army's Project Orbiter would have been given the nod in 1955 over the untried *Vanguard* as the delivery vehicle since as Army officials noted before and after Sputnik, it would have been ready to launch as early as January 1957. By choosing the *Vanguard* the Eisenhower administration made its decision in virtually total disregard for the speed and direction of the Soviet space program. Top priority was given to its goal of establishing the legal precedent that space was international territory and not the property of sovereign states.[28] Given the secrecy of Soviet society, and pending advances in military reconnaissance satellites, this precedent was seen as essential to the continued U.S. ability to gather information on Soviet military capabilities. Eisenhower felt that this would be best realized if the first satellite put into Earth orbit was scientific and not military in nature.

Third, American officials failed to respond to evidence of a growing Soviet satellite program, due to the overriding influence of budgetary considerations. Van Dyke correctly noted that the "general air of conservatism and the stress on economy under Eisenhower and Secretary of Defense Wilson were not propitious to boldness pertaining to space or for that matter imaginative research."[29] The power of this commitment to limit government spending to serve as a restraining force on the U.S. space program comes through even after Sputnik is sent into orbit, as Eisenhower fought doggedly to limit the funds that would be directed at space research. He continued to view excessive government spending and a weak economy as far greater threats to American national security than a nascent Soviet ICBM force.

Finally, an unstated American belief in its presumed technological and scientific superiority over other states seems to have blinded policy makers to Soviet advances. In spite of warnings that the Soviet Union was moving steadily toward launching a satellite, no serious consideration appears to have been given by the Eisenhower administration to the possibility that the United States was in or that it might lose such a race. This sense of technological superiority was the basis for Eisenhower's "New Look" deterrence posture maintained in the face of two jolting challenges: the Soviet A and H bomb tests.

SPUTNIK AS TECHNOLOGICAL SURPRISE

The ability to explain the surprise surrounding the launching of Sputnik is significant because it does not fit neatly into either of the two dominant categories that form the basis for theorizing. Strictly speaking, it is neither a case of military surprise (although Sputnik had great military significance), nor of diplomatic surprise (although it had great propaganda value to the Soviet Union for its Cold War diplomacy).

Sputnik is best seen as a case of technological surprise and its defining characteristics are drawn equally from those Handel uses to define military and diplomatic surprise.

Using Sputnik as a point of departure, technological surprise shares with military surprise the following characteristics. First, surprise is an inherent part of technological activity. Where diplomacy values consistency and predictability, technology values innovation and experimentation. Thinking in terms of technological surprise does not require developing new concepts or new ways of thinking on the part of scientists or engineers. Second, technological surprise poses many of the same types of forecasting challenges as does military surprise. Like military surprise, the scale of activity involved makes it difficult to maintain 100 percent secrecy surrounding one's efforts. At the same time, because of the wide variety of ways in which the surprise might occur, correctly predicting all of the details of a technological breakthrough is difficult. Sputnik, for example, did not surprise U.S. observers so much in terms of the timing of its occurrence as it did in its size. Third, anticipating technological surprise requires an attention to both capabilities and intentions. In the case of Sputnik, the capability existed in both the United States and Soviet Union. A key difference in the two space programs was that where the Soviets relied upon "off-the-shelf" technology readily available to Soviet defense planners in assembling Sputnik, the United States sought to push the technological window further. More significant than this difference in their attitudes toward technology was the differing level of the political commitment to make use of this technology. Writing more generally on the nature of technology and public policy, Dorothy Nelkin notes that "perhaps [the] overriding factor shaping priorities for science and technology is the convergence of technological opportunity...with political readiness to accept technological change.[30] Korolev was very much the policy entrepreneur and he succeeded in convincing Khrushchev of the political and strategic benefits that would follow from being first in space. Neither Wernher von Braun nor any other American scientist operated as freely or successfully within the American political process.

Technological surprise also shares certain characteristics with diplomatic surprise. First, as with diplomatic surprise, the significance of technological surprise does not have to be immediate. The Eisenhower administration correctly judged that Sputnik did not present a short term challenge to American national security interests. Both sides in the Cold War, however, realized that Sputnik did foreshadow the possibility of a fundamental shift in the global balance of power in the years ahead. Second, where military surprise only offers advantages to the initiating side, technological and diplomatic surprise may also offer advantages to the surprised state. In the case of Sputnik, the Eisenhower administration saw in the world's reaction to Sputnik an affirmation of the principle of free skies that it hoped to establish with the launching of a scientific satellite. Finally, technological surprise does not have to be a hostile, negative act. It can also produce positive results. Sputnik produced a mixed bag of consequences in the United States. On the negative side, it set off an often bitter partisan political debate over whose fault it was that the Untied States was surprised and what to do about it. On the positive side it shocked the United States out of a sense of complacency regarding its technological prowess and led to a burst of creative activity (and funding) for science education, and research and development.

SUMMARY

The forces which helped create an environment in which surprise was possible with regard to Sputnik are consistent with those found in other studies of surprise. Sputnik was not a bolt from the blue. There was warning. What was absent on the part of senior officials in the Eisenhower administration was an appreciation of the political significance of Sputnik. Convinced of their own technological superiority, committed to a policy of fiscal conservatism, and focusing on the narrowly defined issue of how to establish the legality of overflights, the Administration felt no sense of urgency in moving ahead with its own space program.

The manner in which the United States was surprised by Sputnik should be of interest to those concerned with the evolving dynamics of the post-Cold War international order. The agenda of the post-Cold War international system gives a prominent place to issues such as the environment, health, and economic growth, and arms control in which technology plays a central role. The continued failure to appreciate the dynamics of technological surprise heightens the possibility of future "Sputniks"—if not in terms of the actual technological feat involved, then certainly in terms of the public's reaction to it.

Technology is also one of the foundations of "soft power" which Joseph Nye and other commentators hold to be the key to the international politics of the next century.[31] And, as in the 1950s, it is assumed by many that the United States is a leading source (if not the leading source) of technological change and innovation. If technology and technological change are among the primary driving forces behind issues and a key ingredient of power that is brought to bear on solving them, then preventing technological surprise ought to be a prime concern of policy makers.

The events leading up to Sputnik suggest two facets of technological surprise which appear to guarantee that technological surprises will continue to be experienced. First, technological surprise is characterized by an abundance of "routine" information. Overwarning rather than deception or secrecy appears to be a dominant motif. Second, policy makers are not well-educated or sensitive to the larger implications of technological breakthroughs; therefore, they do not act on the warning intelligence they receive. As with military surprise, the most prudent course of action open to policy makers may be that of coupling a concern for preventing surprise with an increased capacity for responding to it and minimizing its consequences, something the Eisenhower administration failed at in the case of Sputnik.

1. Robert A. Divine, *The Sputnik Challenge* (New York: Oxford University Press, 1993), pp. xiii-xv.
2. Walter A. McDougall, *...the Heavens and the Earth: A Political History of the Space Age* (New York: Basic Books, 1985), p. 142.
3. Divine, *Sputnik Challenge*, pp. 44–45.
4. McDougall, *...the Heavens and the Earth*, p. 226.
5. Barry B. Hughes, *World Futures: A Critical Analysis of Alternatives* (Baltimore, MD: Johns Hopkins University Press, 1985), p. 145.
6. Ibid., p. 151.
7. Richard Betts, *Surprise Attack* (Washington, DC: The Brookings Institution, 1982).

8.   Ibid., p. 11; Katrina Brodin, "Surprise Attack: The Case of Sweden," *Journal of Strategic Studies* 1 (1978): 99.

9.   Michael I. Handel, *The Diplomacy of Surprise: Hitler, Nixon, Sadat* (Cambridge, MA: Harvard University Press, 1991), p. 1-6.

10.   Michael I. Handel, "Avoiding Political and Technological Surprise in the 1980s," in Roy Godson, ed., *Intelligence Requirements for the 1980's: Analysis and Estimates* (Washington. DC: Consortium for the Study of Intelligence, 1980), pp. 85–111.

11.   Barton Whaley, *Codeword BARBAROSSA* (Cambridge, MA: MIT Press, 1974).

12.   Roberta Wohlstetter, *Pearl Harbor: Warning and Decision* (Stanford, CA: Stanford University Press, 1962).

13.   Shlomo Gazit, "Estimates and Fortunetelling in Intelligence Work," *International Security* 4 (1980): 36–56.

14.   Robert Jervis, *Perception and Misperception in International Politics* (Princeton, NJ: Princeton University Press, 1976), pp. 203–16.

15.   William Ascher, *Forecasting: An Appraisal for Policy Makers and Planners* (Baltimore, MD: Johns Hopkins University Press, 1978).

16.   Handel, *Diplomacy of Surprise*, pp. 13–24.

17.   McDougall, ... *The Heavens and the Earth*, p. 60.

18.   John Prados, *The Soviet Estimate: U.S. Intelligence and Soviet Strategic Forces* (Princeton, NJ: Princeton University Press, 1982), pp. 55–57.

19.   McDougall, ... *The Heavens and the Earth*, p. 27.

20.   Ibid., pp. 52–53.

21.   Lawrence Freedman, *U.S. Intelligence and the Soviet Strategic Threat* (Princeton, NJ: Princeton University Press, 1977), p. 64.

22.   Jeffrey Richelson, *American Espionage and the Soviet Target* (New York: Quill, 1987), pp. 143–48.

23.   Vernon Van Dyke, *Pride and Power: The Rationale of the Space Program* (Urbana: University of Illinois Press, 1964), p. 15.

24.   McDougall, ... *The Heavens and the Earth*, p. 98.

25.   Van Dyke, *Pride and Power*, p. 15.

26.   Freedman, *U.S. Intelligence and the Soviet Strategic Threat*, p. 68.

27.   McDougall, ...*the Heavens and the Earth*, pp. 52–53.

28.   Ibid., p. 123.

29.   Van Dyke, *Pride and Power*, p. 13.

30.   Dorothy Nelkin, "Technology and Public Policy," in Ina Spiegel-Rossing and Derek de Solla Price, eds., *Science Technology and Society* (London: Sage, 1977), p. 399.

31.   Joseph S. Nye, Jr., *Bound to Lead: The Changing Nature of American Power* (New York: Basic Books, 1990).

Contributors

William P. Barry is a lieutenant colonel in the United States Air Force and assistant professor in the Department of Political Science at the U.S. Air Force Academy, Colorado Springs, Colorado, and a recent Ph.D. from Oxford University, United Kingdom.

Rip Bulkeley is an independent scholar in Oxford, England, writing a history of the 1957–1958 International Geophysical Year, a case study of the interaction between international scientific cooperation and international relations. He has written *The Sputniks Crisis and Early U.S. Space Policy* (Indiana University Press, 1991); *The Anti-Ballistic Missile Treaty and World Security*, with H.G. Brauch (AFES-Press, 1998); and *Space Wespons: Deterrence or Delusion?* with G. Spinardi (Polity Press, 1986).

Dwayne A. Day is a research associate at the Space Policy Institute at George Washington University and a recent Guggenheim Fellow. He has written extensively on U.S. military and civilian space policy and history and was the 1995 winner of the Goddard Space History Essay Contest. He has served as an assistant editor for volumes 1 and 2 of the NASA history books *Exploring the Unknown: Selected Documents in the History of the U.S. Civil Space Program* (NASA SP-4407, 1995–1996), and is the primary editor of *Eye in the Sky: The Story of the Corona Spy Satellite* (Smithsonian Institution Press, 1998). He is also the guest editor of several issues of the *Journal of the British Interplanetary Society* on military space history.

John A. Douglass is a Research Fellow at the Center for Studies in Higher Education, and serves as a policy analyst for the Academic Senate of the University of California. He recently completed a study on educational opportunity and the history of University of California admissions policies, and three recent articles in the *History of Higher Education Annual*, the *American Behavioral Scientist*, and

California Politics and Policy. His book, *California and the History of American Higher Education*, was published by Stanford University Press, 1997.

Eilene Galloway participated in the legislative process that led to the enactment of the National Aeronautics and Space Act of 1958. As Senior Specialist in International Relations (National Security), Congressional Research Service, Dr. Galloway was responsible for the analysis of issues combining national defense and foreign relations. Senator Lyndon B. Johnson appointed her to assist with the hearings which began on November 17, 1957, in response to the orbiting of *Sputnik 1* on October 4, 1957. She also served as Special Consultant to the Senate Committee on Aeronautical and Space Sciences, February 6, 1958, which was responsible for outer space legislation, a position she held throughout the life of that committee when jurisdiction was given to the Senate Committee on Science, Commerce, and Transportation, for which she wrote the Senate document on the Moon Agreement. Beginning in 1959, Dr. Galloway was appointed to participate in sessions of the United Nations Committee on the Peaceful Uses of Outer Space (COPUOS) ad its Legal Subcommittee. She is the author of more than 100 published articles and documents concerned with space matters.

Peter A. Gorin is a specialist in political science and international relations, as well as a long-time researcher of aerospace history. He is a graduate of the Moscow University and worked in the USSR as a political analyst. Currently he continues his research in space history in association with the National Air and Space Museum of the Smithsonian Institution. Dr. Gorin has published articles in U.S. and Russian magazines and presented research papers at a number of history forums. He also is the author of the chapter on Soviet photo-reconnaissance satellites in a monograph *Corona: Between the Sun & the Earth* published by the American Society for Photogrammetry and Remote Sensing (ASPRS) in 1997.

James J. Harford is Executive Director-Emeritus, American Institute of Aeronautics and Astronautics, and served as director between 1964 and 1988. The author of numerous articles and papers, he also has published, *Korolev: How One Man Masterminded the Soviet Drive to Beat America to the Moon* (John Wiley, 1997).

Glenn P. Hastedt is professor of political science at James Madison University in northern Virginia. The author of numerous works on the Cold War and intelligence, he has also published *American Foreign Policy: Past Present, Future* (Prentice-Hall, 1997), now in its third edition.

John Krige is Director, Centre de Recherche en Historie des Sciences et des Techniques, Paris, France. He is a coauthor of *The History of CERN* (Amsterdam: North Holland, 1990), and several monographs on the history of ESA, including *The Prehistory of ESRO, 1959/60: From the First Initiatives to the Formations of COPERS* (ESA Publication HSR-1, July 1992); *The Early Activities of the COPERS and the Drafting of the ESRO Convention (1960/61)* (ESA Publication HSR-4, January 1993); *The Launch of ELDO* (ESA Publication HSR-7, March 1993); *Europe into Space: The Auger Years (1959–1967)* (ESA Publication HSR-8, May 1993); and coauthor with Arturo Russo, *Europe in Space 1960–1973* (ESA Publication SP-1172, September 1994).

Sergey N. Krushchev is Senior Fellow at the Thomas J. Watson, Jr., Institute for International Studies, Brown University. Dr. Krushchev has his Soviet Doctoral degree from the Ukranian Academy of Science, a Ph.D. from the Moscow Technical University, and an M.A. with distinction from the Moscow Electric Power Institute. He edited his father's memoirs, Nikita Khrushchev, *Khrushchev Remembers* (1970), *Khrushchev Remembers: Last Testament* (1974) and *Khrushchev Remembers: Tapes of Glasnost* (Little, Brown, 1990). He is also the author of *Khrushchev on Khrushchev: An Inside Account of the Man and His Era, by His Son* (Little, Brown, 1990) and other works.

Roger D. Launius is chief historian of the National Aeronautics and Space Administration, Washington, D.C. He is the author or editor of several books on aerospace history, including *Frontiers of Space Exploration* (Greenwood Press, 1998); *Spaceflight and the Myth of*

Presidential Leadership (University of Illinois Press, 1997); *Organizing for the Use of Space: Historical Perspectives on a Persistent Issue* (Univelt, Inc., AAS History Series, 1995); and *NASA: A History of the U.S. Civil Space Program* (Krieger Publishing Co., 1994).

John M. Logsdon is Director of both the Center for International Science and Technology Policy and the Space Policy Institute of George Washington University's Elliott School of International Affairs, where he is also Professor of Political Science and International Affairs. He holds a B.S. in physics from Xavier University and a Ph.D. in political science from New York University. He has been at George Washington University since 1970, and previously taught at the Catholic University of America. Dr. Logsdon's research interests include space policy, the history of the U.S. space program, the structure and process of government decision-making for research and development programs, and international science and technology policy. He is author of *The Decision to Go to the Moon: Project Apollo and the National Interest* (MIT Press, 1970); general editor of *Exploring the Unknown: Selected Documents in the History of the U.S. Civil Space Program*, Vols. I-III (NASA SP-4407, 1995–1998); and has written numerous articles and reports on space policy and science and technology policy. In addition, he is North American editor for the journal *Space Policy*.

Walter A. McDougall is Alloy-Ansin Professor of International Relations and the Professor of History at the University of Pennsylvania. He is the author of the Pulitzer Prize-winning history, ... *The Heavens and the Earth: A Political History of the Space Age* (Basic Books, 1985). He also has written *Promised Land, Crusader State: The American Encounter with the World Since 1776* (Houghton Mifflin, 1997); *Let the Sea Make a Noise ...: A History of the North Pacific from Magellan to MacArthur* (Basic Books, 1993); and *France's Rhineland Diplomacy, 1914–1918: The Last Bid for a Balance of Power in Europe* (Princeton University Press, 1978).

Michael J. Neufeld is a curator of Aeronautics at the National Air and Space Museum, Smithsonian Institution, Washington, D.C. A specialist in the aerospace technology of Germany, he is the author of the prize-winning history, *The Rocket and the Reich: Peenemünde and the Coming of the Ballistic Missile Era* (Free Press, 1995).

Kenneth A. Osgood is completing his Ph.D. in U.S. diplomatic history from the University of California, Santa Barbara. His dissertation relates to the national security of the United States during the Cold War.

Asif A. Siddiqi received his B.S. and M.S. degrees from Texas A&M University. He has is presently in doctoral program at Carnegie-Mellon University under a National Science Foundation Fellowship to study Cold War science and technology. Mr. Siddiqi is currently a NASA Contract Historian finishing a book on the history of the Soviet human space exploration program that will be published by the Johns Hopkins University Press. He has published extensively in *Spaceflight*, the *Journal of the British Interplanetary Society*, *Quest: The Journal of Spaceflight History*, and *Countdown* on the Soviet/Russian space program. He has also presented papers at meetings of the Society for the History of Technology (1994), the American Association for the Advancement of Slavic Studies (1995), and the Society for History in the Federal Government (1996). He was the 1997 recipient of the Robert H. Goddard Historical Essay Award sponsored by the National Space Club and the 1997 American Institute for Aeronautics and Astronautics (AIAA) historical manuscript prize. He currently lives in Philadelphia, Pennsylvania.

Robert W. Smith is chair of the Space History Department of the National Air and Space Museum, Smithsonian Institution, Washington, D.C. He is the author of the prize-winning history, *The Space Telescope: A Study of NASA, Science, Technology, and Politics* (Cambridge University Press, 1989); *The Expanding Universe* (Cambridge University Press, 1982); and several other works.

Gretchen J. Van Dyke received both her M.A. and Ph.D. in Foreign Affairs from the University of Virginia's Woodrow Wilson Department of Government and Foreign Affairs. Her dissertation and subsequent research projects have concentrated on the Kennedy administration's development of the flexible response strategy; her article on Robert McNamara's and McGeorge Bundy's roles in the flexible response policy making process appeared in the Spring 1997 edition of the *Miller Center Journal*. Dr. Van Dyke is currently an assistant professor of political science at the University of Scranton (Pennsylvania), where she teaches International Relations, American Foreign Policy and Decision Making, and European Foreign Policy and Politics.

Index

Advanced Research Projects
  Agency (ARPA), 185–86, 312
Adzhubey, Aleksey I., 52
Aelita, 20, 267–68
Aerobee Rocket, 235, 242
Aleksandr II, Tsar, 22
Almaz space station, 284–85
Alsop, Joseph, 86
Alsop, Steward, 413
Amaldi, Edoardi, 301–302, 304
American Rocket Society, 83, 165,
  233, 238, 241, 311
Annals of the IGY, 141
Army Ballistic Missile Agency
  (ABMA), 123–24, 165; and
  selection of IGY satellite plan,
  174–79, 231–57; and Stewart
  committee, 240–47
Astronautics, 83
Atkinson, Richard, 357, 359
Atlas Rocket, 161, 168, 210, 245,
  415
Auger, Pierre, 301, 304
Aviation Week, 33, 413

Barber, Richard J., 353, 354
Bardin, Ivan P., 47, 126–27, 128,
  129–30, 131, 137; and satellite
  data exchange, 142–48;
Barmin, V. P., 99
Bedrosian, Alex, 340–41
Beloussov, V. V., 127, 138, 151;
  and satellite data exchange,
  142–48;
Belyayev, Aleksandr R., 21

Beriia, Laventii, 99–100
Berkner, Lloyd, V., 127, 130, 132,
  236; and cooperation with
  Soviets, 133–39; and CSAGI
  Conference on Rockets and
  Satellites, 133; and *Guide to
  IGY World Data Centres*,
  133–39; and Moscow IGY
  Assembly, 139–42; and satellite
  data exchange, 142–48; and
  selection of IGY satellite plan,
  174–79; and WS-117L
  program, 184–90
Bissell, Richard, 164
Blackband, W. T., 130
Black Knight Rocket, 302
Blagonravov, Anatoliy A., and
  *Guide to IGY World Data
  Centres*, 133–39; and decision
  for Sputnik, 33–40, 50, 75–76;
  and Moscow IGY Assembly,
  139–42; and satellite data
  exchange, 142–48
Bloch, Erich, 357
Blue Streak Rocket, 294–99, 302
Bogdanov, Aleksandr A., 13
Bok, Derek, 357
Bolley, William, 234
*Bolshaya Sovetskaya
  entsiklopediya (Great Soviet
  Encyclopedia)*, 36–37
Bolshevik party, 13, 15–16, 26
Bolshevik Revolution, 7, 15, 16,
  22
Bonner, Thomas N., 328, 339–40

Bowles, Chester, 386
Braun, Wernher von, 77, 88, 162;
and American reaction to
Sputnik 1, 83–84, 180–82,
309–25; and selection of IGY
satellite plan, 174–79, 231–57;
and Stewart committee, 240–47
Brezhnev, Leonid, 104, 105
British Committee on Space
Research, 300–302
Brundage, Percival, 167, 183, 184
Buffalo, University of, 241
Bulganin, Nikolay A., 52
Buran, 285
Bush, George G. W., and higher
education, 327–64
Bush, Vannevar, 414–15

California Institute of Technology
(Caltech), 241
California, University of, 351–52,
359
Cape Canaveral, Florida, 55, 179
Castro, Fidel, 407
Central Intelligence Agency (CIA),
150, 166, 170, 180, 203, 375;
and election of 1960, 378–85;
and origins of Orbiter, 233–34;
relationship to scientific satellite
program, 182–84, 203–206,
236–37; and WS-117L
program, 184–90
Centre european de le research
nuclear (CERN), 301–302
Centre national d'études spatiales
(CNES), 303
Chapman, Sydney, 127, 134, 135,
136
Chekunov, Boris S., 63–64
Chelomei, Vladimir, 104,
107–108, 267, 278–87
Chertok, Boris, 278–79

Clement, George, 241, 245
Colliers' magazine, 162
Committee on the Peaceful Uses
of Outer Space (COPUOS),
United States, 187, 318–21
Committee on Space Research
(COSPAR), United Nations,
141–42
Council of Economic Advisors,
358–59

Defense Mobilization, Office
of-Defense Department
(ODM-Defense), 213–14
Denikin, Anton I., 16
DeWitt, Nicholas, 339–40
Douglas Aircraft Co., 75
Dryden, Hugh L., 89
Dulles, Allen, 166–67, 206, 220,
236, 375
Dulles, John Foster, 119, 172, 401
Durant III, Frederick C., 48, 174,
233–34, 236, 247
Dyna-Soar, 285

Edson, James B., 245
Eideman, Robert P., 28
Eisenhower, Dwight D., 37, 48,
53, 77, 122, 162–63, 261;
and American reaction to
Sputnik 1, 83–84, 180–82,
309–25; and CIA relationship
to scientific satellite, 182–84;
and "freedom of space,"
168–69, 172–73, 187–90,
203–205, 242–43; and higher
education, 327–64; and 1960
presidential campaign,
365–400; National Aeronautics
and Space Act of 1958, 309–25;
and NSC 5520, 121, 167–69,

208–10, 212; "peace offensive" of, 199–202; and reconnaissance activities, 163–65, 172–73; and scientific satellite program, 165–67; and selection of IGY satellite plan, 174–79, 231–57; space policy of, 197–229; and space technology development, 212–19; and Technological Capabilities Panel (TCP), 120, 163–65, 203; and TCP recommendations, 169–73, 203; and technological surprise, 401–23; and United Nations, 318–21; and WS-117L program, 184–90

Ellington, Charles G., 178–79
Energia Design Bureau, 29
Esnault-Pelterie, Robert, 20
Experimental Design Bureau No. 1 (OKB-1), and building of *Sputnik 1*, 54–67; and conception of *Sputnik 1*, 44–54, 97–101; and creation of Soviet space industry, 95–115; and decision for Sputnik, 33–40; and development of R-7 rocket, 76–77; and flight of *Sputnik 1*, 62–66, 82–83; and flight of *Sputnik 2*, 86–87; and flight of *Sputnik 3*, 88–91; and IGY, 43–72; and initial Soviet reaction to *Sputnik 1*, 73–75, 309–25; and myths of *Sputnik 1*, 43–44, 101–102; and results of *Sputnik 1*, 65–68, 101–108; scientific objectives of *Sputnik 1*, 59–60; and *Sputnik 1*, 43–72, 78–80, 267–87; triple play of, 73–94; and V-2, 28–33, 97–98

Experimental Design Bureau No. 10 (OKB-10), 105
Experimental Design Bureau No. 23 (OKB-23), 104
Experimental Design Bureau No. 52 (OKB-52), 104
Experimental Design Bureau No. 301 (OKB-301), 104
Experimental Design Bureau No. 586 (OKB-586), 57–58, 105
Explorer 1, 88, 91, 119, 123, 146–47, 150; and American reaction to *Sputnik 1*, 83–84, 180–82, 309–25; and NSC 5520, 121, 167–69, 208–10, 212; and selection of IGY satellite plan, 174–79, 231–57; and Stewart committee, 240–47

Federal Technology Transfer Act of 1986, 356
Fedorov, Aleksandr P., 14, 137, 146, 152
Feedback, Project, 163, 165, 166
Feoktistov, Konstantin, 81
Flammirion, Camille, 12–13
Florianskiy, Mikhail, 82
Flight International, 33
Frau im Mond (Woman on the Moon), 21
From the Earth to the Moon, 12, 13
Furnas, Charles C., 176; and Stewart committee, 240–47

Gagarin, Yuri A., 275
Gaither Committee Report, 367–68
Gardner, Trevor, 217–18

Gas Dynamics Laboratory (GDL), 24–28
Gates, Thomas, 374–75, 381–82
Gaulle, Charles de, 303–304
Gavin, James M., 368–69
Gazit, Shlomo, 406, 417
Geiger, Roger, 357–58
Geshvend, Fedor R., 13, 22
Gibbons, Michael, 356
Gilzin, Karl A., 32–33, 34
GIRD, see OSOAVIAKHIM Reactive Motion Research Groups
Glenn L. Martin Company, 249–50
Glennan, T. Keith, 188
Glushko, Valentin P., 4, 12, 20, 279; and creation of Soviet space industry, 95–115; and decision for Sputnik, 33–40, 50, 52, 62, 99; and purges, 28–30; and MosGIRD, 24–28; and rivalry with Korolev, 278–87; writes *Rakety: ikh ustroistvo i primeneniye (Rockets: Their Design and Applications)*, 27; and V-2, 28–33, 97–98
Goddard, Robert H., 17, 20
Goodpaster, Andrew, 164, 175–76, 178–79, 181–82
Gray, Gordon, 188
Grechko, Georgiy, 7
Greenewalt, Crawford H., 345–46
Griffith, Martha W., 341
Grosse, Aristid V., 121, 202–203, 205
Guide to IGY World Data Centres, 133–39; and Moscow IGY Assembly, 139–42
Gurevich, Mark, 280–81

Hagerty, James C., 48
Halley's Comet, 19
Harvard Observatory, 146
Haviland, Robert P., 75
Heller, Gerhard, 234, 236
Hermes Rocket, 241
Herter, Christian, 220
Higher Education Act of 1965, 350
Higher Education Facilities Act of 1963, 355
Higher Education for American Democracy, 335
Hilsberry, Clarence, 340
Hoover, George, 233, 234
Hoover, Herbert, Jr., 215
Hughes, Barry, 402–403

Interdepartmental Commission on Interplanetary Communications (MKMS), and decision for Sputnik, 33–40
International Astronautical Congress, 48
International Astronautical Federation (IAF), 48, 233
International Council of Scientific Unions (ICSU), 5, 126
International Geophysical Year (IGY), 5, 165, and building of *Sputnik 1*, 54–67; and conception of *Sputnik 1*, 44–54, 97–101; and cooperation with Soviets, 133–39; and CSAGI Conference on Rockets and Satellites, 133; and data exchange, 142–48; and decision for Sputnik, 33–40; and *Guide to IGY World Data Centres*, 133–39; and initial Soviet reaction to *Sputnik 1*, 73–75; and Moscow Assembly,

139– 42; and myths of *Sputnik 1*, 43–44, 101–102; and NSC 5520, 121, 167–69, 208–10, 212; and problems of Soviet committee, 131–33; and race between U.S. and Soviet Union, 179–80; and results of *Sputnik 1*, 65–68, 101–108; and revelation of Soviet satellite at IAF, 77–78; satellite program of, 5–6, 125–59; setting for, 119–24; Soviet response to, 126–30; and *Sputnik 1*, 43–72; and technological surprise, 401–23; and U.S. satellite proposal, 126, 165–67, 174–79; and western European reaction to *Sputnik 1*, 289–307; and world reaction to *Sputnik 1*, 84–86

Iskustvennye sputniki zemli (Artificial Satellites), 33

Issledovaniya mirovykh prostranstv reaktivnymi priborami (Exploration of Space by Reactive Devices), 14, 18

Ivanovskiy, Oleg G., 60, 81, 88

Jackson, Bruce, 340–41
Jackson, C.D., 207
Jackson, Henry, 338–39, 375
Jet Propulsion Laboratory, 88; and Stewart committee, 240–47
Johnson, Lyndon B., 89, 262–63, 283; and higher education, 327–64; and National Aeronautics and Space Act of 1958, 309–25; investigations of, 309–14, 368–69, 371; and 960 presidential campaign, 365–400

Johnson, Roy, 185–86
Journal of Higher Education, 328
Jupiter-C Rocket, 63, 88, 168, 295; and selection of IGY satellite plan, 174–79, 231–57; and Stewart committee, 240–47

Kaliningrad, Soviet Union, 55
Kalmykov, Valeriy D., 62
Kaluga, Soviet Union, 13
Kaplan, Joseph, 74, 206–207, 236; and Stewart committee, 240–47
Kapustin Yar, Soviet Union, 55
Kasatkin, A. M., 129–30, 133–34
Kazantsev, Aleksandr P., 32
Kealey, Terence, 359
Keldysh, Mstislav V., 6, 35–36, 49, 50, 51, 53, 62, 67–68, 78–80
Keller, K. T., 415
Kendrick, J. B., 236
Kennan, George, 86
Kennedy, John F., 204, 265–65, 276, 281–83, 303, 343; and higher education, 327–64; and 1960 presidential campaign, 365–400; and "missile gap" issue, 366–69, 388–92
Kerr, Clark, 353, 357–58
Killian, James A., 89, 166, 339, 344–45; and NSC 5520, 121, 167–69, 208–10, 212; and Technological Capabilities Panel (TCP), 120, 163–65, 203, 236–37, 415–16; and TCP recommendations, 169–73, 203; and WS-117L program, 184–90, 416
Khomyakov, Mikhail S., 60
Kibalchich, Nikolay I., 21–22

Kondratyuk, Yuriy V., 16, 19, 27–28, 30
Korolev, Sergey P., 4, 5, 6–8, 150, 261; and building of *Sputnik 1*, 54–67; and conception of *Sputnik 1*, 44–54, 97–101; and creation of Soviet space industry, 95–115; and decision for Sputnik, 33–40, 98–101; and development of R-7 rocket, 76–77; and flight of *Sputnik 1*, 62–66, 82–83; and flight of *Sputnik 2*, 86–87; and flight of *Sputnik 3*, 88–91; and IGY, 43–72; and initial Soviet reaction to *Sputnik 1*, 73–75; and myths of *Sputnik 1*, 43–44, 101–102; and purges, 28–30; and MosGIRD, 24–28; and results of *Sputnik 1*, 65–68, 101–108; and rivalry with Glushko, 278–87; scientific objectives of *Sputnik 1*, 59–60; and *Sputnik 1*, 43–72, 78–80, 267–87; and technological surprise, 401–23; triple play of, 73–94; writes *Raketniy polet v stratosfere (Rocket Flight in Stratosphere)*, 27; and V-2, 28–33, 97–98;
Krasnaya zvezda (Red Star), 13
Kroshkin, M. G., 147. 151
Krushchev, Nikita A., 7–8, 51, 52, 65–66, 100, 132, 261, 265–66; and flight of *Sputnik 2*, 86–87; and initial Soviet reaction to *Sputnik 1*, 73–75; and results of *Sputnik 1*, 65–68, 101–108; son's recollections of, 267–87; and technological surprise, 401–23; and U-2 incident, 189–90

Krushchev, Sergey, 267–87
Kryukow, Sergey S., 52, 54
Kuznetsov, V. I., 99

Land, Edwin, 163, 165; and WS-117L program, 184–90
Langemak, Georgiy, 27
Lappo, Vyecheslav I., 64–65
Lauritsen, Charles C., 241, 245, 246
Lavrov, Ilya V., 48–49
Layka, 86–87, 274, 291
Le Figaro, 85
Leghorn, Richard, 188–89
LenGIRD, 25, 28
Leningrad, Soviet Union, 24, 27, 38
Leningrad Transportation Institute, 18
Levochkin, Semen, 104, 105–106
Lidorenko, Nikolay S., 55–56
Lindbergh, Charles A., 83
Lockheed Skunk Works, 163
Lovell, A. C. B., 74
LRD-D-1 Rocket, 25

McCormack, John W., 312, 316
McElroy, Neil, 373; and election of 1960, 378–85; and WS-117L program, 184–90
McMath, Robert C., 241
McNamara, Robert S., 354
McQuade, Lawrence, 383–84
Malenkov, Georgiy M., 46
Malyshev, Vyecheslav A., 46
Manchester Guardian, The, 85, 89
Manned Orbiting Laboratory (MOL), 106
Manual on Rockets and Satellites, 138, 141
Martynov, Georgiy, 32

Massachusetts Institute of Technology (MIT), 163, 236
Massevitch, A.G., 135, 136
Massey, Harrie, 129, 293–99, 300, 304
Matkin, Gary, 358
Medaris, John B., and selection of IGY satellite plan, 174–79, 231–57
"Meeting the Threat of Surprise Attack" study, 120–21, 163–65
Menzel, Donald, 146
Merkulov, Igor A., 26
Mezhplanetnye puteshestviya (Interplanetary Travel), 12, 19
Mezhplanetnye soobshcheniy (Interplanetary Communications), 19
Mir Space Station, 15, 81
Mishin, Vasiliy, 76, 281–82
Molotov, Vyecheslav M., 52
Mooney, Ross L., 353
Moscow Air Fleet Academy, 4, 17
MosGIRD, 24–28
Mrykin, Aleksandr G., 6, 49
Murphree, E. V., 177–78
M.V. Keldysh Research Center, 25

N-1 Rocket, 278–79, 284–87
National Academy of Sciences, 165, 166, 236, 313; and selection of IGY satellite plan, 174–79, 231–57; and Stewart committee, 240–47
National Advisory Committee for Aeronautics (NACA), 312
National Aeronautics and Space Act of 1958, 309–25
National Aeronautics and Space Administration (NASA), 301; established, 88–89, 309–25

National Defense Education Act (NDEA) of 1958, 327–64
National Education Act, 352–53
National Review, 119
National Science Foundation (NSF), 166, 179, 207, 236, 239, 313, 356, 357, 359
National Security Council, 83, 121, 122; and NSC 162/2, 200–201; and NSC 5501, 201–202, 214; and NSC 5520, 121, 167–69, 208–10, 212, 221; NSC 5522, 205, 221; NSC 5602, 214; NSC 5707, 214l; NSC 5814, 221; and "peace offensive," 199–202 and selection of IGY satellite plan, 174–79, 231–57; and space technology development, 212–19; and *Sputnik 1*, 220–21; and Technological Capabilities Panel (TCP), 120, 163–65, 203; and TCP recommendations, 169–73, 203; and technological surprise, 401–23; and space technology development, 212–19
Navaho Rocket, 245–46
Naval Research Laboratory, 123–24, 341; and origins of Orbiter, 233–37; and selection of IGY satellite plan, 174–79, 231–57; and Viking rocket, 210–11
Naval Research, Office of, 77, 165
Nedelin, Mitrofan I., 61, 62
Nesmeyanov, Aleksandr N., 62
New Production of Knowledge, The, 356
New Republic, 402
New York Times, 85–86, 388, 391–92, 413

Newell, Homer E., and Moscow
IGY Assembly, 139–42
Newsweek, 162, 401–402
Nickerson, John, 243
Nitze, Paul, 383
Nixon, Richard M., 216, 405; and
1960 presidential campaign,
365–400; and "missile gap"
issue, 366–69, 388–92
North American Aviation, Inc.,
247–48
North Atlantic Treaty
Organization (NATO), 125,
301
Nosov, Aleksandr I., 63
Novatskiy, Pavel, 74
*Novye printsipy vozdukhoplaniya
(New Principles of Air Flight)*,
14
NPO Energiya, 107
NSC 162/2, 200–201
NSC 5501, 201–202, 214
NSC 5520, 121, 167–69, 208–10,
212, 221
NSC 5522, 205, 221
NSC 5602, 214
NSC 5707, 214
NSC 5814, 221

Oberg, James E., 44
Oberth, Hermann, 18, 20
Odessa, Soviet Union, 12
Odishaw, Hugh, 143, 146–47
Ogorodnikov, Kirill F., 48
Orbiter, Project, 77, 122–23, 165,
231–57; and American reaction
to *Sputnik 1*, 83–84, 180–82,
309–25; competition to,
237–40; and NSC 5520, 121,
167–69, 208–10, 212; origins
of, 233–37; and selection of
IGY satellite plan, 174–79,

231–57; and Stewart
committee, 240–47; and
technological surprise, 401–23
OSOAVIAKHIM (Society for the
Support of Aviation and
Chemical Development), 23,
25, 28
OSOAVIAKHIM Reactive
Motion Research Groups
(GIRD), 23–26

Patrick Air Force Base, Florida, 55
Perelman, Yakov I., 12, 14, 19,
25, 28, 33, 268
Pervukhin, Mikhail G., 52
Petrograd, Soviet Union, 15
Pickering, William H., 88; and
Stewart committee, 240–47
Piliugin, N. A., 99
Pioneer 10, 19
Pioneer 11, 19
Pionerskaya Pravda, 34–35
Piore, E. R., 238–39
Poltava, Ukraine, 15
Polyarniy, Aleksandr I., 25–26
Porter, Richard W., 241, 245, 246
Power, Thomas S., 238, 375,
382
Powers, Francis Gary, 189
Pravda, 66, 73–74, 87, 127
"Preliminary Design of an
Experimental World-Circling
Spaceship," 75
President's Science Advisory
Committee (PSAC), 314–15,
344–45
*Pryzhok v nichto (Jump Into
Nowhere)*, 21
Pushkov, N. V., 127, 131
*Putechestviye k dalekim miram
(Journey to the Distant Words)*,
34

Quarles, Donald A., 122, 169; and American reaction to *Sputnik 1*, 83–84, 180–82, 220, 309–25; and competition to Orbiter, 237–40; NSC 5520, 121, 167–69, 208–10, 212; and selection of IGY satellite plan, 174–79, 231–57, 248–49; and Stewart committee, 240–47; and technological surprise, 401–23; and WS-117L program, 184–90

R-1 Rocket, 30
R-7 Rocket, 6, 18, 45–46, 49, 52, 58, 60–62, 100–101, 102–103, 268, 275–76
R-12 Rocket, 57–58
Raketa v kosmicheskoye prostrantstvo (The Rocket into Space), 18
Raketoplan, 285
Raketniy polet v stratosfere (Rocket Flight in Stratosphere), 27
Rakety: ikh ustroistvo i primeneniye (Rockets: Their Design and Applications), 27
Ramsey, Norman, 301
Rand Corp., 75, 163, 202, 237, 241, 415
Randall, Clarence, 401
Raushenbakh, Boris V., and initial Soviet reaction to *Sputnik 1*, 73–75;
Razumov, Vladimir V., 25
Reactive Scientific Research Institute (RNII), 4, 25–28, 33
Rebrov, Mikhail, 82–83
Red Star in Orbit, 44

Redstone Rocket, 174–75, 177; and Stewart committee, 240–47
Reston, James, 388
Ridgway, Matthew, 368–69
Robertson, Reuben B., Jr., 177
Rockefeller, Nelson A., 207–208, 367–68, 388–89
Rosen, Milton, 238; and Stewart committee, 240–47
Rosser, J. Barkley, 241, 245–46
Royal Society for the Protection of Cruelty to Animals, 291–92
Russell, Richard, 374–75
Russian Space Agency, 108–109, 110
Ryabikov, Vasiliy M., 49, 51, 61, 62, 63, 65
Ryazanskiy, Mikhail S., 55–56, 65, 99
Rynin Nikoley A., 4, 17–18, 19–20, 25

St. Petersburg, Russia, 12
Sageev, Roald, 76–77, 90
Salyut space station, 286–87
Samos Project, 186
Sänger, Eugen, 31
Schriever, Bernard A., 161–62, 237–38
Scientific-Research Institute for Current Sources, 55
Seaborg, Glen T., 345–46
Sedov, Leonid I., and decision for Sputnik, 33–40, 47–48, 50, 53; reveals Soviet satellite at IAF, 77–78, 247, 313–14
Seitz, Fred, 301
Shargey, Aleksandr I., 15, 16
Sharp, Dudley C., 374
Shternfeld, Ari A., 33, 268
Silver Spring, Maryland, 46

Simon, Leslie, 210

Singer, S. Fred, 233, 236

Skylark Rocket, 293–94

Smirnov, Leonid, 284

Smithsonian Astrophysical Observatory, 135

Snow, C. P., 340–41

Society for the Study of Interplanetary Communications, 4, 17–18, 26–27

Soviet Academy of Sciences, aerospace R&D, 95–97; and building of *Sputnik 1*, 54–67; and conception of *Sputnik 1*, 44–54; and cooperation with IGY, 133–39; and creation of Soviet space industry, 95–115; and CSAGI Conference on Rockets and Satellites, 133; and decision for Sputnik, 33–40, 46–48, 313–14; and flight of *Sputnik 1*, 62–66, 82–83; and flight of *Sputnik 2*, 86–87; and flight of *Sputnik 3*, 88–91; and *Guide to IGY World Data Centres*, 133–39; and IGY, 43–72, 125–59; and initial Soviet reaction to *Sputnik 1*, 73–75; and Korolev's triple play, 73–94; and Moscow IGY Assembly, 139–42; and problems with IGY committee, 131–33; and results of *Sputnik 1*, 65–68, 101–108; and revelation of Soviet satellite at IAF, 77–78; and satellite data exchange, 142–48; and scientific objectives of *Sputnik 1*, 59–60;

Soviet Union, aerospace R&D in, 95–97; and building of *Sputnik 1*, 54–67; and conception of *Sputnik 1*, 44–54, 97–101; and cooperation with IGY, 133–39; creation of space industry in, 95–115; and CSAGI Conference on Rockets and Satellites, 133; and decision for Sputnik, 33–40; early rocketry in, 23–28; and flight of *Sputnik 1*, 62–66, 78–80, 82–83, 267–87; and flight of *Sputnik 2*, 86–87; and flight of *Sputnik 3*, 88–91; and *Guide to IGY World Data Centres*, 133–39; and IGY, 43–72, 125–59; and initial Soviet reaction to *Sputnik 1*, 73–75; and Korolev's triple play, 73–94; and MosGIRD, 24–28; and problems with IGY committee, 131–33; and myths of *Sputnik 1*, 43–44, 101–102; perceptions of spaceflight in, 11–42; purges in, 28–30; and results of *Sputnik 1*, 65–68, 101–108; and revelation of Soviet satellite at IAF, 77–78; and satellite data exchange, 142–48; scientific objectives of *Sputnik 1*, 59–60; space boom in, 16–23; and spaceflight precursors, 11–16; and technological surprise, 401–23; and UFOs, 31–32; and World War II, 28–30; and V-2, 28–33, 97–98;

Spaceflight, 33

Specialized Design Bureau at the Scientific Research Institute No. 88 (NII-88), 45

Sputnik 1, 3, 7, 11, 18, 101; American reaction to, 83–84, 180–82, 309–25; building of,

54–67, 76–83; conception of, 44–54, 97–101; and creation of Soviet space industry, 95–115; decision for, 33–40; flight of, 62–66, 78–80, 267–87; and *Guide to IGY World Data Centres*, 133–39; and higher education, 327–64; and 1960 presidential campaign, 365–400; Korolev and, 43–72; and Korolev's triple play, 73–94; launch preparations for, 82–83; and Moscow IGY Assembly, 139–42; myths of, 43–44, 101–102; and satellite data exchange, 142–48; scientific objectives of, 59–60; Soviet initial reaction to, 73–75; and technological surprise, 401–23; and U.S. space policy formation, 197–229; western European reaction to, 289–307; world reaction to, 84–86

Sputnik 2, 7, 101; flight of, 86–87; and Korolev's triple play, 73–94;

Sputnik 3, 7, 38, 101, 143; flight of, 88–91; and Korolev's triple play, 73–94

Stalin, Josef, 26, 30, 99, 199

Stewart, Homer J., and selection of IGY satellite plan, 174–79, 210–11, 231–57; and Stewart committee, 240–47

Stuhlinger, Ernst, 88

Summerfield, Martin, 85

Symington, Stuart, 375–77; and election of 1960, 378–85

220 Dney na zvezdolete (220 Days in a Starship), 32

TASS (Soviet News Agency), 58, 66, 73–74

Taylor, Maxwell D., 368–69, 387

Technological Capabilities Panel (TCP), 120, 163–65, 203; and NSC 5520, 121, 167–69, 208–10, 212; recommendations of, 169–73, 203

Teleshov, Nikolay A., 13, 22

Thagard, Norman, 81

Thor Rocket, 295, 299

Todd, Alexander, 300

Tombaugh, Clyde, 235–36

Tikhonravov, Mikhail K., 24, 28, 45, 75, 277; and building of *Sputnik 1*, 54–67; and conception of *Sputnik 1*, 44–54, 97–101; and decision for Sputnik, 33–40; and flight of *Sputnik 1*, 62–66; and IGY, 43–72; and Korolev's triple play, 73–94; and myths of *Sputnik 1*, 43–44, 101–102; and results of *Sputnik 1*, 65–68, 101–108; scientific objectives of *Sputnik 1*, 59–60; and *Sputnik 1*, 43–72; transfer to Korolev design bureau, 78;

Tikhov, Gavriil A., 31

Time, 401–402

Tolstoy, Aleksey N., 20, 267–68

Topchiyev, Gennadiy V., 49–50

Troitskayam V. A., 136

Trow, Martin, 356

Tsander, Frederik, 4, 14–15, 17, 19, 20, 27, 30

Tsesevich, Vladimir P., 31

Tsiolkovskiy, Konstantin E., 3–4, 13, 14–15, 18, 20, 21, 22–23, 26, 30, 34, 53, 75–76, 411

Tukhachevskiy, Mikhail N., 24–25

Tunguska Meteorite, 32
Tyura-Tam, Soviet Union, 58,
  60–66

U-2, 163–64, 172–73, 182–83,
  189, 198, 278, 392, 413
UCLA, 241
Uncertain Trumpet, The, 387
*Une Zemli (Beyond the Planet
  Earth)*, 26–27
Ustinov, Dmitriy F., and building
  of Sputnik 1, 57–58; and
  creation of Soviet space
  industry, 95–115; and decision
  for Sputnik, 33–40, 52, 61,
  98–101; and development of R-
  7 rocket, 76–77; and flight of
  Sputnik 1, 62–66
USIA, 391–92
U.S. News and World Report, 402
"U.S. Scientific Satellite Program"
  (NSC 5520), 121

V-2 Rocket, 4–5, 28–33, 97–98
Valier, Max, 20
Van Allen, James A., 46, 88, 91,
  119, 133
Vanguard, Project, 37, 38, 67–68,
  77–78, 83, 87–88, 120, 121,
  123, 126, 128, 131, 322–32,
  341–42; and American reaction
  to *Sputnik 1*, 83–84, 180–82;
  and CIA, 182–84; and selection
  of IGY satellite plan, 174–79,
  211–12, 231–57; and Stewart
  committee, 240–47; and
  technological surprise, 401–23
Vega 1, 19
Vega 2, 19
Verne, Jules, 12, 13
Vernov, Sergey, 90–91

Vetchinkin, Vladimir P., 19
Vetrov, Georgy S., 20
Viking Rocket, 210–11, 235–36,
  249; and Stewart committee,
  240–47; and technological
  surprise, 401–23
Villard, O.G., Jr., 187
Vorontsev-Velyaminov, Boris A.,
  31
Voskresenskiy, Leonid A., 63–64
Voyager 1, 19
Voyager 2, 19

War and Peace in the Space Age,
  368–69
War of the Worlds, 13
Warsaw Pact, 125
Waterman, Alan T., 142, 166–67,
  179–80, 207
Webb, James E., 343, 344
Wells, H.G., 12–13, 267
Western Development District
  (WDD), 161
Whipple, Fred L., 135, 143, 233,
  235, 236
White, Thomas D., 374–75
Wilson, Charles, 174, 401
Witkin, Richard, 84
WS-117L, 184–90

Yangel, Mikhail K., 57–58, 101,
  103

*Zanimatelnaya fizika
  (Entertaining Physics)*, 12
*Zavoevanie mirovykh prostranstv
  (Conquest of Interplanetary
  Space)*, 19
Zhuravlev V. N., 21
Zvezda K. E. Ts. (Star K.E.Ts.), 21

Other titles in Studies in the History of Science, Technology and Medicine

Volume 7
Cold War, Hot Science: Applied Research in Britain's Defence Laboratories
1945–1990
Edited by Robert Bud and Philip Gummett

Volume 8
Planning Armageddon: Britain, the United States and the Command of Western
Nuclear Forces 1945–1964
Stephen Twigge and Len Scott

Volume 9
Cultures of Control
Miriam Levin

Volume 10
Science, Cold War and the American State: Lloyd V. Berkner and the Balance
of Professional Ideals
Allan A. Needell

Volume 11
Reconsidering Sputnik: Forty Years Since the Soviet Satellite
Edited by Robert D. Launius, John M. Logsdon and Robert W. Smith

This book is part of a series. The publisher will accept continuation orders which may be cancelled
at any time and which provide for automatic billing and shipping of each title in the series upon
publication. Please write for details.